Matrix Algebra
for Applied Economics

Matrix Algebra
for Applied Economics

Shayle R. Searle
Departments of Biometrics and of Statistical Science
Cornell University

Lois Schertz Willett
Food and Resource Economics Department
University of Florida

A Wiley-Interscience Publication
JOHN WILEY & SONS, INC.

Copyright © 2001 by John Wiley & Sons, Inc., New York. All rights reserved.

Published simultaneously in Canada.

For ordering and customer service, call 1-800-CALL-WILEY.

Library of Congress Cataloging-in-Publication Data:

Searle, S. R. (Shayle R.) 1928–
    Matrix algebra for applied economics / Shayle R. Searle, Lois Schertz Willett.
        p. cm. — (Wiley series in probability and statistics. Applied probability and statistics section)
    "A Wiley-Interscience publication."
    Includes bibliographical references and index.
    ISBN 0-471-32207-5 (cloth : alk. paper)
    1. Economics, Mathematical.   2. Economics—Econometric models.   3. Matrices.   I.
Willett, Lois Schertz.   II. Title.   III. Wiley series in probability and statistics. Applied
probability and statistics.

    HB135 .S38 2001
    330'.01'5129434—dc21   2001026759

Printed in the United States of America.

10 9 8 7 6 5 4 3 2 1

List of Chapters

I. BASICS

II. NECESSARY THEORY

III. WORKING WITH MATRICES

Contents

Preface

The great economist John Maynard Keynes, in his 1924 biography of Alfred Marshall, suggests that economics, the study of how we allocate scarce resources among competing needs, is an easy subject in which few excel! He argues that an economist must combine the talents of mathematician, historian, statesman and philosopher.

It is to the first of these talents, mathematics, that this book is directed. It is designed for students of economics and for practicing economists, to help them understand the exceedingly useful tool of matrix algebra. This tool, now commonly used for the mathematical analysis of current problems in economics, is demonstrated throughout each chapter and its exercises. What are called examples demonstrate numerically the foundation operations of matrix algebra; and illustrations demonstrate a variety of economic problems, such as transportation, price transmission, demand analysis, production, game theory, and so on. Many of the illustrations are used recurrently throughout the book to provide the reader with some strands of continuity. All of them illustrate how the matrix formulation of a problem can simplify one's understanding of that problem and lead to its solution.

The book has three parts. Part I, "Basics", is Chapters 1-5, which presents the scope and basic definitions of matrices in Chapter 1; arithmetic and simple operations of matrices constitute Chapter 2, with special matrices and their properties described in Chapter 3. The analog of division is described in Chapter 5, requiring the pre-requisite determinants of Chapter 4.

Part II, "Necessary Theory", is Chapters 6-12. It deals with concepts and operations that are needed for using matrices in real-life situations. Its seven chapters begin with linear dependence and independence, and proceed through rank, canonical forms, generalized inverses, solving equations, eigen roots and vectors, and finally, as Chapter 12, a catch-all of other useful operations.

Part III, "Working with Matrices", has five chapters, 13-17, each dealing with a topic of major interest to economists wherein matrix algebra substantively simplifies the understanding and solution of real questions in economics. These five chapters present linear equations, regression, linear models, linear programming and Markov chains — all with several examples from real-life economics.

Students in their third or fourth year of degrees in economics or applied economics, as well as those early in Master's or Ph.D. programs, will find the mathematical level of this book exactly as they need it: no pre-requisite other than high school algebra (and no calculus).

Practicing economists wanting to add more mathematics to their skills will also find the book to be at the right level for them, no matter what their applied interests are. In fact, it is those economists in applied disciplines such as agricultural, resource and consumer economics who will find the book especially helpful, since it is often these economists who struggle with the mathematical modelings of economic problems facing the world.

This book is suited to a one-semester course in matrix algebra. Students wanting a thorough understanding of matrix algebra and its useful applications in economics will want to work though all of Parts I, II and III of the book. On the other hand, courses emphasizing just basics and applications could be based on Parts I and III; and likewise for emphasizing theory only, Parts I and II are sufficient.

Since the focus of the book is on the techniques of using matrices, and on the application of matrices to economics, no particular computing software is linked to the book. Several software packages are available for matrix calculation (e.g., SHAZAM, Mathematica, and MATLAB) which, in combination with formulating economic problems in terms of matrix algebra, greatly enhances the ability of economists to solve such problems.

It is a great pleasure to thank Norma Phalen for her many successes with LATEX in preparing the manuscript for this book.

Shayle R. Searle
Lois Schertz Willett

Ithaca, NY
August 2001

Matrix Algebra
for Applied Economics

Part I

BASICS

Chapter 1

INTRODUCTION

1.1 The Scope of Matrix Algebra

Economics has become increasingly quantitative. As a result, economists find themselves confronted more and more with large amounts of numerical data, measurements of one form or another gathered from such sources as market research and government agencies, production facilities, and household surveys. But the mere collecting and recording of data achieve nothing; having been collected, data must be investigated to see what information they contain concerning the problem at hand. Any investigation so made, being quantitative in nature, will involve mathematics in some way, even if only the calculating of percentages or averages. However, one often has to submit data to more complex calculations, requiring procedures that involve a higher level of mathematical skill. In order to carry out the mathematics, economists must either learn the procedures or at least learn something of the language of mathematics, so that they can communicate satisfactorily with mathematicians whose aid is enlisted. In either case there are numerous, valuable mathematical tools essential to understand. Matrices and their algebra are the language of many of these tools.

This book deals with the application of matrix algebra to problems in economics. Specific applications range from the allocation of scarce resources such as equipment (assuming conditions of certainty) to the investigation of portfolio decisions under conditions of uncertainty. There are also examples of using matrix algebra in applications such as supply and demand elasticities, international trade, processing production costs and game theory. The reader should appreciate that new applications and theories making use of the mathematical techniques introduced in this book are developing so rapidly that we can only sample from what is a rich population of available applications.

The manipulative skill required for using matrices is neither great nor excessively demanding, whereas in terms of brevity, simplicity and clarity the value of matrix algebra is often appreciable. Furthermore, the almost universal nature of matrix expressions has great appeal. Often the same results can be applied, with only minor changes, to situations involving both small amounts of data and extremely large amounts. Matrix algebra is therefore a vehicle by which mathematical procedures for many problems, both large and small, can be described independently of the size of the problem. Size does not affect the understanding of the procedures, only the amount of calculation involved. This in turn determines time and cost, factors whose importance in today's world of computers is rapidly diminishing. The presence of computers is additional reason for enhancing the value of knowing matrix algebra. Expressing a set of calculations in terms of matrices frequently leads to easy and efficient use of computers; thus matrices and computing go hand in hand.

1.2   Using Computers for Matrix Arithmetic

Several computer software packages are available for the arithmetic of dealing with economic problems using matrix algebra. These packages, under continual development and enhancement, provide simple matrix addition and subtraction as well as some of the more complex matrix functions such as generalized inverses. The ability of a software package to deal effectively with economic analysis is related to the number of specific matrix algebra procedures programmed into the software as well as the software's constraints on the size of the data set it can manipulate. As computer hardware and software are enhanced, the time and cost of using computers for economic analysis decrease. Likewise, the feasibility of expressing economic problems in terms of matrices and solving these problems using the tools of matrix algebra with the help of computers increases. The focus of this book is on the techniques of matrix algebra and the application of matrix algebra to economics; it is not on the use of computers for these techniques. Thus the examples and illustrations we present do not include reference to commands necessary for using computer software.

1.3 A Matrix is an Array

A matrix is a rectangular array of numbers arranged in rows and columns. As such, it is a device frequently used in organizing the presentation of numerical data so that they may be handled with ease mathematically. The use of matrices can condense a wealth of mathematical manipulations into a small set of symbols. As a result, matrices are of great help in describing data analysis procedures suited to quantitative methods in economics. When mathematical analysis is needed, matrices are useful both in organizing the calculations involved and in clarifying an understanding of them.

The aspect of a matrix which makes it an aid in organizing data provides a convenient starting point for its description. We begin with an illustration.

Illustration

The Bureau of Labor Statistics periodically reports the Consumer Price Index (CPI), composed from prices of several categories of consumer goods and services, using 100 for 1982-4 as the base. Table 1.1 shows the CPI (relative to that 100 base) for each of four of the many available categories for three different years.

TABLE 1.1. Four Categories in the CPI, 1995-7,
Based on Taking 1982-4 as 100.

Category	1995	1996	1997
Housing	149	153	157
Apparel	132	132	133
Medical care	220	228	235
Recreation	96	99	100

These numbers mean that apparel, for example, which on average cost $100 in 1982-4 would cost $132 in 1995, and $133 in 1997.

The array of numbers in Table 1.1 can be extracted and written simply as

$$\begin{bmatrix} 149 & 153 & 157 \\ 132 & 132 & 133 \\ 220 & 228 & 235 \\ 96 & 99 & 100 \end{bmatrix}, \tag{1.1}$$

where the position of a number in the array determines its meaning. Each row in the array is for a category and each column is for a year.

Thus, for example, in the third row and first column the entry of 220 is the CPI for medical care in 1995. In this way the position of a number in the array identifies the CPI category and the year to which it applies.

An array of numbers of this sort is a matrix, so identified by enclosing it in square brackets as in (1.1). A matrix is always a rectangular (or square) array, with its entries set out in rows and columns. Those individual entries are called *elements* of the matrix; they can be numbers of any sort or even functions of one or more variables. With few exceptions in this book, we think of elements of a matrix as being positive or negative real numbers, or zero. Matrices with such elements are called *real matrices*: no element involves $\sqrt{-1}$.

Matrix algebra, being the algebra of matrices, is the algebra of arrays as just described, where each array is treated as an entity and denoted by a single symbol. The arithmetic implied by the algebra involves the elements in the arrays, but it is the handling of the arrays themselves, each treated as an entity in its own right, that constitutes matrix algebra.

1.4   Subscript Notation

Algebra is arithmetic with numbers represented by letters of the alphabet. This might seem to restrict a matrix to having no more than 26 elements. A less restrictive notation has each row of a matrix designated by a letter with integer subscripts representing columns. Thus, if the matrix in (1.1) is named **A**, it could be written as

$$\mathbf{A} = \begin{bmatrix} a_1 & a_2 & a_3 \\ b_1 & b_2 & b_3 \\ c_1 & c_2 & c_3 \\ d_1 & d_2 & d_3 \end{bmatrix},$$

using a capital letter in boldface to denote a matrix (a quite customary notation). The elements of the first row, for example, are then read as "*a* one", "*a* two" and "*a* three"; or as "*a* sub one", "*a* sub two" and "*a* sub three". In this instance the subscripts describe the column in which each element belongs and all elements in the same column have a common subscript; for example, the elements in the second column have the subscript 2.

In practice a single letter of the alphabet is commonly used for all elements. Two subscripts are used for each element and written alongside one another, the first specifying the row the element is in and the second the column. Thus **A** could be written as

$$\mathbf{A} = \begin{bmatrix} a_{11} & a_{12} & a_{13} \\ a_{21} & a_{22} & a_{23} \\ a_{31} & a_{32} & a_{33} \\ a_{41} & a_{42} & a_{43} \end{bmatrix}.$$

The elements are read as "a one, one", "a one, two" or as "a sub one, one", "a sub one, two," and so on. In this way an element's subscripts uniquely locate its position in the matrix.

Just as we have used the letter a to denote the value of an element of a matrix, so we can use other letters to denote other things; for example, we might refer to a row as "row i", and if i were equal to 2, "row i" would be the second row of the matrix. In the same way we can use j to reference columns, and "column j" or the "jth column" is, for $j = 3$, the third column. By this means, we refer to the element in the ith row and jth column of a matrix as a_{ij}, using i and j as subscripts to the letter a to denote the row and column the element is in; a_{12}, for example, is in the first row and second column. In general, then, a_{ij} is in the ith row and jth column, e.g., a_{ij} for $i = 2$ and $j = 3$ is a_{23}, the element in the second row and third column. No comma is placed between the two subscripts unless it is needed to avoid confusion. For example, were there twelve columns in a matrix, the element in row 1 and column 12 could be written as $a_{1,12}$ to distinguish it from $a_{11,2}$, the element in the eleventh row and second column.

This notation provides ready opportunity to detail the elements of a row, a column or even a complete matrix, in a very compact manner. Thus the first row of **A**, with elements a_{12}, a_{12} and a_{13}, can be described as a_{1j} for $j = 1, 2, 3$. This representation means that those elements are a_{1j} with j taking in turn each one of the values $1, 2$ and 3. Columns can be described in like fashion. The whole matrix is accounted for by enclosing a_{ij} in curly brackets and writing

$$\mathbf{A} = \{a_{ij}\} \quad \text{for} \quad i = 1, 2 \quad \text{and} \quad j = 1, 2, 3.$$

1.5 Summation Notation

The simplest arithmetic operation is that of adding numbers together. Subscript notation enables this operation to be presented very succinctly. Suppose we wish to add five numbers represented as

$$a_1, \ a_2, \ a_3, \ a_4 \quad \text{and} \quad a_5.$$

Their sum is

$$a_1 + a_2 + a_3 + a_4 + a_5.$$

These symbols can be expressed in words as

"the sum of all values of a_i for $i = 1, 2, \ldots, 5$".

The phrase "the sum of all values of" is customarily represented by Σ, the capital Greek letter sigma. Accordingly, the sum of the as is written as

$$\sum_{i=1}^{5} a_i = a_1 + a_2 + a_3 + a_4 + a_5.$$

The lower limit of i does not have to be 1; for example,

$$\sum_{i=3}^{7} y_i = y_3 + y_4 + y_5 + y_6 + y_7.$$

And, although subscripts in a summation are usually consecutive integers, omissions can be indicated:

$$\sum_{\substack{i=3 \\ i \neq 4}}^{7} y_i = y_3 + y_5 + y_6 + y_7.$$

This notation applies equally well to cases involving two subscripts: in summing over either subscript the other remains unchanged. For example,

$$\sum_{j=1}^{3} a_{ij} = a_{i1} + a_{i2} + a_{i3} \quad \text{and} \quad \sum_{i=1}^{2} a_{ij} = a_{1j} + a_{2j}.$$

Summing the first of these with respect to i leads to the operation of double summation:

$$\sum_{i=1}^{2} \left(\sum_{j=1}^{3} a_{ij} \right) = \sum_{i=1}^{2} (a_{i1} + a_{i2} + a_{i3}) \tag{1.2}$$

$$= (a_{11} + a_{12} + a_{13}) + (a_{21} + a_{22} + a_{23}).$$

Similarly,

$$\sum_{j=1}^{3} \left(\sum_{i=1}^{2} a_{ij} \right) = \sum_{j=1}^{3} (a_{1j} + a_{2j})$$

$$= (a_{11} + a_{21}) + (a_{12} + a_{22}) + (a_{13} + a_{23}),$$

which is seen to be the same as $\sum_{i=1}^{2}\left(\sum_{j=1}^{3} a_{ij}\right)$ in (1.2). Removing the brackets gives the important result

$$\sum_{i=1}^{2}\sum_{j=1}^{3} a_{ij} = \sum_{j=1}^{3}\sum_{i=1}^{2} a_{ij}.$$

In general, the order of summation in double summation is of no consequence:

$$\sum_{i=1}^{m}\sum_{j=1}^{n} a_{ij} = \sum_{j=1}^{n}\sum_{i=1}^{m} a_{ij}.$$

In terms of a matrix of m rows and n columns the left-hand side of the above expression is the sum of row totals and the right-hand side is the sum of column totals, both sums equaling the total of all elements.

Summation notation has so far been described in terms of simple sums, but it also encompasses sums of squares and sums of products, and indeed sums of any series of expressions that can be identified by subscript notation. Thus

$$\sum_{j=1}^{4} a_{ij}^{2} = a_{i1}^{2} + a_{i2}^{2} + a_{i3}^{2} + a_{i4}^{2},$$

$$\sum_{i=1}^{3} a_{ij}b_{ij} = a_{1j}b_{1j} + a_{2j}b_{2j} + a_{3j}b_{3j}$$

and

$$\sum_{j=1}^{3} a_{1j}b_{j1} = a_{11}b_{11} + a_{12}b_{21} + a_{13}b_{31}.$$

Some readers may question the use of j in b_{j1} in this last expression since until now we have used i as the first subscript. There is, however, nothing sacrosanct about i and j as subscripts; any letter of the alphabet may be so used. While i and j are commonly found in this role, both in this book and elsewhere, they are by no means the only letters used as subscripts; for example,

$$\sum_{i=1}^{2}\sum_{j=1}^{3} a_{ij} = \sum_{p=1}^{2}\sum_{q=1}^{3} a_{pq} = \sum_{k=1}^{2}\sum_{t=1}^{3} a_{kt}$$
$$= a_{11} + a_{12} + a_{13} + a_{21} + a_{22} + a_{23}.$$

The expression $\sum_{j=1}^{3} a_{1j}b_{j1}$ used above is an example of the more general form

$$\sum_{j=1}^{n} a_{ij}b_{jk} = a_{i1}b_{1k} + a_{i2}b_{2k} + a_{i3}b_{3k} + \cdots + a_{in}b_{nk},$$

an expression used extensively in multiplying two matrices together. It is discussed in detail in the next chapter (Section 2.6). But to anticipate for a moment we may note that if $\mathbf{A}$ and $\mathbf{B}$ are the matrices

$$\mathbf{A} = \begin{bmatrix} a_{11} & a_{12} & a_{13} \\ a_{21} & a_{22} & a_{23} \end{bmatrix} \quad \text{and} \quad \mathbf{B} = \begin{bmatrix} b_{11} & b_{12} & b_{13} \\ b_{21} & b_{22} & b_{23} \\ b_{31} & b_{32} & b_{33} \end{bmatrix},$$

then

$$\sum_{j=1}^{3} a_{1j} b_{j1} = a_{11} b_{11} + a_{12} b_{21} + a_{13} b_{31}$$

is the sum of the term-by-term products of the elements of the first row of $\mathbf{A}$ and the first column of $\mathbf{B}$. Similarly

$$\sum_{j=1}^{3} a_{ij} b_{jk} = a_{i1} b_{1k} + a_{i2} b_{2k} + a_{i3} b_{3k}$$

is the sum of the term-by-term products of the elements of the ith row of $\mathbf{A}$ and the kth column of $\mathbf{B}$.

Two other attributes of summation operations may be noted. One involves summation of terms that do not have subscripts; for example,

$$\sum_{i=1}^{4} x = x + x + x + x = 4x.$$

The other involves summation of several terms each of which is multiplied by a common constant; for example,

$$\sum_{i=1}^{3} k y_i = k y_1 + k y_2 + k y_3 = k \left(\sum_{i=1}^{3} y_i \right).$$

The generalizations of these results are

$$\sum_{i=1}^{n} x = nx \quad \text{and} \quad \sum_{i=1}^{n} k y_i = k \left(\sum_{i=1}^{n} y_i \right).$$

Thus, summation from 1 to n of a constant is n times the constant; and a constant multiplier can be factored outside a summation sign.

There is also a multiplication procedure analogous to Σ. It is denoted by the Greek capital pi, Π, and involves multiplying together all the terms to which it applies. Thus, whereas

$$\sum_{i=1}^{5} b_i = b_1 + b_2 + b_3 + b_4 + b_5,$$

$$\prod_{i=1}^{5} b_i = b_1 b_2 b_3 b_4 b_5,$$

the product of all the bs. Operationally, Π is equivalent to Σ except that it denotes multiplication instead of addition.

1.6 Dot Notation

A further abbreviation often used is

$$\sum_{i=1}^{m} a_{ij} = a_{.j}$$

where the dot subscript in place of i denotes that summation has taken place over the i subscript. Since the notation $a_{.j}$ shows no indication of the limits of i over which summation has occurred, it is used only when these limits are clear from the context of its use. In line with $a_{.j}$ we also have

$$a_{i.} = \sum_{j=1}^{n} a_{ij}$$

and

$$a_{..} = \sum_{i=1}^{m} a_{i.} = \sum_{j=1}^{n} a_{.j} = \sum_{i=1}^{m} \sum_{j=1}^{n} a_{ij}.$$

1.7 Definition of a Matrix

A matrix is a rectangular (or square) array of numbers arranged in rows and columns. All rows are of equal length, as are all columns. In terms of the notation of Section 1.4 we will let a_{ij} denote the element in the ith row and jth column of a matrix $\mathbf{A}$. If $\mathbf{A}$ has r rows and c columns it is written

$$\mathbf{A} = \begin{bmatrix} a_{11} & a_{12} & a_{13} & \cdots & a_{1j} & \cdots & a_{1c} \\ a_{21} & a_{22} & a_{23} & \cdots & a_{2j} & \cdots & a_{2c} \\ \vdots & & & & \vdots & & \vdots \\ a_{i1} & a_{i2} & a_{i3} & \cdots & a_{ij} & \cdots & a_{ic} \\ \vdots & & & & \vdots & & \vdots \\ a_{r1} & a_{r2} & a_{r3} & \cdots & a_{rj} & \cdots & a_{rc} \end{bmatrix}.$$

The three dots indicate, in the first row for example, that the elements a_{11}, a_{12} and a_{13} continue in sequence up to a_{1j} and on up to a_{1c}; likewise, in the first column the elements a_{11}, a_{21}, continue in sequence up to a_{r1}. The use of three dots to represent omitted values of a long sequence in this manner is standard and used extensively. This form of writing a matrix clearly specifies its terms, and also its size, namely the number of rows and columns. An alternative and briefer form is

$$\mathbf{A} = \{a_{ij}\} \quad \text{for} \quad i = 1, 2, \ldots, r, \quad \text{and} \quad j = 1, 2, \ldots, c,$$

the curly brackets indicating that a_{ij} is a typical element, the limits of i and j being r and c, respectively.

We call a_{ij} the (i, j)th *element*, the first subscript referring to the row and the second to the column in which it is located. Thus a_{23} is the element in the second row and third column. The size of the matrix, i.e., the number of rows and columns, is referred to as its *order* (or sometimes as its *dimension*). Thus $\mathbf{A}$ with r rows and c columns has order $r \times c$ (read as "r by c") and, to emphasize its dimensions, the matrix can be written $\mathbf{A}_{r \times c}$. In connection with order we often describe a matrix as small or large, meaning the magnitude of its order. For example, $\mathbf{A}_{2 \times 3}$ would be a small matrix relative to $\mathbf{B}_{60 \times 100}$, which might be called large. The first term in the first row of a matrix, a_{11} in this case, is called the *leading element* of the matrix. An example of a 2×3 matrix (a matrix of order 2×3) is

$$\mathbf{A} = \begin{bmatrix} 4 & 0 & -3 \\ -7 & 2.73 & 1 \end{bmatrix}.$$

Notice that zero is legitimate as an element, that the elements need not all have the same sign, and that integers and decimal numbers can both be elements of the same matrix.

1.8   Some Basic Special Forms

a.   Square matrices

Many matrices have names which originate from their appearance. The simplest is the *square matrix*, which has the same number of rows as it does columns, r, say. Its elements a_{11}, a_{22}, $a_{33}, \ldots, a_{rr}$ are referred to as the *diagonal elements* of the matrix, and the sum of them is the *trace* of the matrix; that is, when $\mathbf{A}$ is square the trace of $\mathbf{A}$ equals $\sum_{i=1}^{r} a_{ii}$. The trace of a *rectangular* (i.e., not square) matrix is not defined.

b. Diagonal matrices

When all the non-diagonal elements of a square matrix are zero, the matrix is described as a *diagonal matrix*; for example,

$$\mathbf{A} = \begin{bmatrix} 3 & 0 & 0 \\ 0 & -17 & 0 \\ 0 & 0 & 0 \end{bmatrix}$$

is a diagonal matrix. In many cases, but not always, a diagonal matrix has every diagonal element nonzero.

c. Identity matrices

A particularly important form of a diagonal matrix is that having all diagonal elements as unity: e.g.,

$$\mathbf{I} = \begin{bmatrix} 1 & 0 & 0 & 0 \\ 0 & 1 & 0 & 0 \\ 0 & 0 & 1 & 0 \\ 0 & 0 & 0 & 1 \end{bmatrix}.$$

The reason for the name identity becomes apparent when the multiplication of matrices is considered (in Section 2.6c).

d. Triangular matrices

If a square matrix has zero for all elements above (below) its diagonal it is called a lower (upper) *triangular* matrix. Examples are, respectively,

$$\mathbf{L} = \begin{bmatrix} 2 & 0 & 0 \\ 7 & 1 & 0 \\ 4 & 8 & 6 \end{bmatrix} \quad \text{and} \quad \mathbf{U} = \begin{bmatrix} 9 & 5 & 3 \\ 0 & 6 & 2 \\ 0 & 0 & 8 \end{bmatrix}.$$

e. Null matrices

Every matrix, whatever its order, which has zero for every element is called a null matrix. Examples are

$$\begin{bmatrix} 0 & 0 & 0 \\ 0 & 0 & 0 \end{bmatrix}, \quad \begin{bmatrix} 0 & 0 & 0 \\ 0 & 0 & 0 \\ 0 & 0 & 0 \\ 0 & 0 & 0 \end{bmatrix} \quad \text{and} \quad \begin{bmatrix} 0 & 0 \\ 0 & 0 \end{bmatrix}.$$

Thus, in contrast to scalars, which have only one zero, matrices have an infinite number of zeros. They are denoted by $\mathbf{0}$, with order determined by context, or shown as a subscript; e.g., $\mathbf{0}_{2\times 3}$ for the first of the preceding three matrices.

f. Equal matrices

Two matrices are equal when they have the same order and the same elements, element by element; i.e., $\mathbf{A}$ and $\mathbf{B}$ are equal only if $a_{ij} = b_{ij}$ for every i, j pair. Thus for

$$\mathbf{A} = \begin{bmatrix} 2 & 3 & 1 \\ 4 & 7 & 0 \end{bmatrix}, \quad \mathbf{B} = \begin{bmatrix} 2 & 3 & 1 \\ 4 & 7 & 1 \end{bmatrix}, \quad \mathbf{C} = \begin{bmatrix} 2 & 3 & 1 & 5 \\ 4 & 7 & 1 & 6 \end{bmatrix}$$

and

$$\mathbf{D} = \begin{bmatrix} 2 & 3 & 1 \\ 4 & 7 & 0 \end{bmatrix},$$

$\mathbf{A}$ and $\mathbf{D}$ are equal but no other two are.

g. Vectors

A matrix which is just a single column is called a *column vector* (and sometimes just a *vector*); e.g.,

$$\mathbf{x} = \begin{bmatrix} 7 \\ 8 \\ 0 \\ 4 \end{bmatrix}$$

is a column vector of order 4, i.e., a matrix of order 4×1. Similarly a row vector is a matrix which is just a single row; for example,

$$\mathbf{y}' = \begin{bmatrix} 2 & -6 & 9 \end{bmatrix}$$

is a row vector of order 3, a matrix of order 1×3.

h. Scalars

Single numbers, whether integers, functions, fractions, decimals, positive or negative, are called *scalars*. Occasionally it is convenient to think of a scalar as a 1×1 matrix.

1.9 Description by Elements

An abbreviation of the description of a $p \times q$ matrix as

$$\{a_{ij}\} \quad \text{for} \quad i = 1,\ldots,p \quad \text{and} \quad j = 1,\ldots,q$$

is each of the following:

$$\left\{{}_m a_{ij}\right\}_{i=1\ j=1}^{p\ \ q} = \left\{{}_m a_{ij}\right\}_{i,j} = \left\{{}_m a_{ij}\right\}. \qquad (1.3)$$

Although the last two expressions have no indication of the $p \times q$ order, they can be used when order is clear from context.

In each case in (1.3) the pre-script m indicates that the elements a_{ij} are to be arrayed as a matrix. Similar uses of r and c indicate a row and column vector, respectively:

$$\left\{{}_r t_j\right\}_{j=1}^{4} = [t_1 \ t_2 \ t_3 \ t_4] \quad \text{and} \quad \left\{{}_c v_i\right\}_{i=1}^{3} = \begin{bmatrix} v_1 \\ v_2 \\ v_3 \end{bmatrix}.$$

Also, d can denote a diagonal matrix:

$$\left\{{}_d w_s\right\}_{s=1}^{3} = \begin{bmatrix} w_1 & 0 & 0 \\ 0 & w_2 & 0 \\ 0 & 0 & w_3 \end{bmatrix}.$$

This notation can be useful when wanting to describe the consequences of matrix algebra in terms of the elements of the matrices involved.

1.10 Notation

Matrices in this book are denoted by upper case letters and their elements by the lower case counterparts with appropriate subscripts. Vectors are denoted by lower case letters, usually from the end of the alphabet, using the prime superscript to distinguish a row vector from a column vector. Thus $\mathbf{x}$ is a column vector and $\mathbf{x}'$ is a row vector ($\mathbf{x}^\top$ is an alternative used by some authors). Square brackets are used for displaying a matrix, as has already been done in this chapter; other forms of brackets are to be found in the literature. Single vertical lines are seldom used since they are usually reserved for determinants (see Chapter 4). These notation conventions are widely accepted but are far from universal. Likewise, although many books have matrices and vectors printed in boldface type as is done here, not all of them do.

1.11 Examples and Illustrations

Matrix algebra books for mathematicians deal primarily with the operations of matrices and mostly have few numerical examples or illustrations of using matrices for solving real-life problems. The approach in this book is to have simple examples demonstrating algebra, and to have illustrations (and a number of exercises, the solutions to) which show how matrix algebra can be used to solve problems in economics. Indeed, we have already used the illustration of the Consumer Price Index being represented as a matrix. And we have also used simple numerical examples to clarify the form of certain special kinds of (square) matrices.

These uses of examples and illustrations continue throughout the book. Most of our examples involve trite numbers designed to be helpful for understanding matrix operations. Illustrations come from real-life problems in economics involving such topics as transportation, price transmission, demand analysis, time series, game theory, and several others. They are all considered in subsequent chapters, to illustrate how the matrix formulation of a problem can often simplify one's understanding of that problem and straightforwardly lead to its solution. Moreover, many of the illustrations reoccur throughout the book, so providing the reader with a helpful strand of continuity.

1.12 Exercises

E 1.1. For

$$
\begin{array}{llll}
a_{11} = 17 & a_{12} = 31 & a_{13} = 26 & a_{14} = 11 \\
a_{21} = 19 & a_{22} = 27 & a_{23} = 16 & a_{24} = 14 \\
a_{31} = 21 & a_{32} = 23 & a_{33} = 15 & a_{34} = 16
\end{array}
$$

show that

(a) $a_{1.} = 85$, $a_{2.} = 76$ and $a_{3.} = 75$,

(b) $\left\{ {}_{r} a_{.j} \right\}_{j=1}^{4} = [57 \quad 81 \quad 57 \quad 41]$, (c) $a_{..} = 236$,

(d) $\sum_{\substack{i=1 \\ i \neq 2}}^{3} \sum_{\substack{j=1 \\ j \neq 3}}^{4} a_{ij} = 119$,

(e) $\left\{ {}_{m} a_{ij} \right\}_{i,j=1}^{3}$ has trace 59, and

(f) $\left\{ {}_{m} a_{ij} \right\}_{i=1\ j=2}^{3\quad 4}$ has trace 63.

E 1.2. Show that for

$$A = \begin{bmatrix} -1 & 17 & 9 & -2 & 3 \\ 3 & 13 & 10 & 2 & 6 \\ 11 & -9 & 0 & -3 & 2 \\ -6 & -8 & 1 & 4 & 5 \end{bmatrix} = \{a_{ij}\},$$

(a) $\displaystyle\sum_{i=3}^{4} a_{i\cdot}^2 = 17,$

(b) $\displaystyle\prod_{j=1}^{3} a_{4j} = \prod_{i=1}^{4} a_{i4},$

(c) $\displaystyle\sum_{i=1}^{4} \sum_{j=1}^{5} (-1)^j a_{ij} = -29,$

(d) $\displaystyle\prod_{j=1}^{5} a_{2j} a_{3j} = 0,$

(e) $\displaystyle\prod_{j=1}^{5} 2^{(-1)^j a_{2j}} = .0625.$

E 1.3. Which of the following matrices are diagonal matrices, upper triangular matrices, lower triangular matrices?

(a) $\begin{bmatrix} 10 & 0 & 4 \\ 0 & -7 & 0 \\ 0 & 0 & 6 \end{bmatrix}$

(b) $\begin{bmatrix} 1 & 0 & 0 \\ 0 & 1 & 0 \\ 0 & 0 & 1 \end{bmatrix}$

(c) $\begin{bmatrix} 7 & 0 & 0 \\ 0 & .38 & 0 \\ 0 & 0 & 0 \end{bmatrix}$

(d) $\begin{bmatrix} 6 & 0 & 0 \\ 0 & \sqrt{3} & 0 \\ 0 & 0 & -\sqrt{2} \end{bmatrix}$

(e) $\begin{bmatrix} 1 & 0 & 0 \\ 4 & 2 & 0 \\ 6 & 5 & 3 \end{bmatrix}$

(f) $\begin{bmatrix} 1 & 0 & 0 \\ 0 & 1 & 0 \\ 0 & 0 & 0 \end{bmatrix}$

E 1.4. Show that

(a) $\displaystyle\sum_{i=3}^{5} 3^i = 351,$ 　　　　　　　　　(b) $\displaystyle\sum_{k=2}^{7} 2^k = 252,$

(c) $\displaystyle\sum_{r=1}^{5} r = 15,$ 　　　　　　　　　(d) $\displaystyle\sum_{\substack{s=1 \\ s\neq 2}}^{6} s(s+1) = 106,$

(e) $\displaystyle\prod_{i=1}^{4} 2^i = 1024,$ 　　　　　　　　(f) $\displaystyle\prod_{i=1}^{3}\sum_{j=1}^{4}(i+j) = 5544.$

E 1.5. For **A** of Exercise 2 write down

$$\mathbf{B} = \Big\{ _{m}\ a_{i+1,j+2} \Big\}_{i=1\ j=1}^{3\quad 3}, \qquad \mathbf{E} = \Big\{ _{d}\ a_{3j} \Big\}_{j=2}^{5},$$

$$\mathbf{C} = \Big\{ _{m}\ a_{2i,2j-1} \Big\}_{i=1\ j=1}^{2\quad 3}, \qquad \mathbf{F} = \Big\{ _{m}\ a_{ij} + i - j \Big\}_{i=2\ j=1}^{3\quad 4}.$$

E 1.6. Write the matrices

$$\Big\{ _{m}\ i+j \Big\}_{i=1\ j=2}^{2\quad 4}, \qquad \Big\{ _{m}\ k^{t-1} \Big\}_{k=1\ t=1}^{4\quad 3}$$

and

$$\Big\{ _{m}\ 3p + 2(s-1) \Big\}_{p=1\ s=1}^{4\quad 5}.$$

E 1.7. Show that $\displaystyle\sum_{i=1}^{2} a_i b_i \neq \left(\sum_{i=1}^{2} a_i\right)\left(\sum_{i=1}^{2} b_i\right).$

E 1.8. Show that the following identities hold by computing both sides of each expression for the numerical example in Exercise E 1.1. Note that the square of $a_i.$ is $a_{i.}^2 = \left(\displaystyle\sum_{j=1}^{n} a_{ij}\right)^2.$

(a) $\displaystyle\sum_{i=1}^{m}\sum_{j=1}^{n} a_{ij}^2 = \sum_{j=1}^{n}\sum_{i=1}^{m} a_{ij}^2$

(b) $\displaystyle\sum_{i=1}^{m}\left(\sum_{j=1}^{n}a_{ij}\right)^{2}=\sum_{i=1}^{m}a_{i.}^{2}.$

(c) $\displaystyle\sum_{j=1}^{n}\sum_{k=1}^{n}a_{ij}a_{hk}=a_{i.}a_{h.}$ for $i\neq h$

(d) $\displaystyle a_{..}^{2}=\sum_{i=1}^{m}a_{i.}^{2}+2\sum_{i=1}^{m-1}\sum_{h=i+1}^{m}a_{i.}a_{h.}$

$\displaystyle\qquad=\sum_{i=1}^{m}a_{i.}^{2}+2\sum_{i=1}^{m-1}\sum_{h>i}^{m}a_{i.}a_{h.}$

$\displaystyle\qquad=\sum_{i=1}^{m}a_{i.}^{2}+\sum_{i=1}^{m}\sum_{\substack{h=1\\h\neq i}}^{m}a_{i.}a_{h.}$

(e) $\displaystyle\sum_{\substack{i=1\\i\neq p}}^{m}\sum_{\substack{j=1\\j\neq q}}^{n}a_{ij}=a_{..}-a_{p.}-a_{.q}+a_{pq}$ for $p=2$ and $j=3$

(f) $\displaystyle\sum_{i=1}^{m}\sum_{j=1}^{n}(a_{ij}-1)=a_{..}-mn$

(g) $\displaystyle\sum_{j=1}^{n}\sum_{\substack{k=1\\k\neq j}}^{n}a_{ij}a_{ik}=a_{i.}^{2}-\sum_{j=1}^{n}a_{ij}^{2}$ for $i=3$

(h) $\displaystyle\left(\sum_{j=1}^{n}a_{ij}\right)^{2}=\sum_{j=1}^{n}a_{ij}^{2}+2\sum_{j=1}^{n-1}\sum_{p=j+1}^{n}a_{ij}a_{ip}$

$\displaystyle\qquad=\sum_{j=1}^{n}a_{ij}^{2}+2\sum_{j=1}^{n-1}\sum_{p>j}^{n}a_{ij}a_{ip}$

$\displaystyle\qquad=\sum_{j=1}^{n}a_{ij}^{2}+\sum_{j=1}^{n}\sum_{\substack{p=1\\p\neq j}}^{n}a_{ij}a_{ip}$

(i) $\displaystyle\sum_{i=1}^{m}\sum_{j=1}^{n}4a_{ij}=4a_{..}$

E 1.9. Prove the identities of Exercise 1.8.

E 1.10. (Profit matrix) A ticket seller for an out-of-town show wants
to decide how many tickets he should buy. He can place only
one order. Each ticket costs $5 and can be sold for $8; left-
over tickets have no value. The number of tickets he is able to
sell is known to be between one and four. Prepare a matrix
of profits associated with his different decisions and the possible outcomes, letting rows represent decisions and columns
represent outcomes.

E 1.11. (Agricultural input matrix) A midwest farmer who has 250
acres of corn and 125 acres of soybeans is interested in comparing the labor and capital inputs for each crop, as well as
the total inputs required for production. Each acre of corn requires three units of labor and four units of capital. Soybeans
require two units of labor per acre and five units of capital
per acre. Prepare a matrix of inputs summarizing the input
requirements. Let rows represent inputs and columns represent the crop. Identify the total amount of labor and capital
needed by the farmer.

Chapter 2

BASIC MATRIX OPERATIONS

This chapter presents two simple operations on matrices which arise from their being rectangular arrays of numbers; it also describes the arithmetic of adding, subtracting, and multiplying matrices. The concept of division is considerably more difficult for matrices than scalars and so is given a chapter to itself, Chapter 5.

2.1 Transposing a Matrix

The 4×3 matrix in (1.1) is of the Consumer Price Index for four categories in three successive years, each row being for a category and each column for a year. Let us refer to that matrix as $\mathbf{A}$. Choosing its rows to represent categories and its columns to represent years is quite arbitrary, and they could well be interchanged without affecting the meaning of individual elements. After making the interchange, rows would be years and columns would be categories. This would give a matrix which shall be denoted as $\mathbf{A}'$ ("$\mathbf{A}$ prime" or "$\mathbf{A}$ dash"). Thus from the matrix of (1.1)

$$\mathbf{A} = \begin{bmatrix} 149 & 153 & 157 \\ 132 & 132 & 133 \\ 220 & 228 & 235 \\ 96 & 99 & 100 \end{bmatrix}, \quad \text{we get} \quad \mathbf{A}' = \begin{bmatrix} 149 & 132 & 220 & 96 \\ 153 & 132 & 228 & 99 \\ 157 & 133 & 235 & 100 \end{bmatrix}.$$

(2.1)

The matrix $\mathbf{A}'$ is the *transpose* of $\mathbf{A}$. It has the same elements as $\mathbf{A}$, but rows (columns) of $\mathbf{A}$ are columns (rows) of $\mathbf{A}'$. An obvious consequence of this is that with $\mathbf{A}$ being 4×3, the order of $\mathbf{A}'$ is 3×4. In general, the transpose of $\mathbf{A}$ of order $r \times c$ is $\mathbf{A}'$ of order $c \times r$. Transposing also affects the location of elements. With a_{ij} being the element in the ith row and jth column of $\mathbf{A}$ it becomes the element in the jth row and ith column of $\mathbf{A}'$. For example, on defining a'_{st} as the element in the sth row and tth column of $\mathbf{A}'$, we see in (2.1) that 220 is a_{31} in $\mathbf{A}$ but a'_{13} in $\mathbf{A}'$. In general, we thus have a_{ij} of $\mathbf{A}$ being a'_{ji} of $\mathbf{A}'$; i.e.,

$$\mathbf{A} = \left\{_m\ a_{ij}\right\}_{i=1\ j=1}^{r\quad c} \quad \text{and} \quad \mathbf{A}' = \left\{_m\ a'_{ji}\right\}_{j=1\ i=1}^{c\quad r}$$

with

$$a'_{ji} = a_{ij}. \tag{2.2}$$

Transposing can incur a minor notational difficulty. $\mathbf{A}_{r \times c}$ after transposing is $\mathbf{A}'$ of order $c \times r$; for this the most careful notation involving its order as a subscript is either $(\mathbf{A}_{r \times c})'$ or $(\mathbf{A}')_{c \times r}$. Fortunately, this clumsy notation is seldom needed: the simple $\mathbf{A}'$ usually suffices. (Some writers use $\mathbf{A}^\top$ where we use $\mathbf{A}'$.)

Four important consequences of transposing merit attention.

a.   A reflexive operation

This is simple: the transpose of the transpose of a matrix is the matrix itself; i.e., $(\mathbf{A}')' = \mathbf{A}$.

b.   Vectors

Since a column vector of order p is a $p \times 1$ matrix, its transpose is $1 \times p$, a row vector; e.g., with

$$\mathbf{x} = \begin{bmatrix} 7 \\ 6 \\ 5 \end{bmatrix} \qquad \mathbf{x}' = [7\ \ 6\ \ 5].$$

This is why the row vector in Section 1.8g is $\mathbf{y}'$ and not $\mathbf{y}$.

c.   Symmetric matrices

The transpose of a square matrix has the same order as the matrix itself — because there are as many rows as columns. But even more than this, if in a square matrix $\mathbf{A} = \{a_{ij}\}$ every element a_{ij} is the same as a_{ji}, that is, $a_{ij} = a_{ji}$, then $\mathbf{A}$ is said to be a *symmetric matrix*; and $\mathbf{A} = \mathbf{A}'$. For example,

$$\mathbf{A} = \begin{bmatrix} 1 & 2 & 7 \\ 2 & 4 & 9 \\ 7 & 9 & 3 \end{bmatrix} = \mathbf{A}'$$

is symmetric. The nature of the symmetry is that elements on either side of the diagonal are "reflected" through to the other side; e.g., the $a_{13} = 7$ in the preceding $\mathbf{A}$ is also $a_{31} = 7$. Thus the definition of a square matrix being symmetric is $\mathbf{A} = \mathbf{A}'$; and, for all i and j,

$$\text{if } a_{ij} = a_{ji} \text{ then } \mathbf{A} = \mathbf{A}'; \quad \text{and if } \mathbf{A} = \mathbf{A}' \text{ then } a_{ij} = a_{ji}. \quad (2.3)$$

Another description of a symmetric matrix (which is square) is that the elements in the ith row are the same as (and in the same sequence as) the elements in the ith column. And for a symmetric matrix of order $n \times n$ this is true for every $i = 1, 2, \ldots, n$. Matrices which are not symmetric can be called *asymmetric*.

Symmetric matrices are a very important class of matrices. They have many important properties (as shall become evident throughout this book), and they occur in many cases of using matrix algebra to solve real-life problems. A simple example is the correlation matrix such as the following.

Illustration (Citrus fruit)

From ten years of citrus fruit production throughout the United States, the correlation of the U.S.A. annual yield of oranges with that of lemons is 0.05; of oranges with tangelos, -0.15, and of lemons with tangelos, 0.48. Since, for example, the correlation of lemons with oranges is the same as that of oranges with lemons, namely 0.05; and because the correlation of each fruit with itself is 1.0, we can summarize all this information in a matrix:

$$\begin{bmatrix} 1 & 0.05 & -0.15 \\ 0.05 & 1 & 0.48 \\ -0.15 & 0.48 & 1 \end{bmatrix}.$$

It is an example of a correlation matrix, a matrix which is always symmetric, with diagonal elements of 1, and off-diagonal elements between -1 and +1.

d.   Notation: iff and NSC

The two statements in (2.3) are often combined in mathematical writing into a single statement:

$$\text{if and only if (iff) } a_{ij} = a_{ji} \text{ then } \mathbf{A} = \mathbf{A}',$$

often abbreviated to

$$\text{iff } a_{ij} = a_{ji} \text{ then } \mathbf{A} = \mathbf{A}'.$$

An alternative to this is

$$\text{a necessary and sufficient condition (NSC) for } a_{ij} = a_{ji} \text{ is } \mathbf{A} = \mathbf{A}'.$$

This in turn gets abbreviated to

$$\text{a NSC for } a_{ij} = a_{ji} \text{ is } \mathbf{A} = \mathbf{A}'.$$

2.2   Partitioned Matrices

A matrix consists of rows and columns of numbers. As such it can be thought of as rows and columns of matrices of smaller order than its own.

a.   An example

Suppose we draw a dashed line between columns 4 and 5 and between rows 3 and 4 of

$$\mathbf{A} = \begin{bmatrix} 1 & 6 & 8 & 9 & 3 & 8 \\ 2 & 4 & 1 & 6 & 1 & 1 \\ 4 & 3 & 6 & 1 & 2 & 1 \\ 9 & 1 & 4 & 6 & 8 & 7 \\ 6 & 8 & 1 & 4 & 3 & 2 \end{bmatrix}$$

to get

$$\mathbf{B} = \left[\begin{array}{cccc:cc} 1 & 6 & 8 & 9 & 3 & 8 \\ 2 & 4 & 1 & 6 & 1 & 1 \\ 4 & 3 & 6 & 1 & 2 & 1 \\ \hdashline 9 & 1 & 4 & 6 & 8 & 7 \\ 6 & 8 & 1 & 4 & 3 & 2 \end{array} \right]. \tag{2.4}$$

Each array of numbers in the four sections of $\mathbf{B}$ engendered by the dashed lines is a matrix:

$$\mathbf{B}_{11} = \begin{bmatrix} 1 & 6 & 8 & 9 \\ 2 & 4 & 1 & 6 \\ 4 & 3 & 6 & 1 \end{bmatrix}, \qquad \mathbf{B}_{12} = \begin{bmatrix} 3 & 8 \\ 1 & 1 \\ 2 & 1 \end{bmatrix},$$

$$\mathbf{B}_{21} = \begin{bmatrix} 9 & 1 & 4 & 6 \\ 6 & 8 & 1 & 4 \end{bmatrix}, \qquad \mathbf{B}_{22} = \begin{bmatrix} 8 & 7 \\ 3 & 2 \end{bmatrix}. \qquad (2.5)$$

Using these four matrices, $\mathbf{B}$ can be written as a matrix of matrices:

$$\mathbf{B} = \begin{bmatrix} \mathbf{B}_{11} & \mathbf{B}_{12} \\ \mathbf{B}_{21} & \mathbf{B}_{22} \end{bmatrix}. \qquad (2.6)$$

This specification of $\mathbf{B}$ is called a *partitioning*; the subscripted matrices in (2.6) are *sub-matrices* of $\mathbf{B}$, and $\mathbf{B}$ of (2.6) is a *partitioned* matrix.

Note that $\mathbf{B}_{11}$ and $\mathbf{B}_{21}$ have the same number of rows as do $\mathbf{B}_{12}$ and $\mathbf{B}_{22}$, respectively. Likewise, $\mathbf{B}_{11}$ and $\mathbf{B}_{12}$ have the same number of columns as do $\mathbf{B}_{21}$ and $\mathbf{B}_{22}$, respectively. This is the usual method of partitioning, as expressed in the general case for an $r \times c$ matrix:

$$\mathbf{A}_{r \times c} = \begin{bmatrix} \mathbf{K}_{p \times q} & \mathbf{L}_{p \times (c-q)} \\ \mathbf{M}_{(r-p) \times q} & \mathbf{N}_{(r-p) \times (c-q)} \end{bmatrix}$$

where $\mathbf{K}$, $\mathbf{L}$, $\mathbf{M}$ and $\mathbf{N}$ are the sub-matrices with their orders shown as subscripts.

Partitioning is not restricted to dividing a matrix into just four sub-matrices; it can be divided into numerous rows and columns of matrices. Thus if

$$\mathbf{B}_{11,1} = \begin{bmatrix} 1 & 6 & 8 & 9 \\ 2 & 4 & 1 & 6 \end{bmatrix}, \quad \mathbf{B}_{12,1} = \begin{bmatrix} 3 & 8 \\ 1 & 1 \end{bmatrix},$$

$$\mathbf{B}_{11,2} = \begin{bmatrix} 4 & 3 & 6 & 1 \end{bmatrix} \quad \text{and} \quad \mathbf{B}_{12,2} = \begin{bmatrix} 2 & 1 \end{bmatrix}$$

with $\mathbf{B}_{21}$ and $\mathbf{B}_{22}$ unchanged, then $\mathbf{B}$ has the partitioned form

$$\mathbf{B} = \begin{bmatrix} 1 & 6 & 8 & 9 & 3 & 8 \\ 2 & 4 & 1 & 6 & 1 & 1 \\ 4 & 3 & 6 & 1 & 2 & 1 \\ 9 & 1 & 4 & 6 & 8 & 7 \\ 6 & 8 & 1 & 4 & 3 & 2 \end{bmatrix} = \begin{bmatrix} \mathbf{B}_{11,1} & \mathbf{B}_{12,1} \\ \mathbf{B}_{11,2} & \mathbf{B}_{12,2} \\ \mathbf{B}_{21} & \mathbf{B}_{22} \end{bmatrix}.$$

It goes without saying that each such line must always go the full length (or breadth) of the matrix. Partitioning in any staggered manner such as

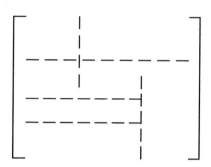

is not allowed.

b. General specification

In general, a matrix $\mathbf{A}$ of order $p \times q$ can be partitioned into r rows and c columns of sub-matrices as

$$
\mathbf{A} = \begin{bmatrix}
\mathbf{A}_{11} & \mathbf{A}_{12} & \cdots & \mathbf{A}_{1c} \\
\mathbf{A}_{21} & \mathbf{A}_{22} & \cdots & \mathbf{A}_{2c} \\
\vdots & \vdots & & \vdots \\
\mathbf{A}_{r1} & \mathbf{A}_{r2} & \cdots & \mathbf{A}_{rc}
\end{bmatrix},
$$

where $\mathbf{A}_{ij}$ is the sub-matrix in the ith row and jth column of sub-matrices. If the ith row of sub-matrices has p_i rows of elements and the jth column of sub-matrices has q_j columns, then $\mathbf{A}_{ij}$ has order $p_i \times q_j$, where

$$
\sum_{i=1}^{r} p_i = p \quad \text{and} \quad \sum_{j=1}^{c} q_j = q.
$$

c. Transposing a partitioned matrix

The transpose of a partitioned matrix is the transposed matrix of transposed sub-matrices:

$$
[\mathbf{X} \ \ \mathbf{Y}]' = \begin{bmatrix} \mathbf{X}' \\ \mathbf{Y}' \end{bmatrix} \quad \text{and} \quad \begin{bmatrix} \mathbf{A} & \mathbf{B} & \mathbf{C} \\ \mathbf{D} & \mathbf{E} & \mathbf{F} \end{bmatrix}' = \begin{bmatrix} \mathbf{A}' & \mathbf{D}' \\ \mathbf{B}' & \mathbf{E}' \\ \mathbf{C}' & \mathbf{F}' \end{bmatrix}.
$$

The reader should note that in general the transpose of $[\mathbf{X} \quad \mathbf{Y}]$ is neither $\begin{bmatrix} \mathbf{X} \\ \mathbf{Y} \end{bmatrix}$ nor $[\mathbf{X}' \quad \mathbf{Y}']$.

Example

$$\begin{bmatrix} 2 & 8 & 9 \\ 3 & 7 & 4 \end{bmatrix}' = \begin{bmatrix} 2 & 3 \\ 8 & 7 \\ 9 & 4 \end{bmatrix}.$$

d. Partitioning into vectors

Suppose $\mathbf{a}_j$ is the jth column of $\mathbf{A}_{r \times c}$. Then

$$\mathbf{A} = [\mathbf{a}_1 \quad \mathbf{a}_2 \cdots \mathbf{a}_j \cdots \mathbf{a}_c] \tag{2.7}$$

is $\mathbf{A}$ partitioned into its c columns. Similarly,

$$\mathbf{A} = \begin{bmatrix} \boldsymbol{\alpha}_1' \\ \boldsymbol{\alpha}_2' \\ \vdots \\ \boldsymbol{\alpha}_i' \\ \vdots \\ \boldsymbol{\alpha}_r' \end{bmatrix} \tag{2.8}$$

is $\mathbf{A}$ partitioned into its r rows $\boldsymbol{\alpha}_i'$ for $i = 1, \ldots, r$.

2.3 Trace of a Square Matrix

The sum of the diagonal elements of a square matrix is called the *trace* of the matrix, written $\text{tr}(\mathbf{A})$; i.e., for $\mathbf{A} = \{a_{ij}\}$ for $i, j = 1, \ldots, n$

$$\text{tr}(\mathbf{A}) = a_{11} + a_{22} + \cdots + a_{nn} = \sum_{i=1}^{n} a_{ii}.$$

For example,

$$\text{tr} \begin{bmatrix} 1 & 7 & 6 \\ 8 & 3 & 9 \\ 4 & -2 & -8 \end{bmatrix} = 1 + 3 - 8 = -4.$$

When $\mathbf{A}$ is not square, the trace is not defined; i.e., it does not exist.

The trace of a transposed matrix is the same as the trace of the matrix itself:

$$\mathrm{tr}(\mathbf{A}') = \mathrm{tr}(\mathbf{A}).$$

This equality is exemplified by

$$\mathrm{tr} \begin{bmatrix} 1 & 8 & 4 \\ 7 & 3 & -2 \\ 6 & 9 & -8 \end{bmatrix} = \mathrm{tr} \begin{bmatrix} 1 & 7 & 6 \\ 8 & 3 & 9 \\ 4 & -2 & -8 \end{bmatrix} = 1 + 3 - 8 = -4.$$

Also, by treating a scalar as a 1×1 matrix we have

$$\mathrm{tr}(\text{scalar}) = \text{scalar}; \ \text{e.g.}, \ \mathrm{tr}(13) = 13.$$

2.4 Matrix Addition

We introduce addition of matrices by means of a simple illustration.

Illustration (Sales)

Suppose a computer software company summarizes its annual sales by product (basic, enhanced, and complete) and by client type (professional and student). Table 2.1 shows what the sales might be both domestic and overseas.

Table 2.1.   Product Sales (in 1000s),
Domestic and Overseas

Product	Domestic		Overseas	
	Professional	Student	Professional	Student
Basic	22	45	17	49
Enhanced	30	39	38	32
Complete	40	8	48	6

Let us write these sales figures as two matrices $\mathbf{S}_1$ for domestic and $\mathbf{S}_2$ for overseas:

$$\mathbf{S}_1 = \begin{bmatrix} 22 & 45 \\ 30 & 39 \\ 40 & 8 \end{bmatrix} \quad \text{and} \quad \mathbf{S}_2 = \begin{bmatrix} 17 & 49 \\ 38 & 32 \\ 48 & 6 \end{bmatrix}. \tag{2.9}$$

Then the total sales (domestic and overseas combined) of basic product to professionals is $22 + 17 = 39$. And this is the basis for adding $\mathbf{S}_1$ and $\mathbf{S}_2$, to get the total sales for each of the three products to the two types of client:

$$\mathbf{S}_1 + \mathbf{S}_2 = \begin{bmatrix} 22 + 17 & 45 + 49 \\ 30 + 38 & 39 + 32 \\ 40 + 48 & 8 + 6 \end{bmatrix} = \begin{bmatrix} 39 & 94 \\ 68 & 71 \\ 88 & 14 \end{bmatrix}. \tag{2.10}$$

Thus adding the matrices element by element gives the matrix sum.

The general result, for two matrices

$$\mathbf{A}_{r \times c} = \{a_{ij}\} \quad \text{and} \quad \mathbf{B}_{r \times c} = \{b_{ij}\},$$

is that their sum is

$$\mathbf{A}_{r \times c} + \mathbf{B}_{r \times c} = \left\{ {}_m a_{ij} + b_{ij} \right\}_{i=1\,j=1}^{r\quad c}. \tag{2.11}$$

An exceedingly important consequence of this is that matrices can be added only if they are of the same order. When they do have the same order they are said to be *conformable for addition*. (The "for addition" is necessary as distinction from "conformable for multiplication", which is discussed in Section 2.6c.)

Note the succinctness of the matrix sum $\mathbf{S}_1 + \mathbf{S}_2$ in (2.10). Relating that matrix to the columns labeled domestic and overseas in Table 2.1, we can immediately compare total sales to professionals to sales to students, product by product (row by row). For example, the basic product sales to professionals are far less than to students (39 vs. 94), for enhanced product they are about the same (68 vs. 71), and complete product sales to professionals far exceed those to students (88 vs. 14). Reinforcing the succinctness of (2.10) is the fact that although it is only a small matrix, of order 2×3, the rule for summing two matrices applies no matter what order the matrices have (so long as their orders are the same). It is this aspect of matrix algebra that makes it so useful: the rules of the algebra are matrix rules, applicable (with few exceptions) regardless of size.

Three consequences of matrix addition are important and useful.

a. Transposing a sum

The transpose of a sum of matrices is the sum of the transposed matrices; i.e.,

$$(\mathbf{A} + \mathbf{B})' = \mathbf{A}' + \mathbf{B}'. \tag{2.12}$$

b. Trace of a sum

The trace of a sum is the sum of the traces; i.e.,

$$\text{tr}(\mathbf{A} + \mathbf{B}) = \text{tr}(\mathbf{A}) + \text{tr}(\mathbf{B}). \tag{2.13}$$

It is left to the reader to create numerical examples and proofs of (2.12) and (2.13) (see E 2.2).

c. Scalar multiplication

It is clear that $\mathbf{A} + \mathbf{A} = 2\mathbf{A} = \{2a_{ij}\}$. Extension to the sum of a number of $\mathbf{A}$s, λ of them, say, where λ is a positive integer, is

$$\lambda\mathbf{A} = \{\lambda a_{ij}\}. \tag{2.14}$$

This result extended to λ being any scalar is the definition of *scalar multiplication* of a matrix. An example is

$$2.1\begin{bmatrix} 1 & 0 & -2 \\ -3 & 4 & -1 \end{bmatrix} = \begin{bmatrix} 2.1 & 0 & -4.2 \\ -6.3 & 8.4 & -2.1 \end{bmatrix}.$$

2.5 Matrix Subtraction

Reverting to Table 2.1 and the matrices $\mathbf{S}_1$ and $\mathbf{S}_2$ which follow it, $\mathbf{S}_1$ is the sales for domestic clients and $\mathbf{S}_2$ is for overseas clients. Then, just as the total sales for basic product to professionals is $22 + 17$ so is $22 - 17$ the extent by which domestic sales exceed overseas sales. This leads to

$$\mathbf{S}_1 - \mathbf{S}_2 = \begin{bmatrix} 22 - 17 & 45 - 49 \\ 30 - 38 & 39 - 32 \\ 40 - 48 & 8 - 6 \end{bmatrix} = \begin{bmatrix} 5 & -4 \\ -8 & 7 \\ -8 & 2 \end{bmatrix}. \tag{2.15}$$

The matrix difference is therefore much like the sum: it is formed by subtracting (rather than adding) the matrices element by element. The general result for $\mathbf{A}_{r \times c} = \{a_{ij}\}$ and $\mathbf{B}_{r \times c} = \{b_{ij}\}$ is that

$$\mathbf{A}_{r \times c} - \mathbf{B}_{r \times c} = \left\{_m\ a_{ij} - b_{ij}\right\}_{i=1\ j=1}^{r\quad c}.$$

This is just like (2.10) only with its plus signs being minuses. And the same important consequence holds: $\mathbf{A}$ and $\mathbf{B}$ can be subtracted from one another only if they have the same order, in which case they are said to be *conformable for subtraction*. And this is identical to being conformable for addition.

Null matrices and equal matrices, defined in Chapter 1, are connected very simply: the difference between two equal matrices is a null matrix. For example,

$$
\begin{bmatrix} 2 & 7 & 4 \\ 3 & 5 & 8 \end{bmatrix} - \begin{bmatrix} 2 & 7 & 4 \\ 3 & 5 & 8 \end{bmatrix} = \begin{bmatrix} 2-2 & 7-7 & 4-4 \\ 3-3 & 5-5 & 8-8 \end{bmatrix} = \begin{bmatrix} 0 & 0 & 0 \\ 0 & 0 & 0 \end{bmatrix}.
$$

The paragraph following that which contains (2.11) has comments about the succinctness of $\mathbf{S}_1 + \mathbf{S}_2$ in (2.10) and about the applicability of the matrix addition rules to matrices of any order. Exactly the same comments apply to subtraction, exemplified by $\mathbf{S}_1 - \mathbf{S}_2$ of (2.15). For example, from (2.15) we find that whenever domestic sales exceed overseas sales for professionals the reverse is true for students, and vice versa.

Illustration (One-person games)

The theory of one-person games against nature uses what are called *payoff matrices*, in which element i, j is the payoff to the player for his action i in the face of outcome j of the game. Thus rows correspond to possible actions by a player and columns correspond to possible outcomes of the game. As an example, suppose that a consulting firm in seeking employees has three possible actions: hire individuals with a B.S. degree, or with a graduate degree, or with a graduate degree and experience. The possible outcomes of the firm's activities relate to the contracts it secures: they require either a low, medium or high level of technical expertise. Suppose the payoff (in units of $1000) matrix related to these actions (rows) and outcomes (columns) is

$$
\mathbf{P} = \begin{bmatrix} 350 & 300 & 150 \\ 100 & 275 & 500 \\ -200 & 175 & 900 \end{bmatrix}.
$$

Thus $p_{12} = 300$ is the payoff for hiring employees with B.S. degrees and getting contracts requiring medium technical expertise.

The firm now wants to develop a strategy for using $\mathbf{P}$ so as to maximize its overall revenue. An optimal strategy, known as the minimax regret criterion, requires using $\mathbf{P}$, the payoff matrix, to form a new matrix, $\mathbf{R}$, known as the *regret matrix*. Column j of the regret matrix has as elements each element of column j of the payoff matrix subtracted from the maximum element of column j of the payoff matrix. Thus elements of column j of the regret matrix represent, for outcome j, the extent to which each action has a payoff less than the maximum payoff for outcome j. And we think of the payoff being "less than" as a "regret". It is certainly regrettable.

In **P** for the consulting firm we see that the largest element in the three columns is, respectively, 350, 300 and 900. Therefore, from **P**, the regret matrix is

$$\mathbf{R} = \begin{bmatrix} 350 - 350 & 300 - 300 & 900 - 150 \\ 350 - 100 & 300 - 275 & 900 - 500 \\ 350 - (-200) & 300 - 175 & 900 - 900 \end{bmatrix}$$

$$= \begin{bmatrix} 0 & 0 & 750 \\ 250 & 25 & 400 \\ 550 & 125 & 0 \end{bmatrix}.$$

In decision theory, as applied in economics, elements of the regret matrix are often described as being (maximum) losses incurred by not choosing the best (largest) payoff for each particular outcome. But in reality this use of the word "loss" is not correct: it is a diminution of the payoff, not an actual loss.

A regret matrix provides information for a player to choose a strategy using the minimax regret criterion. This strategy is the action corresponding to the row of the regret matrix which minimizes the maximum regret which may occur, i.e., the row which contains the minimum of the maximum entries of all rows.

To do this the consulting firm (the player) must first identify in **R** the maximum regret for each action, i.e., the largest element in each row of **R**. This gives a vector,

$$\mathbf{r}_m = \begin{bmatrix} 750 \\ 400 \\ 550 \end{bmatrix}$$

which we call the *maximum regret vector*. It contains the information which is the basis of the minimax-regret criterion: choose the action corresponding to the smallest element in the maximum regret vector. For the consulting firm the smallest element in r_m is 400 of row 2, and so the firm will choose action 2, that of hiring individuals with a graduate degree. This action minimizes the maximum regret.

2.6 Multiplication

Multiplication of matrices is approached step-wise by first considering a product of two vectors, then the product of a matrix and a vector, and finally the product of two matrices. Each of these products is introduced using the computing software illustration already described.

a. Inner product of two vectors

Suppose for the preceding illustration that the selling price of the software is $5 for basic, $10 for enhanced and $20 for complete. Then the sales revenue from the domestic sales to professionals who (from Table 2.1) buy 22, 30 and 40 of these products, respectively, is

$$5(22) + 10(30) + 20(40) = 1210 \qquad (2.16)$$

thousand dollars. Suppose the prices are written as a row vector $\mathbf{a}'$ and the numbers of product sold as a column vector $\mathbf{x}$:

$$\mathbf{a}' = [5 \ \ 10 \ \ 20] \qquad \text{and} \qquad \mathbf{x} = \begin{bmatrix} 22 \\ 30 \\ 40 \end{bmatrix}.$$

Then the total revenue (in $1000) is 1210, the sum of products of the elements of $\mathbf{a}'$ each multiplied by the corresponding element of $\mathbf{x}$. This is the definition of the product $\mathbf{a}'\mathbf{x}$. The example is written as

$$\mathbf{a}'\mathbf{x} = [5 \ \ 10 \ \ 20] \begin{bmatrix} 22 \\ 30 \\ 40 \end{bmatrix} = 5(22) + 10(30) + 20(40) = 1210. \quad (2.17)$$

This illustrates the general method for obtaining $\mathbf{a}'\mathbf{x}$: multiply each element of the row vector $\mathbf{a}'$ by the corresponding element of the column vector $\mathbf{x}$ and add the products. The sum is $\mathbf{a}'\mathbf{x}$. Thus if

$$\mathbf{a}' = [a_1 \ a_2 \ \cdots \ a_n] \qquad \text{and} \qquad \mathbf{x} = \begin{bmatrix} x_1 \\ x_2 \\ \vdots \\ x_n \end{bmatrix},$$

their product $\mathbf{a}'\mathbf{x}$ is defined as

$$\mathbf{a}'\mathbf{x} = a_1 x_1 + a_2 x_2 + \cdots + a_n x_n = \sum_{i=1}^{n} a_i x_i.$$

It is called the *inner product* of the vectors $\mathbf{a}$ and $\mathbf{x}$. It exists only when $\mathbf{a}$ and $\mathbf{x}$ have the same order; when they are not of the same order the product $\mathbf{a}'\mathbf{x}$ is undefined.

b.   A matrix-vector product

Continuing with the software numbers, suppose at an end-of-year meeting of the company it was suggested that the three selling prices of 5, 10 and 20 be raised for the forthcoming year to 7, 12 and 23. Then, had those higher selling prices been used for the current year, the sales revenue from domestic professionals would have been

$$[7 \quad 12 \quad 23] \begin{bmatrix} 22 \\ 30 \\ 40 \end{bmatrix} = 7(22) + 12(30) + 23(40) = 1434 \qquad (2.18)$$

thousand dollars, calculated just like (2.17).

Now put the two sets of prices as the rows of a matrix,

$$\mathbf{A} = \begin{bmatrix} 5 & 10 & 20 \\ 7 & 12 & 23 \end{bmatrix}.$$

Then the products (2.17) and (2.18) can be represented simultaneously as a single product of the matrix $\mathbf{A}$ and the vector $\mathbf{x}$:

$$\mathbf{A}\mathbf{x} = \begin{bmatrix} 5 & 10 & 20 \\ 7 & 12 & 23 \end{bmatrix} \begin{bmatrix} 22 \\ 30 \\ 40 \end{bmatrix} = \begin{bmatrix} 5(22) + 10(30) + 20(40) \\ 7(22) + 12(30) + 23(40) \end{bmatrix} = \begin{bmatrix} 1210 \\ 1434 \end{bmatrix}.$$

$$(2.19)$$

The result is a vector, its elements being the inner products (2.17) and (2.18). In terms of the rows of $\mathbf{A}$, this means that the elements of the vector $\mathbf{A}\mathbf{x}$ are derived in exactly the same way as the product $\mathbf{a}'\mathbf{x}$ developed earlier, using the successive rows of $\mathbf{A}$ as the vector $\mathbf{a}'$. The result is the product $\mathbf{A}\mathbf{x}$: that is, $\mathbf{A}\mathbf{x}$ is obtained by repetitions of the product $\mathbf{a}'\mathbf{x}$ using the rows of $\mathbf{A}$ successively for $\mathbf{a}'$ and writing the results as a column vector. Hence, on using the notation of (2.8), with $\mathbf{a}_1'$ and $\mathbf{a}_2'$ being the rows of $\mathbf{A}$, we see that (2.19) is

$$\mathbf{Ax} = \left[\begin{array}{c} \alpha_1' \\ \alpha_2' \end{array}\right] \mathbf{x} = \left[\begin{array}{c} \alpha_1'\mathbf{x} \\ \alpha_2'\mathbf{x} \end{array}\right].$$

This generalizes at once to $\mathbf{A}$ having r rows:

$$\mathbf{Ax} = \left[\begin{array}{c} \alpha_1' \\ \alpha_2' \\ \vdots \\ \alpha_i' \\ \vdots \\ \alpha_r' \end{array}\right] \mathbf{x} = \left[\begin{array}{c} \alpha_1'\mathbf{x} \\ \alpha_2'\mathbf{x} \\ \vdots \\ \alpha_i'\mathbf{x} \\ \vdots \\ \alpha_r'\mathbf{x} \end{array}\right]. \tag{2.20}$$

Thus $\mathbf{Ax}$ is a column vector, with its ith element being the inner product of the ith row of $\mathbf{A}$ with the column vector $\mathbf{x}$. Providing neither $\mathbf{A}$ nor $\mathbf{x}$ is a scalar, it is clear from this definition and from the example that $\mathbf{Ax}$ is defined only when the number of elements in each row of $\mathbf{A}$ (i.e., number of columns) is the same as the number of elements in the column vector $\mathbf{x}$. And when this occurs $\mathbf{Ax}$ is a column vector having the same number of elements as there are rows in $\mathbf{A}$. Therefore, when $\mathbf{A}$ has r rows and c columns and $\mathbf{x}$ is of order c, $\mathbf{Ax}$ is a column vector of order r; its ith element is $\Sigma_{k=1}^{c} a_{ik}x_k$ for $i = 1, 2, \ldots, r$. More formally, when

$$\mathbf{A} = \{a_{ij}\} \quad \text{and} \quad \mathbf{x} = \{x_j\} \quad \text{for} \quad i = 1, 2, \ldots, r \quad \text{and} \quad j = 1, 2, \ldots, c,$$

then

$$\mathbf{Ax} = \left\{\sum_{j=1}^{c} a_{ij}x_j\right\} \quad \text{for} \quad i = 1, 2, \ldots, r.$$

Illustration (Investment)

Suppose a stockbroker has three clients who invest various amounts of capital in four different money market funds connected to industries such as minerals, health care, communications and aeronautics. The amounts invested, in thousands of dollars, are represented by the 3×4 matrix $\mathbf{A}$; and the funds' annual interest rates are shown in the 4×1

vector **x**:

$$
A = \begin{bmatrix} 50 & 100 & 350 & 30 \\ 700 & 1000 & 20 & 6 \\ 20 & 8 & 2 & 12 \end{bmatrix} \quad \text{and} \quad x = \begin{bmatrix} 5\% \\ 12\% \\ 10\% \\ 8\% \end{bmatrix},
$$

then the three clients' annual incomes (per \$1000 invested) is the 3×1 vector

$$
Ax = \begin{bmatrix} 51.90 \\ 157.48 \\ 3.12 \end{bmatrix}.
$$

c. A matrix-matrix product

Multiplying two matrices is just a simple repetitive extension of multiplying a matrix by a vector.

Continuing the illustration of the software company, equation (2.19) shows the calculation of total revenue from sales to domestic professionals at prices 5, 10 and 20; and at prices 7, 12 and 23. Similar calculation for the other columns in Table 2.1, namely from domestic students, overseas professionals and overseas students are, respectively,

$$
\begin{bmatrix} 5 & 10 & 20 \\ 7 & 12 & 23 \end{bmatrix} \begin{bmatrix} 45 \\ 39 \\ 8 \end{bmatrix} = \begin{bmatrix} 775 \\ 967 \end{bmatrix}, \quad \begin{bmatrix} 5 & 10 & 20 \\ 7 & 12 & 23 \end{bmatrix} \begin{bmatrix} 17 \\ 38 \\ 48 \end{bmatrix} = \begin{bmatrix} 1425 \\ 1679 \end{bmatrix}
$$

$$(2.21)$$

and

$$
\begin{bmatrix} 5 & 10 & 20 \\ 7 & 12 & 23 \end{bmatrix} \begin{bmatrix} 49 \\ 32 \\ 6 \end{bmatrix} = \begin{bmatrix} 685 \\ 865 \end{bmatrix}. \qquad (2.22)
$$

Now, by writing alongside one another, as a matrix, the four column vectors of sales figures in Table 2.1, we have the matrix

$$
B = \begin{bmatrix} 22 & 45 & 17 & 49 \\ 30 & 39 & 38 & 32 \\ 40 & 8 & 48 & 6 \end{bmatrix}.
$$

Then the products in (2.19), (2.21) and (2.22) can be represented as the single matrix product

$$
\begin{bmatrix} 5 & 10 & 20 \\ 7 & 12 & 23 \end{bmatrix} \begin{bmatrix} 22 & 45 & 17 & 49 \\ 30 & 39 & 38 & 32 \\ 40 & 8 & 48 & 6 \end{bmatrix} = \begin{bmatrix} 1210 & 775 & 1425 & 685 \\ 1434 & 967 & 1679 & 865 \end{bmatrix}.
$$

This example illustrates how the product $\mathbf{AB}$ is simply a case of obtaining the product of $\mathbf{A}$ with each column of $\mathbf{B}$ and setting the products alongside one another. Thus for

$$\mathbf{B} = [\mathbf{b}_1 \quad \mathbf{b}_2 \quad \mathbf{b}_3 \quad \mathbf{b}_4]$$

partitioned into columns in the manner of (2.7),

$$\mathbf{AB} = [\mathbf{Ab}_1 \quad \mathbf{Ab}_2 \quad \mathbf{Ab}_3 \quad \mathbf{Ab}_4].$$

Then, with $\mathbf{A}$ partitioned into its rows

$$\mathbf{A} = \begin{bmatrix} \alpha_1' \\ \alpha_2' \end{bmatrix} \quad \text{we have} \quad \mathbf{AB} = \begin{bmatrix} \alpha_1'\mathbf{b}_1 & \alpha_1'\mathbf{b}_2 & \alpha_1'\mathbf{b}_3 & \alpha_1'\mathbf{b}_4 \\ \alpha_2'\mathbf{b}_1 & \alpha_2'\mathbf{b}_2 & \alpha_2'\mathbf{b}_3 & \alpha_2'\mathbf{b}_4 \end{bmatrix}.$$

Hence the element of $\mathbf{AB}$ in its ith row and jth column is the inner product $\alpha_i'\mathbf{b}_j$ of the ith row of $\mathbf{A}$ and the jth column of $\mathbf{B}$. And this is true in general:

$$\begin{aligned}
\mathbf{A}_{r \times c}\mathbf{B}_{c \times s} &= \left\{_m \ \alpha_i'\mathbf{b}_j\right\}_{i=1 \ j=1}^{r \quad s} \\
&= \left\{_m \ a_{i1}b_{1j} + a_{i2}b_{2j} + \cdots + a_{ic}b_{cj}\right\}_{i=1 \ i=j}^{r \quad s} \quad (2.23) \\
&= \left\{_m \ \sum_{k=1}^{c} a_{ik}b_{kj}\right\}_{i=1 \ j=1}^{r \quad s}.
\end{aligned}$$

The (i,j)th element of $\mathbf{AB}$ can therefore be obtained by thinking of moving from element to element along the ith row of $\mathbf{A}$ and simultaneously down the jth column of $\mathbf{B}$, summing the products of corresponding elements. The resulting sum is the (i,j)th element of $\mathbf{AB}$. Schematically the operation to obtain $\mathbf{AB}$ can be represented as follows.

$$\begin{bmatrix} \text{row } i \text{ of } \mathbf{A} \\ \longrightarrow \end{bmatrix}_{r \times c} \begin{bmatrix} \downarrow & \text{column } j \\ & \text{of } \mathbf{B} \end{bmatrix}_{c \times s} = \begin{bmatrix} \text{element} \\ i,j \text{ of } \mathbf{AB} \end{bmatrix}_{r \times s}$$

for $i = 1, 2, \ldots, r$ and $j = 1, 2, \ldots, s$. The arrows indicate moving along the ith row and simultaneously down the jth column, summing the products of corresponding elements to get the (i,j)th element of the product.

Once again, this is a matrix operation defined only if a certain condition is met: the ith row of $\mathbf{A}$ (and hence all rows) must have the same number of elements as does the jth column of $\mathbf{B}$ (and hence all columns). Since the number of elements in a row of a matrix is the number of columns in the matrix (and the number of elements in a column is the number of rows), this means that there must be exactly as many columns in $\mathbf{A}$ as there are rows in $\mathbf{B}$. Thus the matrix product $\mathbf{AB}$ is defined only if the number of columns in $\mathbf{A}$ equals the number of rows of $\mathbf{B}$. Note also, particularly in the numerical examples, that the product $\mathbf{AB}$ has the same number of rows as $\mathbf{A}$ and the same number of columns as $\mathbf{B}$. This is true in general.

The important consequences of the definition of matrix multiplication are therefore as follows. The product $\mathbf{AB}$ of two matrices $\mathbf{A}$ and $\mathbf{B}$ is defined and therefore exists only if the number of columns in $\mathbf{A}$ equals the number of rows in $\mathbf{B}$; the matrices are then said to be *conformable for multiplication for the product* $\mathbf{AB}$, and $\mathbf{AB}$ has the same number of rows as $\mathbf{A}$ and the same number of columns as $\mathbf{B}$. And the (i, j)th element of $\mathbf{AB}$ is the inner product of the ith row of $\mathbf{A}$ and the jth column of $\mathbf{B}$.

Example

For

$$\mathbf{A} = \begin{bmatrix} 1 & 0 & 2 \\ -1 & 4 & 3 \end{bmatrix} \quad \text{and} \quad \mathbf{B} = \begin{bmatrix} 0 & 6 & 1 & 5 \\ 1 & 1 & 0 & 7 \\ 3 & 4 & 4 & 3 \end{bmatrix}$$

the element in the first row and first column of the product $\mathbf{AB}$ is the inner product of the first row of $\mathbf{A}$ and the first column of $\mathbf{B}$ and is

$$1(0) + 0(1) + 2(3) = 6;$$

the element in the first row and second column is

$$1(6) + 0(1) + 2(4) = 14;$$

and the element of $\mathbf{AB}$ in the second row and third column is

$$-1(1) + 4(0) + 3(4) = 11.$$

In this way $\mathbf{AB}$ is obtained as

$$\mathbf{AB} = \begin{bmatrix} 6 & 14 & 9 & 11 \\ 13 & 10 & 11 & 32 \end{bmatrix}.$$

d.   Existence of matrix products

As has been noted, subscript notation can denote the order of a matrix; thus $\mathbf{A}_{r \times c}$ is a matrix of order r by c (r rows and c columns). A product $\mathbf{AB}$ can be written in this notation as

$$\mathbf{A}_{r \times c}\mathbf{B}_{c \times s} = \mathbf{P}_{r \times s},$$

a form which provides both for checking the conformability of $\mathbf{A}$ and $\mathbf{B}$ and for ascertaining the order of their product. Repeated use of this also simplifies determining the order of a matrix derived by multiplying several matrices together. Adjacent subscripts (which must be equal for conformability) "cancel out," leaving the first and last subscripts as the order of the product. For example, the product

$$\mathbf{A}_{r \times c}\mathbf{B}_{c \times s}\mathbf{C}_{s \times t}\mathbf{D}_{t \times u}$$

is a matrix of order $r \times u$.

This notation also demonstrates what is by now readily apparent from the definition of matrix multiplication, namely that the product $\mathbf{BA}$ does not necessarily exist, even if $\mathbf{AB}$ does. For $\mathbf{BA}$ can be written as $\mathbf{B}_{c \times s}\mathbf{A}_{r \times t}$, which we see at once is a legitimate product only if $s = r$. Otherwise $\mathbf{BA}$ is not defined. There are therefore three situations regarding the product of two matrices $\mathbf{A}$ and $\mathbf{B}$. If $\mathbf{A}$ is of order $r \times c$

(i) $\mathbf{AB}$ exists only if $\mathbf{B}$ has c rows,

(ii) $\mathbf{BA}$ exists only if $\mathbf{B}$ has r columns,

(iii) $\mathbf{AB}$ and $\mathbf{BA}$ both exist only if $\mathbf{B}$ is $c \times r$.

A consequence of (iii) is that $\mathbf{A}^2$ exists only when $\mathbf{A}$ is square. Another consequence is that both $\mathbf{AB}$ and $\mathbf{BA}$ always exist and are of the same order only when $\mathbf{A}$ and $\mathbf{B}$ are square and of the same order. But as is shown subsequently, the two products are not necessarily equal. Their inequality is discussed when considering the commutative law of multiplication, but meanwhile we simply state that they are not in general equal.

As a means of distinction , $\mathbf{AB}$ is described as $\mathbf{A}$ *post-multiplied* by $\mathbf{B}$, or as $\mathbf{A}$ *multiplied on the right* by $\mathbf{B}$; and $\mathbf{BA}$ is either $\mathbf{A}$ *pre-multiplied* by $\mathbf{B}$, or $\mathbf{A}$ *multiplied on the left* by $\mathbf{B}$.

e. Products with vectors

Both the inner product of two vectors and the product of a matrix post-multiplied by a column vector (see Sections 2.6a and b) are special cases of the general matrix product $\mathbf{A}_{r \times c} \mathbf{B}_{c \times s} = \mathbf{P}_{r \times s}$. For the inner product of two vectors when $r = 1$ and $s = 1$, which means that $\mathbf{A}_{r \times c}$ becomes $\mathbf{A}_{1 \times c}$, a row vector $(\mathbf{a})'_{1 \times c}$, say; and $\mathbf{B}_{c \times s}$ becomes $\mathbf{B}_{c \times 1}$, a column vector $\mathbf{b}_{c \times 1}$; we then have the inner product

$$(\mathbf{a}')_{1 \times c} \mathbf{b}_{c \times 1} = \mathbf{p}_{1 \times 1},$$

a scalar. And the product in reverse order, where $\mathbf{a}$ and $\mathbf{b}$ can now be of different orders, is

$$\mathbf{b}_{c \times 1} (\mathbf{a}')_{1 \times r} = \mathbf{P}_{c \times r},$$

a matrix called the *outer product of* $\mathbf{b}$ *and* $\mathbf{a}'$. For the matrix-vector product of Section 2.6b, we have

$$\mathbf{A}_{r \times c} \mathbf{b}_{c \times 1} = \mathbf{p}_{r \times 1},$$

the product being a column vector; similarly, a row vector post-multiplied by a matrix is a row vector:

$$(\mathbf{a}')_{1 \times c} \mathbf{B}_{c \times r} = \mathbf{p}'_{1 \times r}.$$

In words these four results (for conformable products) are as follows:

(i) A row vector post-multiplied by a column vector is a scalar.

(ii) A column vector post-multiplied by a row vector is a matrix.

(iii) A matrix post-multiplied by a column vector is a column vector.

(iv) A row vector post-multiplied by a matrix is a row vector.

Examples

Given

$$\mathbf{A} = \begin{bmatrix} 5 & 1 \\ 3 & 6 \end{bmatrix}, \quad \mathbf{B} = \begin{bmatrix} 2 & 0 & 7 \\ 1 & 2 & -5 \end{bmatrix}, \quad \mathbf{a}' = [2 \ \ 3], \quad \mathbf{b} = \begin{bmatrix} 1 \\ 4 \end{bmatrix},$$

then

(i) $\mathbf{a'b} = \begin{bmatrix} 2 & 3 \end{bmatrix} \begin{bmatrix} 1 \\ 4 \end{bmatrix} = 14,$

(ii) $\mathbf{ba'} = \begin{bmatrix} 1 \\ 4 \end{bmatrix} \begin{bmatrix} 2 & 3 \end{bmatrix} = \begin{bmatrix} 2 & 3 \\ 8 & 12 \end{bmatrix},$

(iii) $\mathbf{Ab} = \begin{bmatrix} 5 & 1 \\ 3 & 6 \end{bmatrix} \begin{bmatrix} 1 \\ 4 \end{bmatrix} = \begin{bmatrix} 9 \\ 27 \end{bmatrix},$

(iv) $\mathbf{a'B} = \begin{bmatrix} 2 & 3 \end{bmatrix} \begin{bmatrix} 2 & 0 & 7 \\ 1 & 2 & -5 \end{bmatrix} = \begin{bmatrix} 7 & 6 & -1 \end{bmatrix}.$

Illustration (Taxicab fares)

Suppose an experienced taxicab driver has ascertained that when in Town 1 there is a probability of .2 that the next fare will be within Town 1 and a probability of .8 that it will be to Town 2. But when in Town 2 the probabilities are .4 of going to Town 1 and .6 of staying in Town 2. These probabilities can be assembled in a matrix

$$\mathbf{P} = \begin{bmatrix} .2 & .8 \\ .4 & .6 \end{bmatrix} = \{p_{ij}\} \quad \text{for} \quad i, j = 1, 2. \tag{2.24}$$

p_{ij} is the probability when in Town i of the next fare being to Town j.

The matrix in (2.24) is an example of a transition probability matrix. It is an array of probabilities of making a transition from what is called state i to state j. (The states in the example are the two towns.) Thus p_{ij} is a transition probability, and hence the name: *transition probability matrix*.

In (2.24) the sum of the elements in each row (the *row sum*) is 1.00. This is so because whenever the taxi is in Town i it must, with its next fare, either stay in that town or go to the other one. This is a feature of all transition probability matrices. In general, for such a matrix

$$\mathbf{P} = \begin{bmatrix} p_{11} & p_{12} & \cdots & p_{1j} & \cdots & p_{1m} \\ p_{21} & p_{22} & \cdots & p_{2j} & \cdots & p_{2m} \\ \vdots & \vdots & & \vdots & & \vdots \\ p_{i1} & p_{i2} & \cdots & p_{ij} & \cdots & p_{im} \\ \vdots & \vdots & & \vdots & & \vdots \\ p_{m1} & p_{m2} & \cdots & p_{mj} & \cdots & p_{mm} \end{bmatrix}$$

$$= \{p_{ij}\} \quad \text{for} \quad i, j = 1, 2 \ldots, m, \tag{2.25}$$

corresponding to m different states, the sum of the probabilities of going from state i to any one of the m states (including staying in state i) must be 1. Hence

$$p_{i1} + p_{i2} + \cdots + p_{ij} + \cdots + p_{im} = 1, \qquad (2.26)$$

i.e.,

$$\sum_{j=1}^{m} p_{ij} = 1 \quad \text{for all} \quad i = 1, 2, \ldots, m.$$

Situations like the taxicab example form a class of probability models known as *Markov chains*. They arise in a variety of ways: for example, in considering the probabilities of changes in the prime interest rate from week to week, or the probabilities of certain types of mating in genetics, or in studying the yearly mobility of labor where p_{ij} can be the probability of moving during a calendar year from category i in the labor force to category j. Assembling probabilities of this nature into a matrix then enables matrix algebra to be used as a tool for answering questions of interest; for example, the probabilities of moving from category i to category j in two years (assuming the probabilities are the same from year to year) are given by $\mathbf{P}^2$; in three years, by $\mathbf{P}^3$; and so on.

Now suppose the driver lives in Town 1 and starts work from there each day. Denote this by a vector $\mathbf{x}_0' = [1 \ 0]$. In general $\mathbf{x}'$ is called a *state probability vector* (or simply *state vector*). The subscript 0 represents the beginning of the day, and the elements 1 and 0 are probabilities of starting in Town 1 and in Town 2, respectively. This being so, the probabilities of being in Town 1 or Town 2 after the morning's first fare are, using the transition probability matrix $\mathbf{P}$,

$$\mathbf{x}_1' = \mathbf{x}_0'\mathbf{P} = [1 \ 0] \begin{bmatrix} .2 & .8 \\ .4 & .6 \end{bmatrix} = [.2 \ .8].$$

And after the second fare the probabilities are

$$\mathbf{x}_2' = \mathbf{x}_1'\mathbf{P} = [.2 \ .8] \begin{bmatrix} .2 & .8 \\ .4 & .6 \end{bmatrix} = [.36 \ .64].$$

Markov chain situations such as this one are considered more fully in Chapter 17.

f. Products with scalars

To the extent that a scalar can be considered a 1×1 matrix, certain cases of the scalar multiplication of a matrix in Section 2.4c are included in the general matrix product $\mathbf{A}_{r \times c} \mathbf{B}_{c \times s} = \mathbf{P}_{r \times s}$: to wit $a_{1 \times 1} \mathbf{b}_{1 \times c} = a\mathbf{b}'$ and $(\mathbf{a}')_{r \times 1} b_{1 \times 1} = b\mathbf{a}'$.

g. Products with null matrices

For any matrix $\mathbf{A}_{r \times s}$, pre- or post-multiplication by a null matrix of appropriate order results in a null matrix. Thus if $\mathbf{0}_{c \times r}$ is a null matrix of order $c \times r$,

$$\mathbf{0}_{c \times r} \mathbf{A}_{r \times s} = \mathbf{0}_{c \times s} \quad \text{and} \quad \mathbf{A}_{r \times s} \mathbf{0}_{s \times p} = \mathbf{0}_{r \times p}.$$

For example,

$$[0 \ \ 0] \begin{bmatrix} 1 & 2 & -4 \\ 9 & 7 & 2 \end{bmatrix} = [0 \ \ 0 \ \ 0].$$

In writing this equality as $\mathbf{0A} = \mathbf{0}$, it is important to notice that the two $\mathbf{0}$s are not of the same order.

h. Products with diagonal matrices

A diagonal matrix is defined in Section 1.8b as a square matrix having all off-diagonal elements zero. Multiplication by a diagonal matrix is particularly easy: premultiplication of the matrix $\mathbf{A}$ by a diagonal matrix $\mathbf{D}$ gives a matrix whose rows are those of $\mathbf{A}$ multiplied by the respective diagonal elements of $\mathbf{D}$.

Illustration (Sales, *continued*)

In the matrix

$$\mathbf{A} = \begin{bmatrix} 5 & 10 & 20 \\ 7 & 12 & 23 \end{bmatrix}$$

of (2.19), rows represent current and future prices. Suppose these prices were to increase by 5% and 20%, respectively. Then with

$$\mathbf{D} = \begin{bmatrix} 1.05 & 0 \\ 0 & 1.2 \end{bmatrix}$$

the matrix of new prices is

$$\mathbf{DA} = \begin{bmatrix} 1.05 & 0 \\ 0 & 1.2 \end{bmatrix} \begin{bmatrix} 5 & 10 & 20 \\ 7 & 12 & 23 \end{bmatrix} = \begin{bmatrix} 5.25 & 10.5 & 21.0 \\ 8.40 & 14.4 & 27.6 \end{bmatrix}.$$

In contrast, suppose the prices for basic product were to increase by
20%, those for enhanced product were to remain as is, and those for
complete product were to decline 10%. Then with Δ being the diagonal
matrix representing these price changes the new prices would be

$$\mathbf{A\Delta} = \begin{bmatrix} 5 & 10 & 20 \\ 7 & 12 & 23 \end{bmatrix} \begin{bmatrix} 1.2 & 0 & 0 \\ 0 & 1 & 0 \\ 0 & 0 & .9 \end{bmatrix} = \begin{bmatrix} 6.0 & 10 & 18.0 \\ 8.4 & 12 & 20.7 \end{bmatrix}.$$

i.   Products with identity matrices

As in Section 1.8c, an identity matrix is a diagonal matrix with all
its diagonal elements being 1.0. Thus multiplication of a matrix $\mathbf{A}$ by
an identity matrix does not alter $\mathbf{A}$: e.g.,

$$\begin{bmatrix} 1 & 0 \\ 0 & 1 \end{bmatrix} \begin{bmatrix} 2 & 3 & -1 \\ 4 & 8 & 7 \end{bmatrix} = \begin{bmatrix} 2 & 3 & -1 \\ 4 & 8 & 7 \end{bmatrix}.$$

Thus just as a null matrix is a zero of matrix algebra, so is an identity
matrix a "one" or unit of the algebra. A distinction from scalar algebra
is that whereas with scalar algebra there is only one zero and one "one",
with matrix algebra there are many zeros (null matrices of every order)
and many "ones" (identity matrices of every order).

Identity matrices are usually denoted by the letter $\mathbf{I}$ with order shown
as a subscript. And although we can think of an $\mathbf{I}$ as a "one" of matrix
algebra, in the products $\mathbf{IA}$ and $\mathbf{AI}$, which both equal $\mathbf{A}$, the two $\mathbf{I}$s
are not necessarily the same. This is because for $\mathbf{A}_{p \times q}$

$$\mathbf{I}_p \mathbf{A}_{p \times q} = \mathbf{A}_{p \times q} = \mathbf{A}_{p \times q} \mathbf{I}_q.$$

Sometimes a matrix of the form $\lambda \mathbf{I}$ where λ is a scalar is called a
scalar matrix; e.g.,

$$4\mathbf{I} = \begin{bmatrix} 4 & 0 \\ 0 & 4 \end{bmatrix}.$$

It is simply a diagonal matrix with all diagonal elements the same.
Multiplication by a scalar matrix has the same effect as scalar multiplication; e.g., $4\mathbf{IA} = 4\mathbf{A}$.

j.   Transpose of a product

The transpose of a product matrix is the product of the transposed
matrices taken in reverse sequence, i.e., $(\mathbf{AB})' = \mathbf{B}'\mathbf{A}'$.

Example

$$\mathbf{AB} = \begin{bmatrix} 1 & 0 & -1 \\ 2 & -1 & 3 \end{bmatrix} \begin{bmatrix} 1 & 1 & 1 \\ 0 & 2 & 4 \\ 3 & 0 & 7 \end{bmatrix} = \begin{bmatrix} -2 & 1 & -6 \\ 11 & 0 & 19 \end{bmatrix}$$

$$\mathbf{B'A'} = \begin{bmatrix} 1 & 0 & 3 \\ 1 & 2 & 0 \\ 1 & 4 & 7 \end{bmatrix} \begin{bmatrix} 1 & 2 \\ 0 & -1 \\ -1 & 3 \end{bmatrix} = \begin{bmatrix} -2 & 11 \\ 1 & 0 \\ -6 & 19 \end{bmatrix} = (\mathbf{AB})'.$$

Consideration of order and conformability for multiplication confirms this result. If $\mathbf{A}$ is $r \times s$ and $\mathbf{B}$ is $s \times t$, the product $\mathbf{P} = \mathbf{AB}$ is $r \times t$; i.e., $\mathbf{A}_{r \times s}\mathbf{B}_{s \times t} = \mathbf{P}_{r \times t}$. But $\mathbf{A}'$ is $s \times r$ and $\mathbf{B}'$ is $t \times s$ and the only product to be derived from these is $(\mathbf{B}')_{t \times s}(\mathbf{A}')_{s \times r} = \mathbf{Q}_{t \times r}$, say. That $\mathbf{Q} = \mathbf{B'A'}$ is indeed the transpose of $\mathbf{P} = \mathbf{AB}$ is apparent from the definition of multiplication: the (i, j)th term of $\mathbf{Q}$ is the inner product of the ith row of $\mathbf{B}'$ and the jth column of $\mathbf{A}'$, which in turn is the inner product of the ith column of $\mathbf{B}$ and the jth row of $\mathbf{A}$, and this by the definition of multiplication is the (j, i)th term of $\mathbf{P}$. Hence $\mathbf{Q} = \mathbf{P}'$, or $\mathbf{B'A'} = (\mathbf{AB})'$.

This result for the transpose of the product of two matrices extends directly to the product of more than two. For example, $(\mathbf{ABC})' = \mathbf{C'B'A'}$ and $(\mathbf{ABCD})' = \mathbf{D'C'B'A'}$. Proof is left as exercise E 2.10 for the reader.

k. Trace of a product

The trace of a matrix is defined in Section 2.3 for $\mathbf{A} = \{a_{ij}\}_{i,j=1}^{n}$. Its trace is $\text{tr}(\mathbf{A}) = \Sigma_{i=1}^{n} a_{ii}$, the sum of diagonal elements. We now show that $\mathbf{AB}$ and $\mathbf{BA}$ have the same trace, i.e.,

$$\text{tr}(\mathbf{AB}) = \text{tr}(\mathbf{BA}) \tag{2.27}$$

and hence

$$\text{tr}(\mathbf{ABC}) = \text{tr}(\mathbf{BCA}) = \text{tr}(\mathbf{CAB}). \tag{2.28}$$

To develop (2.27) note first that $\text{tr}(\mathbf{AB})$ exists only if $\mathbf{AB}$ is square, which occurs only when $\mathbf{A}$ is $r \times c$ and $\mathbf{B}$ is $c \times r$. Then if $\mathbf{AB} = \mathbf{P} = \{p_{ij}\}$ and $\mathbf{BA} = \mathbf{T} = \{t_{ij}\}$,

$$\text{tr}(\mathbf{AB}) = \sum_{i=1}^{r} p_{ii} = \sum_{i=1}^{r} \left(\sum_{j=1}^{c} a_{ij} b_{ji} \right)$$

$$= \sum_{j=1}^{c} \left(\sum_{i=1}^{r} b_{ji} a_{ij} \right) = \sum_{j=1}^{c} t_{jj} = \text{tr}(\mathbf{BA}).$$

Thus (2.27) is established and using it leads to (2.28): $\text{tr}(\mathbf{ABC}) = \text{tr}[(\mathbf{AB})\mathbf{C}] = \text{tr}(\mathbf{CAB})$, and so on. A useful special case of (2.27) is to have $\mathbf{B}$ be $\mathbf{A}'$ and get

$$\text{tr}(\mathbf{AA}') = \text{tr}(\mathbf{A}'\mathbf{A}) = \sum_{i=1}^{r} \sum_{j=1}^{c} a_{ij}^2,$$

that the trace of $\mathbf{AA}'$ (and of $\mathbf{A}'\mathbf{A}$) is the sum of squares of elements of $\mathbf{A}$. This result is so because

$$
\begin{aligned}
\text{tr}(\mathbf{AA}') &= \sum_{i=1}^{r} i\text{th diagonal element of } \mathbf{AA}' \\
&= \sum_{i=1}^{r} (i\text{th row of } \mathbf{A})(i\text{th column of } \mathbf{A}') \\
&= \sum_{i=1}^{r} (i\text{th row of } \mathbf{A})(i\text{th row of } \mathbf{A} \text{ written as a column}) \\
&= \sum_{i=1}^{r} \left(\sum_{j=1}^{c} a_{ij}^2 \right).
\end{aligned}
$$

1.   Powers of a matrix

Since $\mathbf{A}_{r \times c} \mathbf{A}_{r \times c}$ exists only if $r = c$, i.e., only if $\mathbf{A}$ is square, we see that $\mathbf{A}^2$ exists only when $\mathbf{A}$ is square; and then $\mathbf{A}^k$ exists for all positive integers k. And, in keeping with scalar arithmetic where $x^0 = 1$, we take $\mathbf{A}^0 = \mathbf{I}$ for $\mathbf{A}$ square.

Illustration   (Taxicab, *continued*)

In the taxicab illustration, it is clear from the equations at the end of Section 2.6e that

$$
\begin{aligned}
\mathbf{x}_2' &= \mathbf{x}_0' \mathbf{P}^2 = [1 \ \ 0] \begin{bmatrix} .2 & .8 \\ .4 & .6 \end{bmatrix}^2 \\
&= [1 \ \ 0] \begin{bmatrix} .2 & .8 \\ .4 & .6 \end{bmatrix} \begin{bmatrix} .2 & .8 \\ .4 & .6 \end{bmatrix} = [1 \ \ 0] \begin{bmatrix} .36 & .64 \\ .32 & .68 \end{bmatrix} = [.36 \ \ .64].
\end{aligned}
$$

Similarly, $\mathbf{x}_3' = \mathbf{x}_0' \mathbf{P}^3$, and in general $\mathbf{x}_n' = \mathbf{x}_0' \mathbf{P}^n$.

Illustration (Graph theory)

Suppose in a communications network of five stations that messages can be sent only in the directions of the arrows of the following diagram.

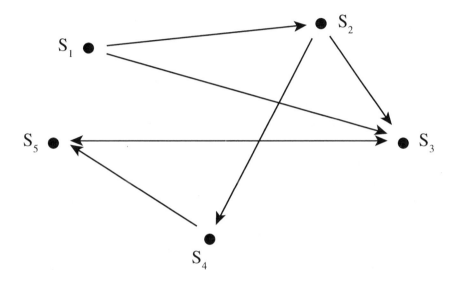

This chart of the possible message routes can also be represented by a matrix $\mathbf{T} = \{t_{ij}\}$ say, where $t_{ij} = 0$ except $t_{ij} = 1$ if a message can be sent from S_i and S_j. Hence

$$\mathbf{T} = \begin{bmatrix} 0 & 1 & 1 & 0 & 0 \\ 0 & 0 & 1 & 1 & 0 \\ 0 & 0 & 0 & 0 & 1 \\ 0 & 0 & 0 & 0 & 1 \\ 0 & 0 & 1 & 0 & 0 \end{bmatrix}.$$

In the rth power of $\mathbf{T}$, say $\mathbf{T}^r = \left\{ t_{ij}^{(r)} \right\}$, the element $t_{ij}^{(r)}$ is then the number of ways of getting a message from station i to station j in exactly r steps. Thus

$$\mathbf{T}^2 = \begin{bmatrix} 0 & 0 & 1 & 1 & 1 \\ 0 & 0 & 0 & 0 & 2 \\ 0 & 0 & 1 & 0 & 0 \\ 0 & 0 & 1 & 0 & 0 \\ 0 & 0 & 0 & 0 & 1 \end{bmatrix} \quad \text{and} \quad \mathbf{T}^3 = \begin{bmatrix} 0 & 0 & 1 & 0 & 2 \\ 0 & 0 & 2 & 0 & 0 \\ 0 & 0 & 0 & 0 & 1 \\ 0 & 0 & 0 & 0 & 1 \\ 0 & 0 & 1 & 0 & 0 \end{bmatrix}$$

show that messages can be transmitted in two ways from S_2 to S_5 in two steps and in two ways from S_2 to S_3 in three steps. In this manner the pathways through a network can be counted simply by looking at powers of $\mathbf{T}$. More than that, if there is no direct path from S_i to S_j, i.e., $t_{ij} = 0$, the powers of $\mathbf{T}$ can be used to ascertain if there are any indirect paths. For example, $t_{25} = 0$ but $t_{25}^{(2)} = 2$, showing that the path from S_2 to S_5 cannot be traversed directly but there are two indirect routes of two steps each. Thus only if $\Sigma_{r=0}^{\infty} t_{ij}^{(r)} = 0$ is there no path at all from S_i to S_j.

Applications of matrices like $\mathbf{T}$ arise in many varied circumstances. For example, instead of the Ss being stations in a communications network they could be people in a social or business group, whereupon each arrow of the figure could represent the dominance of one person by another. Then $\mathbf{T}^2$ represents the two-stage dominances existing within the group of people; and so on. Or the figure could represent the spreading of a rumor between people or groups of people; or it could be the winning of sports contests; and so on. In this way, graph theory, supported by matrix representations and attendant matrix algebra, can provide insight to a host of real-life problems.

m.   Multiplying partitioned matrices

When matrices $\mathbf{A}$ and $\mathbf{B}$ are partitioned so that their sub-matrices are appropriately conformable for multiplication, the product $\mathbf{AB}$ can be expressed in partitioned form having sub-matrices that are functions of the sub-matrices of $\mathbf{A}$ and $\mathbf{B}$. For example, if

$$\mathbf{A} = \begin{bmatrix} \mathbf{A}_{11} & \mathbf{A}_{12} \\ \mathbf{A}_{21} & \mathbf{A}_{22} \end{bmatrix} \quad \text{and} \quad \mathbf{B} = \begin{bmatrix} \mathbf{B}_{11} \\ \mathbf{B}_{21} \end{bmatrix},$$

then

$$\mathbf{AB} = \begin{bmatrix} \mathbf{A}_{11} & \mathbf{A}_{12} \\ \mathbf{A}_{21} & \mathbf{A}_{22} \end{bmatrix} \begin{bmatrix} \mathbf{B}_{11} \\ \mathbf{B}_{21} \end{bmatrix} = \begin{bmatrix} \mathbf{A}_{11}\mathbf{B}_{11} + \mathbf{A}_{12}\mathbf{B}_{21} \\ \mathbf{A}_{21}\mathbf{B}_{11} + \mathbf{A}_{22}\mathbf{B}_{21} \end{bmatrix}.$$

This means that the partitioning of $\mathbf{A}$ along its columns must be the same as that of $\mathbf{B}$ along its rows. Then $\mathbf{A}_{11}$ (and $\mathbf{A}_{21}$) has the same number of columns as $\mathbf{B}_{11}$ has rows, and $\mathbf{A}_{12}$ (and $\mathbf{A}_{22}$) has the same number of columns as $\mathbf{B}_{21}$ has rows. We see at once that when two matrices are appropriately partitioned the sub-matrices of their product are obtained by treating the sub-matrices of each of them as elements in a normal matrix product, and the individual elements of the product are derived in the usual way from the products of the sub-matrices.

The preceding sentence explains the product of two partitioned matrices in a manner which is much easier to understand than is the formal description which, for the sake of completeness, is given as follows. When $\mathbf{A}$ and $\mathbf{B}$ are two partitioned matrices conformable for the product $\mathbf{AB}$, where

$$\mathbf{A}_{p \times q} = \left\{_m \mathbf{A}_{p_i \times q_i}\right\}_{i=1}^{r} {}_{j=1}^{c}$$

with $p = \Sigma_{i=1}^{r} p_i$ and $q = \Sigma_{j=1}^{c} q_j$, and

$$\mathbf{B}_{q \times s} = \left\{_m \mathbf{B}_{q_j \times s_k}\right\}_{j=1}^{c} {}_{k=1}^{t}$$

with $s = \Sigma_{k=1}^{t} s_k$, then the product $\mathbf{AB}$ is

$$(\mathbf{AB})_{p \times s} = \left\{_m \sum_{j=1}^{c} \mathbf{A}_{p_i \times q_j} \mathbf{B}_{q_j \times s_k}\right\}_{i=1}^{r} {}_{k=1}^{t}.$$

n. Hadamard products

The definition of a product in the preceding pages is the one most generally used. But because matrices are arrays of numbers, they provide opportunity for defining products in several different ways. One of these is described here and another in Section 12.7b.

The Hadamard product of matrices $\mathbf{A} = \{a_{ij}\}$ and $\mathbf{B} = \{b_{ij}\}$ is defined only when $\mathbf{A}$ and $\mathbf{B}$ have the same order. It is the matrix of the element-by-element products of corresponding elements in $\mathbf{A}$ and $\mathbf{B}$:

$$\mathbf{A} \bullet \mathbf{B} = \{a_{ij} b_{ij}\}.$$

Note the dot between $\mathbf{A}$ and $\mathbf{B}$. It is the symbol for a matrix having elements $a_{ij} b_{ij}$; indeed, the name *dot product* is often used.

Illustration (Taxicab, *continued*).

Suppose in the taxicab illustration that the average fare between the two towns is \$10 and within Town 1 it is \$7 and within Town 2 it is \$9. Then, corresponding to the transition probability matrix, the reward matrix is

$$\mathbf{R} = \begin{bmatrix} 7 & 10 \\ 10 & 9 \end{bmatrix}.$$

Therefore, since the expected reward for a fare from Town i to Town j is $p_{ij}r_{ij}$, the matrix of expected rewards is the Hadamard product of $\mathbf{P}$ and $\mathbf{R}$:

$$\mathbf{P} \bullet \mathbf{P} = \begin{bmatrix} .2 & .8 \\ .4 & .6 \end{bmatrix} \bullet \begin{bmatrix} 7 & 10 \\ 10 & 9 \end{bmatrix} = \begin{bmatrix} .2(7) & .8(10) \\ .4(10) & .6(9) \end{bmatrix} = \begin{bmatrix} 1.4 & 8.0 \\ 4.0 & 5.4 \end{bmatrix}.$$

2.7 The Laws of Algebra

We now give formal consideration to the associative, commutative and distributive laws of algebra as they relate to the addition and multiplication of matrices.

a. Associative laws

The addition of matrices is associative provided the matrices are conformable for addition. For with $\mathbf{A}$, $\mathbf{B}$ and $\mathbf{C}$ having the same order,

$$(\mathbf{A} + \mathbf{B}) + \mathbf{C} = \{a_{ij} + b_{ij}\} + \{c_{ij}\} = \{a_{ij} + b_{ij} + c_{ij}\} = \mathbf{A} + \mathbf{B} + \mathbf{C}.$$

Also,

$$\{a_{ij} + b_{ij} + c_{ij}\} = \{a_{ij}\} + \{b_{ij} + c_{ij}\} = \mathbf{A} + (\mathbf{B} + \mathbf{C}),$$

so proving the associative law of addition.

In general, the laws of algebra that hold for matrices do so because matrix results follow directly from corresponding scalar results for their elements — as illustrated here. Further proofs in this section are therefore omitted.

The associative law is also true for multiplication, provided the matrices are conformable for multiplication. For if $\mathbf{A}$ is $p \times q$, $\mathbf{B}$ is $q \times r$ and $\mathbf{C}$ is $r \times s$, then $(\mathbf{AB})\mathbf{C} = \mathbf{A}(\mathbf{BC}) = \mathbf{ABC}$.

b. Distributive law

The distributive law holds true. For example,

$$\mathbf{A}(\mathbf{B} + \mathbf{C}) = \mathbf{A}\mathbf{B} + \mathbf{A}\mathbf{C}$$

provided both $\mathbf{B}$ and $\mathbf{C}$ are conformable for addition (necessarily of the same order) and $\mathbf{A}$ and $\mathbf{B}$ are conformable for multiplication (and hence $\mathbf{A}$ and $\mathbf{C}$ also).

c. Commutative laws

Addition of matrices is commutative (provided the matrices are conformable for addition). If $\mathbf{A}$ and $\mathbf{B}$ are of the same order

$$\mathbf{A} + \mathbf{B} = \{a_{ij} + b_{ij}\} = \{b_{ij} + a_{ij}\} = \mathbf{B} + \mathbf{A}.$$

Multiplication of matrices is not in general commutative. As seen earlier, there are two possible products that can be derived from matrices $\mathbf{A}$ and $\mathbf{B}$, $\mathbf{A}\mathbf{B}$ and $\mathbf{B}\mathbf{A}$, and if $\mathbf{A}$ is of order $r \times c$, both products exist only if $\mathbf{B}$ is of order $c \times r$. $\mathbf{A}\mathbf{B}$ is then square, of order r, and $\mathbf{B}\mathbf{A}$ is also square, of order c. Possible equality of $\mathbf{A}\mathbf{B}$ and $\mathbf{B}\mathbf{A}$ can therefore be considered only where $r = c$, in which case $\mathbf{A}$ and $\mathbf{B}$ are both square and have the same order r. The products are then

$$\mathbf{A}\mathbf{B} = \left\{ \sum_{k=1}^{r} a_{ik}b_{kj} \right\}_{i,j=1}^{r} \quad \text{and} \quad \mathbf{B}\mathbf{A} = \left\{ \sum_{k=1}^{r} b_{ik}a_{kj} \right\}_{i,j=1}^{r}.$$

It can be seen that the (i, j)th elements of these products do not necessarily have even a single term in common in their sums of products, let alone be equal. Therefore, even when $\mathbf{A}\mathbf{B}$ and $\mathbf{B}\mathbf{A}$ both exist and are of the same order, they are not in general equal; for example,

$$\begin{bmatrix} 1 & 2 \\ 3 & 4 \end{bmatrix} \begin{bmatrix} 0 & -1 \\ 1 & -1 \end{bmatrix} = \begin{bmatrix} 2 & -3 \\ 4 & -7 \end{bmatrix}$$

$$\neq \begin{bmatrix} 0 & -1 \\ 1 & -1 \end{bmatrix} \begin{bmatrix} 1 & 2 \\ 3 & 4 \end{bmatrix} = \begin{bmatrix} -3 & -4 \\ -2 & -2 \end{bmatrix}.$$

But in certain cases $\mathbf{A}\mathbf{B}$ and $\mathbf{B}\mathbf{A}$ *are* equal; for example,

$$\begin{bmatrix} 1 & 2 \\ 3 & 4 \end{bmatrix} \begin{bmatrix} 0 & 2 \\ 3 & 3 \end{bmatrix} = \begin{bmatrix} 6 & 8 \\ 12 & 18 \end{bmatrix} = \begin{bmatrix} 0 & 2 \\ 3 & 3 \end{bmatrix} \begin{bmatrix} 1 & 2 \\ 3 & 4 \end{bmatrix}.$$

Two special cases of matrix multiplication being commutative are $\mathbf{IA} = \mathbf{AI} = \mathbf{A}$ and $\mathbf{0A} = \mathbf{A0} = \mathbf{0}$ for $\mathbf{A}$ being square. When $\mathbf{A}$ is rectangular the former holds, with identity matrices of different orders, i.e., $\mathbf{I}_r\mathbf{A}_{r\times c} = \mathbf{A}_{r\times c}\mathbf{I}_c = \mathbf{A}_{r\times c}$; the latter can be expressed more generally as two separate statements: $\mathbf{0}_{p\times r}\mathbf{A}_{r\times c} = \mathbf{0}_{p\times c}$ and $\mathbf{A}_{r\times c}\mathbf{0}_{c\times s}$ and $\mathbf{A}_{r\times c}\mathbf{0}_{c\times s} = \mathbf{0}_{r\times s}$.

2.8   Contrasts with Scalar Algebra

The definition of matrix multiplication leads to results in matrix algebra that have no counterpart in scalar algebra. In fact, some matrix results contradict their scalar analogues. We give some examples.

First, although in scalar algebra $ax + bx$ can be factored either as $x(a + b)$ or as $(a + b)x$, this duality is not generally possible with matrices:

$$\mathbf{AX} + \mathbf{BX} = (\mathbf{A} + \mathbf{B})\mathbf{X} \quad \text{and} \quad \mathbf{XA} + \mathbf{XB} = \mathbf{X}(\mathbf{A} + \mathbf{B}),$$

but

$$\mathbf{XP} + \mathbf{QX} \text{ generally does not have } \mathbf{X} \text{ as a factor.}$$

Another example of factoring is that, similar to $xy - x = x(y - 1)$ in scalar algebra, $\mathbf{XY} - \mathbf{X} = \mathbf{X}(\mathbf{Y} - \mathbf{I})$ in matrix algebra, but with the second term inside the parentheses being $\mathbf{I}$ and not the scalar 1. (It cannot be 1 because 1 is not conformable for subtraction from $\mathbf{Y}$.) Furthermore, the fact that it is $\mathbf{I}$ emphasizes, because of conformability for subtraction from $\mathbf{Y}$, that $\mathbf{Y}$ must be square. Of course this can also be gleaned directly from $\mathbf{XY} - \mathbf{X}$ itself: if $\mathbf{X}$ is $r \times c$, then $\mathbf{Y}$ must have c rows in order for $\mathbf{XY}$ to be defined, and $\mathbf{Y}$ must have c columns in order for $\mathbf{XY}$ and $\mathbf{X}$ to be conformable for subtraction. This simple example illustrates the need for constantly keeping conformability in mind, especially when matrix symbols do not have their orders attached.

Another consequence of matrix multiplication is that even when $\mathbf{AB}$ and $\mathbf{BA}$ both exist, they may not be equal. Thus for

$$\mathbf{A} = \begin{bmatrix} 4 & 4 \\ 4 & 4 \end{bmatrix} \quad \text{and} \quad \mathbf{B} = \begin{bmatrix} 1 & 1 \\ -1 & -1 \end{bmatrix}$$

we have

$$\mathbf{AB} = \begin{bmatrix} 0 & 0 \\ 0 & 0 \end{bmatrix} \neq \mathbf{BA} = \begin{bmatrix} 8 & 8 \\ -8 & -8 \end{bmatrix}.$$

Notice here that even though $\mathbf{AB} = \mathbf{0}$, neither $\mathbf{A}$ nor $\mathbf{B}$ is $\mathbf{0}$. This illustrates an extremely important feature of matrices: the equation $\mathbf{AB} = \mathbf{0}$ does *not* always lead to the conclusion that $\mathbf{A}$ or $\mathbf{B}$ is $\mathbf{0}$, as would be the case with scalars. A further illustration arises from observing that for $\mathbf{A}$ and $\mathbf{B}$, we have $\mathbf{BA} = 8\mathbf{B}$. This can be rewritten as $\mathbf{BA} - 8\mathbf{B} = \mathbf{0}$, equivalent to $\mathbf{B}(\mathbf{A} - 8\mathbf{I}) = \mathbf{0}$. But from this we cannot conclude either that $\mathbf{A} - 8\mathbf{I}$ is $\mathbf{0}$ or that $\mathbf{B} = \mathbf{0}$. Similarly, not even the equation $\mathbf{X}^2 = \mathbf{0}$ means that $\mathbf{X} = \mathbf{0}$; for example, with

$$\mathbf{X} = \begin{bmatrix} 1 & 2 & 5 \\ 2 & 4 & 10 \\ -1 & -2 & -5 \end{bmatrix} \quad \text{we have} \quad \mathbf{X}^2 = \mathbf{0}.$$

Likewise $\mathbf{Y}^2 = \mathbf{I}$ implies neither $\mathbf{Y} = \mathbf{I}$ nor $\mathbf{Y} = -\mathbf{I}$; e.g.,

$$\mathbf{Y} = \begin{bmatrix} 1 & 0 \\ 4 & -1 \end{bmatrix}, \quad \text{but} \quad \mathbf{Y}^2 = \mathbf{I} = \begin{bmatrix} 1 & 0 \\ 0 & 1 \end{bmatrix}.$$

Similarly, we can have $\mathbf{M}^2 = \mathbf{M}$ with both $\mathbf{M} \neq \mathbf{I}$ and $\mathbf{M} \neq \mathbf{0}$; e.g.,

$$\mathbf{M} = \begin{bmatrix} 3 & -2 \\ 3 & -2 \end{bmatrix} = \mathbf{M}^2.$$

Such a matrix $\mathbf{M}$ with $\mathbf{M} = \mathbf{M}^2$ is said to be *idempotent*, a property which is discussed more fully in Section 3.4.

2.9 Exercises

E 2.1. For $\mathbf{A} = \begin{bmatrix} 3 & 6 \\ 2 & 1 \end{bmatrix}$, $\mathbf{B} = \begin{bmatrix} 1 & 0 & 3 & 2 \\ 0 & -1 & -1 & 1 \end{bmatrix}$,

$$\mathbf{x} = \begin{bmatrix} 1 \\ 1 \\ 0 \\ -1 \end{bmatrix} \quad \text{and} \quad \mathbf{y} = \begin{bmatrix} 1 \\ -1 \end{bmatrix}$$

show that

(a) $\mathbf{AB} = \begin{bmatrix} 3 & -6 & 3 & 12 \\ 2 & -1 & 5 & 5 \end{bmatrix}$ and $\mathbf{A'B} = \begin{bmatrix} 3 & -2 & 7 & 8 \\ 6 & -1 & 17 & 13 \end{bmatrix}$,

(b) $(\mathbf{A} + \mathbf{A'})\mathbf{B} = \begin{bmatrix} 6 & 8 \\ 8 & 2 \end{bmatrix} \mathbf{B} = \begin{bmatrix} 6 & -8 & 10 & 20 \\ 8 & -2 & 22 & 18 \end{bmatrix}$

$$= \mathbf{AB} + \mathbf{A}'\mathbf{B},$$

(c) $\mathbf{BB}' = \begin{bmatrix} 14 & -1 \\ -1 & 3 \end{bmatrix}$ and $\mathbf{B}'\mathbf{B} = \begin{bmatrix} 1 & 0 & 3 & 2 \\ 0 & 1 & 1 & -1 \\ 3 & 1 & 10 & 5 \\ 2 & -1 & 5 & 5 \end{bmatrix}$,

(d) trace $\mathbf{BB}' = $ trace $\mathbf{B}'\mathbf{B} = 17$,

(e) $\mathbf{Bx} = \begin{bmatrix} -1 \\ -2 \end{bmatrix}$, $\quad \mathbf{B}'\mathbf{Bx} = \begin{bmatrix} -1 \\ 2 \\ -1 \\ -4 \end{bmatrix}$, $\quad \mathbf{x}'\mathbf{B}'\mathbf{Bx} = 5$

$$= (\mathbf{Bx})'\mathbf{Bx},$$

(f) $\mathbf{Ay} = \begin{bmatrix} -3 \\ 1 \end{bmatrix}$, $\quad \mathbf{A}'\mathbf{Ay} \begin{bmatrix} -7 \\ -17 \end{bmatrix}$, $\quad \mathbf{y}'\mathbf{A}'\mathbf{Ay} = 10$

$$= (\mathbf{Ay})'\mathbf{Ay},$$

(g) $\mathbf{A}^2 - 4\mathbf{A} - 9\mathbf{I} = \mathbf{0}$,

(h) $\frac{1}{9}\mathbf{A} \begin{bmatrix} -1 & 6 \\ 2 & -3 \end{bmatrix} = \frac{1}{9} \begin{bmatrix} -1 & 6 \\ 2 & -3 \end{bmatrix} \mathbf{A} = \begin{bmatrix} 1 & 0 \\ 0 & 1 \end{bmatrix}$.

E 2.2. Confirm each of the following:

(a) $\begin{bmatrix} 1 & 1 \\ 1 & 1 \end{bmatrix} \begin{bmatrix} 2 & 3 \\ 3 & 2 \end{bmatrix} = \begin{bmatrix} 2 & 3 \\ 3 & 2 \end{bmatrix} \begin{bmatrix} 1 & 1 \\ 1 & 1 \end{bmatrix}$.

(b) If $\mathbf{A} = \frac{1}{3} \begin{bmatrix} 1 & 1 & 1 \\ 1 & 1 & 1 \\ 1 & 1 & 1 \end{bmatrix}$, then $\mathbf{A}^2 = \mathbf{A}$.

(c) If $\mathbf{B} = \begin{bmatrix} 1/\sqrt{6} & 1/\sqrt{3} & 1/\sqrt{2} \\ -2/\sqrt{6} & 1/\sqrt{3} & 0 \\ 1/\sqrt{6} & 1/\sqrt{3} & -1/\sqrt{2} \end{bmatrix}$, $\quad \mathbf{BB}' = \mathbf{B}'\mathbf{B} = \mathbf{I}_3$.

(d) If $\mathbf{C} = \begin{bmatrix} 6 & -4 \\ 9 & -6 \end{bmatrix}$, then $\mathbf{C}^2$ is null.

(e) $\frac{1}{9}$ $\begin{bmatrix} 4 & -5 & -1 \\ 1 & 1 & 2 \\ 4 & 4 & -1 \end{bmatrix}$ $\begin{bmatrix} 1 & 1 & 1 \\ -1 & 0 & 1 \\ 0 & 4 & -1 \end{bmatrix}$ is an identity matrix.

E 2.3. For $\mathbf{X} = \begin{bmatrix} 1 & 2 & 3 \\ 0 & -1 & -2 \\ -1 & 0 & 7 \end{bmatrix}$ and $\mathbf{Y} = \begin{bmatrix} 6 & 0 & 0 \\ -3 & 4 & 0 \\ 0 & -5 & 2 \end{bmatrix}$:

(a) Verify that $(\mathbf{X} + \mathbf{Y})' = \mathbf{X}' + \mathbf{Y}'$.
Also verify $\text{tr}(\mathbf{X} + \mathbf{Y}) = \text{tr}(\mathbf{X}) + \text{tr}(\mathbf{Y})$.

(b) Find $\mathbf{X}^2, \mathbf{Y}^2, \mathbf{XY}$ and $\mathbf{YX}$, and show that

$$(\mathbf{X} + \mathbf{Y})^2 = \mathbf{X}^2 + \mathbf{XY} + \mathbf{YX} + \mathbf{Y}^2 = \begin{bmatrix} 40 & 5 & 44 \\ -28 & 13 & -33 \\ -1 & -62 & 88 \end{bmatrix}.$$

E 2.4. Given $\mathbf{A} = \begin{bmatrix} 1 & 0 & 2 \\ 0 & 1 & 1 \\ 2 & 0 & 2 \end{bmatrix}$, $\mathbf{B} = \begin{bmatrix} 1 & 3 & 0 \\ 0 & 4 & -1 \\ 2 & 3 & 0 \end{bmatrix}$, $\mathbf{X} = \begin{bmatrix} 6 & 5 & 7 \\ 2 & 2 & 4 \\ 3 & 3 & 6 \end{bmatrix}$,

show that $\mathbf{AX} = \mathbf{BX}$ even though $\mathbf{A} \neq \mathbf{B}$.

E 2.5. Show that

(a) $\begin{bmatrix} 6 & 3 \\ 0 & 7 \\ -5 & 1 \end{bmatrix} + \begin{bmatrix} 3 & 8 \\ 2 & -1 \\ 6 & -4 \end{bmatrix} = \begin{bmatrix} 12 & 17 \\ -11 & 2 \\ 3 & 9 \end{bmatrix} - \begin{bmatrix} 3 & 6 \\ -13 & -4 \\ 2 & 12 \end{bmatrix}$,

(b) $4 \begin{bmatrix} 3 & 8 \\ 1 & 9 \end{bmatrix} - 3 \begin{bmatrix} 1 & -2 \\ 0 & 1 \end{bmatrix} = 2 \begin{bmatrix} 2 & -1 \\ 7 & 4 \end{bmatrix} + 5 \begin{bmatrix} 1 & 8 \\ -2 & 5 \end{bmatrix}$,

(c) if $\mathbf{A} = \begin{bmatrix} 3 & 8 & 4 \\ 8 & 7 & -1 \\ 4 & -1 & 2 \end{bmatrix} + \begin{bmatrix} 1 & -1 & 3 \\ -1 & 2 & 4 \\ 3 & 4 & 6 \end{bmatrix}$, then $\mathbf{A} = \mathbf{A}'$.

E 2.6. For $\mathbf{A} = \begin{bmatrix} 1 & 2 & 3 \\ 6 & 8 & 4 \end{bmatrix}$ and $\mathbf{B} = \begin{bmatrix} 1 & 6 \\ 8 & 9 \\ 2 & -7 \end{bmatrix}$,

(a) Write down the transpose of $\mathbf{A}$ and of $\mathbf{B}$.

(b) Calculate $\mathbf{A} + \mathbf{B}'$ and $\mathbf{A}' + \mathbf{B}$.

(c) What relationship exists between the two sums obtained in (b)?

E 2.7. (a) If, in general, $\mathbf{A}$ is $r \times c$, what must be the order of $\mathbf{B}$ so that $\mathbf{A} + \mathbf{B}'$ exists? Why? For $\mathbf{A}_{r \times c}$, prove the relationship between $\mathbf{A} + \mathbf{B}'$ and $\mathbf{A}' + \mathbf{B}$ illustrated in Exercise 2.6.

(b) Explain why $\mathbf{AVA}' = \mathbf{BVB}'$ implies that $\mathbf{A}$ and $\mathbf{B}$ have the same order and $\mathbf{V}$ is square.

E 2.8. (Free-trade agreements) Free-trade agreements promote international trade, often leading to changes in employment statistics in the countries concerned. Suppose employment (thousands of jobs) before, and five years after, a trade agreement is represented, respectively, by the two matrices

$$\begin{bmatrix} 12 & 9 & 8 & 4 \\ 10 & 2 & 4 & 3 \\ 6 & 1 & 10 & 1 \end{bmatrix} \quad \text{and} \quad \begin{bmatrix} 8 & 10 & 2 & 8 \\ 9 & 2 & 3 & 5 \\ 10 & 0 & 18 & 0 \end{bmatrix}.$$

In these matrices, rows represent countries and columns represent the agricultural, automobile, apparel, and finance industries. What are the shifts in employment over the five years?

E 2.9. Prove that $(\mathbf{ABCD})' = \mathbf{D}'\mathbf{C}'\mathbf{B}'\mathbf{A}'$.

Note: Whenever asked to prove a result such as this, one can assume that the matrices involved are conformable for the expressions given; e.g., $\mathbf{A}$, $\mathbf{B}$, $\mathbf{C}$ and $\mathbf{D}$ are here conformable for the product $\mathbf{ABCD}$.

E 2.10. Under what conditions do both $\mathrm{tr}(\mathbf{ABC})$ and $\mathrm{tr}(\mathbf{BAC})$ exist? When they exist, do you expect them to be equal? Why?

Calculate tr(**ABC**) and tr(**BAC**) for

$$\mathbf{A} = \begin{bmatrix} 1 & 2 \\ 2 & 3 \end{bmatrix}, \quad \mathbf{B} = \begin{bmatrix} 1 & 0 \\ 2 & -1 \end{bmatrix} \quad \text{and} \quad \mathbf{C} = \begin{bmatrix} 1 & 3 \\ 4 & 5 \end{bmatrix}.$$

E 2.11. Explain why each of the following statements is true.

(a) $(\mathbf{A} + \mathbf{A'})\mathbf{B} = \mathbf{AB} + \mathbf{A'B}$.

(b) $\text{tr}(\mathbf{B'B}) = \text{tr}(\mathbf{BB'})$.

(c) $\mathbf{x'B'Bx} = (\mathbf{Bx})'\mathbf{Bx} = \text{tr}(\mathbf{B'Bxx'})$.

Verify each statement for the following values of **A**, **B** and **x**:

$$\mathbf{A} = \begin{bmatrix} -1 & 2 \\ 7 & 3 \end{bmatrix}, \quad \mathbf{B} = \begin{bmatrix} 6 & 8 & -1 & 0 \\ 2 & 3 & 1 & 4 \end{bmatrix}, \quad \mathbf{x'} = \begin{bmatrix} 1 & 1 & 2 & 3 \end{bmatrix}.$$

E 2.12. For **A** and **B** having the same order, explain why

$$\begin{aligned} (\mathbf{A} + \mathbf{B})(\mathbf{A} + \mathbf{B})' &= (\mathbf{A} + \mathbf{B})(\mathbf{A'} + \mathbf{B'}) \\ &= \mathbf{AA'} + \mathbf{AB'} + \mathbf{BA'} + \mathbf{BB'}. \end{aligned}$$

Will these expressions also generally equal $(\mathbf{A} + \mathbf{B})'(\mathbf{A} + \mathbf{B})$? Why must **A** and **B** have the same order here? Verify these equalities for

$$\mathbf{A} = \begin{bmatrix} 2 & 3 & 7 \\ 6 & 1 & -2 \end{bmatrix} \quad \text{and} \quad \mathbf{B} = \begin{bmatrix} -1 & 0 & 1 \\ 0 & 1 & 7 \end{bmatrix}.$$

E 2.13. For a matrix **A** from any preceding exercise, calculate **AA'** and **A'A** and verify that the trace of each is the sum of squares of elements of **A**.

E 2.14. (a) When does $(\mathbf{A} + \mathbf{B})(\mathbf{A} - \mathbf{B}) = \mathbf{A}^2 - \mathbf{B}^2$?

(b) When $\mathbf{A} = \mathbf{A'}$, prove that $\text{tr}(\mathbf{AB}) = \text{tr}(\mathbf{AB'})$.

(c) When $\mathbf{X'XGX'X} = \mathbf{X'X}$, prove that $\mathbf{X'XG'X'X} = \mathbf{X'X}$.

E 2.15. With $\mathbf{A} = \begin{bmatrix} 1 & 0 & 3 \\ 2 & -1 & 0 \\ 0 & 4 & 1 \end{bmatrix}$ and $\mathbf{B} = \begin{bmatrix} 3 & 3 & 1 \\ -4 & 2 & -1 \\ -2 & 0 & 0 \end{bmatrix}$:

(a) Partition **A** and **B** as

$$\mathbf{A} = \begin{bmatrix} \mathbf{A}_{11} & \mathbf{A}_{12} \\ \mathbf{A}_{21} & \mathbf{A}_{22} \end{bmatrix}, \quad \mathbf{B} = \begin{bmatrix} \mathbf{B}_{11} & \mathbf{B}_{12} \\ \mathbf{B}_{21} & \mathbf{B}_{22} \end{bmatrix},$$

where both $\mathbf{A}_{21}$ and $\mathbf{B}_{21}$ have order 1×2.

(b) Calculate **AB** both with and without the partitioning, to demonstrate the validity of multiplication of partitioned matrices.

(c) Calculate $\mathbf{AB}'$, showing that

$$\mathbf{B}' = \begin{bmatrix} \mathbf{B}'_{11} & \mathbf{B}'_{21} \\ \mathbf{B}'_{12} & \mathbf{B}'_{22} \end{bmatrix}.$$

E 2.16. Partition the following matrices in any manner conformable for the products indicated. Calculate the products both with and without using the partitioned forms, to test their validity.

(a) $\mathbf{A} = \begin{bmatrix} 1 & 1 & 0 \\ 0 & 1 & 1 \\ 1 & 1 & 1 \end{bmatrix}$, $\mathbf{B} = \begin{bmatrix} 2 & 1 \\ 1 & 3 \\ 0 & 1 \end{bmatrix}$, and the product **AB**.

(b) $\mathbf{C} = \begin{bmatrix} 4 & 1 \\ -1 & 1 \end{bmatrix}$, $\mathbf{D} = \begin{bmatrix} 1 & 0 & 0 & 4 \\ 0 & 2 & 3 & -5 \end{bmatrix}$, and **CD**.

(c) $\mathbf{E} = \begin{bmatrix} 1 & 0 \\ 3 & -1 \end{bmatrix}$, $\mathbf{F} = \begin{bmatrix} 0 & 1 & 3 \\ 4 & 4 & -2 \end{bmatrix}$,

$$\mathbf{G} = \begin{bmatrix} 1 & 2 & 3 \\ 4 & -3 & -2 \\ -1 & 0 & -1 \end{bmatrix}, \quad \text{and the product } \mathbf{EFG}$$

E 2.17. (a) Prove that for any **P**, **Q**, **R** and **S**

$$\begin{bmatrix} \mathbf{I} \\ \mathbf{A} \end{bmatrix} \mathbf{B} \begin{bmatrix} \mathbf{I} & \mathbf{C} \end{bmatrix} = \begin{bmatrix} \mathbf{I} & \mathbf{P} \\ \mathbf{A} & \mathbf{Q} \end{bmatrix} \begin{bmatrix} \mathbf{B} & 0 \\ 0 & 0 \end{bmatrix} \begin{bmatrix} \mathbf{I} & \mathbf{C} \\ \mathbf{R} & \mathbf{S} \end{bmatrix}.$$

(b) For **A** of order $p \times q$, **B** of order $r \times s$ and $\mathbf{C}_{t \times w}$ in (a), what relationships exist among p, q, r, s, t and w?

(c) If, in addition to (b), the lower right null matrix in the matrix containing **B** is $k \times m$, what are the orders of all the matrices in (a)?

E 2.18. (Payoff matrix) A construction company must decide whether to build condominiums, townhouses, single-family homes or apartment rental units. The outcomes they face relate to the growth of the company's locality being low, medium or high. Suppose the payoff matrix (in units of $1000) related to these actions and outcomes is

$$\mathbf{P} = \begin{bmatrix} 550 & 350 & 450 \\ 400 & 525 & 650 \\ -100 & 475 & 425 \\ 750 & 700 & 500 \end{bmatrix}.$$

Using the minimax regret criterion, determine the regret matrix, the maximum regret vector and the optimal strategy for the construction company.

E 2.19. (Game theory) Another aspect of game theory is the two-person game, in which there are two payoff matrices, one for each player. Although the rows and columns of payoff matrices in one-person games represent actions and outcomes, in two-person games rows represent one person's actions and columns represent the other person's actions. Within this framework consider the payoff matrices for players p and q as being

$$\mathbf{P} = \begin{bmatrix} 10 & 1 & 8 \\ 6 & 7 & 17 \\ 22 & 4 & 10 \end{bmatrix} \quad \text{and} \quad \mathbf{Q} = \begin{bmatrix} 4 & 18 & 12 \\ 8 & 8 & 14 \\ 1 & 10 & 13 \end{bmatrix}.$$

(a) What actions by p and q maximize p's payoff?

(b) What actions by p and q maximize q's payoff?

(c) If p and q combined as a team, what is the team's payoff matrix; and what action maximizes the team's payoff?

(d) Which actions would p prefer if p wanted to maximize the difference between her payoff and q's payoff?

(e) Answer (d) again, after interchanging p and q.

E 2.20. (Manufacturing costs) Shipping costs (in dollars) from three manufacturing plants to four destinations are shown in the following matrix:

$$\begin{bmatrix} 10 & 15 & 9 & 7 \\ 14 & 8 & 12 & 8 \\ 6 & 14 & 22 & 17 \end{bmatrix}.$$

Production costs are $40, $38 and $41 at the three plants. What is the matrix of total costs, production plus shipping?

E 2.21. (Graph theory) After several meetings of a four-person committee, its secretary realizes that A dominates C, B dominates A, C dominates D and B, and D dominates A.

(a) Draw a graph showing these dominations.

(b) Write a matrix representing them.

(c) Calculate the first five powers of the matrix.

(d) What do you notice about these powers?

E 2.22. (Transition probabilities) Suppose a college senior who is interviewing for jobs in banking, in marketing and in consulting, determines that if her first offer is in banking then the next offer she receives has a probability of .6 of being in banking, .2 of being in marketing and .2 of being in consulting. But if the first offer is in marketing, the probabilities are .3, 0, and .7 that the next offer will be in banking, marketing, and consulting, respectively. Finally, for a first offer in consulting those three probabilities are .2, .1 and .7, respectively.

(a) Write down the transition probability matrix for the college senior.

(b) Assuming the senior has received an offer in banking, what are the job offer probabilities (i).   for a second offer?

(ii).   for a third offer?

(iii).   for a fourth offer?

(c) Repeat (b) for the first offer being in marketing.

(d) Repeat (b) for the first offer being in consulting.

(e) Do you notice anything significant about the results in (b), (c) and (d)?

E 2.23.   (Steady-state vector) Suppose for some large positive integer s and a transition probability matrix $\mathbf{P}$ that the sth and $(s+1)$th powers of $\mathbf{P}$ are effectively equal. Then

$$\mathbf{x}_{s+1} = \mathbf{x}_0'\mathbf{P}^{s+1} = \mathbf{x}_0'\mathbf{P}^s = \mathbf{x}_s$$

and so the standard relationship $\mathbf{x}_{s+1}' = \mathbf{x}_s'\mathbf{P}$ becomes $\mathbf{x}_s' = \mathbf{x}_s'\mathbf{P}$. Then $\mathbf{x}_s'$ is known as the *steady-state vector* and can be determined (so long as s exists, i.e., so long as $\mathbf{P}^n$ tends to a limit as n tends to infinity) from the equations $\mathbf{x}_s' = \mathbf{x}_s'P$. The equation $\mathbf{x}_s'1 = 1$ must also be used. Find $\mathbf{x}_s$ for $\mathbf{P}$ of (2.24).

Chapter 3

SPECIAL MATRICES

Numerous matrices with particular properties have attracted special names. Historically, each of them probably originated in a specific mathematical problem or application, and was later found to have properties of broader interest. Of the vast number of such matrices, this chapter presents but a small selection of general use and interest. Others appear in subsequent chapters.

3.1 Linear Transformations

Section 2.6b shows that the product $\mathbf{Ax}$ is a vector. Call it $\mathbf{y}$, so that $\mathbf{Ax} = \mathbf{y}$; e.g.,

$$\mathbf{y} = \begin{bmatrix} y_1 \\ y_2 \end{bmatrix} = \mathbf{Ax} = \begin{bmatrix} a_{11} & a_{12} & a_{13} \\ a_{21} & a_{22} & a_{23} \end{bmatrix} \begin{bmatrix} x_1 \\ x_2 \\ x_3 \end{bmatrix}$$

gives

$$y_1 = a_{11}x_1 + a_{12}x_2 + a_{13}x_3$$

and

$$y_2 = a_{21}x_1 + a_{22}x_2 + a_{23}x_3.$$

Thus (so long as each row of $\mathbf{A}$ has at least one non-zero element) each of y_1 and y_2 is a weighted sum of the elements of $\mathbf{x}$, the weights being elements of $\mathbf{A}$. Such a weighted sum is called a *linear combination* of the elements of $\mathbf{x}$; it is linear because the elements of $\mathbf{x}$ are not raised to powers other than unity (and with only one element, $y = ax$ represents a straight line), and no term has more than one element of $\mathbf{x}$. In this situation the matrix $\mathbf{A}$ of $\mathbf{y} = \mathbf{Ax}$ is said to represent the *linear transformation* of **x** into **y**. It is the means by which **x** is transformed (linearly) into **y**. And, if further, **x** = **Bw**, then **y** = **Ax** = **ABw** and **AB** represents the linear transformation of **w** into **y**.

Example

In the example (2.19),

$$\mathbf{Ax} = \begin{bmatrix} 5 & 10 & 20 \\ 7 & 12 & 23 \end{bmatrix} \begin{bmatrix} 22 \\ 30 \\ 40 \end{bmatrix} = \begin{bmatrix} 1210 \\ 1434 \end{bmatrix} = \mathbf{y},$$

A is the matrix of prices which transforms **x**, the vector of quantities, into **y**, the vector of sales revenue.

3.2   Symmetry: Some Basic Outcomes

Section 2.1c describes a square matrix as being symmetric, namely

$$\mathbf{A} = \mathbf{A}', \quad \text{iff} \quad \alpha_{ij} = a_{ji} \,\forall\, i \neq j.$$

We here consider some basic outcomes of symmetry.

a.   Verifying symmetry

Symmetric matrices have many special properties which are helpful in practical problems. Therefore, when we have a matrix which we think might be symmetric, it is useful to be able to verify that. Because $\mathbf{A}' = \mathbf{A}$ is the only definition of **A** being symmetric, the only way of establishing whether some matrix **X** is symmetric or not is to transpose **X** and ascertain whether **X**′ equals **X** or not. If **X**′ equals **X** then, of course, **X** is symmetric.

b.   Product of symmetric matrices

In general, a product of symmetric matrices is not symmetric. First, from Section 2.6j

$$(\mathbf{AB})' = \mathbf{B}'\mathbf{A}';$$

then for symmetric **A** and **B**

$$(\mathbf{AB})' = \mathbf{B}'\mathbf{A}' = \mathbf{BA}, \quad \text{for} \quad \mathbf{A} = \mathbf{A}' \quad \text{and} \quad \mathbf{B} = \mathbf{B}'.$$

But even though $\mathbf{AB}$ and $\mathbf{BA}$ will here be square and of the same order (due to $\mathbf{A} = \mathbf{A}'$ and $\mathbf{B} = \mathbf{B}'$), the product $\mathbf{BA}$ is generally different from $\mathbf{AB}$, and so $\mathbf{AB}'$ is not $\mathbf{AB}$; hence $\mathbf{AB}$ is not symmetric. The reader should demonstrate the lack of symmetry using two simple matrices of order two.

c. Properties of $\mathbf{AA}'$ and $\mathbf{A}'\mathbf{A}$

Products of a matrix and its transpose always exist and are symmetric:

$$(\mathbf{AA}')' = (\mathbf{A}')'\mathbf{A}' = \mathbf{AA}' \quad \text{and} \quad (\mathbf{A}'\mathbf{A})' = \mathbf{A}'(\mathbf{A}')' = \mathbf{A}'\mathbf{A}.$$

Although both products $\mathbf{AA}'$ and $\mathbf{A}'\mathbf{A}$ are symmetric, they are not necessarily equal. In fact, only when $\mathbf{A}$ is square might they be equal, because it is only then that both products have the same order. Thus for $\mathbf{A}_{r \times c}$, the product $\mathbf{AA}' = \mathbf{A}_{r \times c}(\mathbf{A}')_{c \times r}$ has order $r \times r$, whereas $\mathbf{A}'\mathbf{A} = (\mathbf{A}')_{c \times r}\mathbf{A}_{r \times c}$ has order $c \times c$.

The nature of matrix multiplication ensures that elements of $\mathbf{AA}'$ and of $\mathbf{A}'\mathbf{A}$ are inner products of rows and of columns, respectively, of $\mathbf{A}$. Thus for $\mathbf{A}_{r \times c}$

$$\mathbf{AA}' = \left\{_m \text{ inner product of } i\text{th and } k\text{th rows of } \mathbf{A} \right\}_{i,k=1}^{r} \tag{3.1}$$

and

$$\mathbf{A}'\mathbf{A} = \left\{_m \text{ inner product of } j\text{th and } t\text{th columns of } \mathbf{A} \right\}_{j,t=1}^{c}.$$

In particular, the ith diagonal element of $\mathbf{AA}'$ is a sum of squares of the elements of the ith row of $\mathbf{A}$:

$$i\text{th diagonal element of } \mathbf{AA}' \text{ is } \sum_{j=1}^{c} a_{ij}^2. \tag{3.2}$$

Confining attention to real matrices (Section 1.3), we can use the fact that a sum of squares of real numbers is positive (unless they are all zero), and so observe that for real matrices $\mathbf{A}$, the matrix $\mathbf{AA}'$ has diagonal elements that are positive (or zero). Similar results hold for $\mathbf{A}'\mathbf{A}$ in terms of columns of $\mathbf{A}$. This fact is also the basis for proving that for any real matrix $\mathbf{A}$

$$\mathbf{AA}' = 0 \quad \text{implies} \quad \mathbf{A} = 0 \tag{3.3}$$

and

$$\text{tr}(\mathbf{AA}') = 0 \quad \text{implies} \quad \mathbf{A} = \mathbf{0}. \tag{3.4}$$

Result (3.3) is true because $\mathbf{AA}' = \mathbf{0}$ implies that every diagonal element of $\mathbf{AA}'$ is zero; and for the ith such element this means, from (3.2), that $\Sigma_{j=1}^c a_{ij}^2 = 0$. Hence $a_{ij} = 0$ for $j = 1, \ldots, c$; and this is so for every i. Therefore $\mathbf{A} = \mathbf{0}$. Proof of (3.4) is similar,

$$\text{tr}(\mathbf{AA}') = \sum_{i=1}^r (i\text{th diagonal element of } \mathbf{AA}') = \sum_{i=1}^c \sum_{j=1}^c a_{ij}^2$$

and so $\text{tr}(\mathbf{AA}') = 0$ implies that every a_{ij} is zero; i.e., that every element of $\mathbf{A}$ is zero. Hence $\mathbf{A} = \mathbf{0}$.

Results (3.3) and (3.4) are seldom useful for the sake of some particular matrix $\mathbf{A}$, but they can be helpful in developing results in matrix algebra when $\mathbf{A}$ is a function of other matrices. For example, by means of (3.3) we can prove, for real matrices, $\mathbf{P}$, $\mathbf{Q}$ and $\mathbf{X}$, that

$$\mathbf{PXX}' = \mathbf{QXX}' \quad \text{implies} \quad \mathbf{PX} = \mathbf{QX}. \tag{3.5}$$

The proof consists of observing that

$$(\mathbf{PXX}' - \mathbf{QXX}')(\mathbf{P}' - \mathbf{Q}') \equiv (\mathbf{PX} - \mathbf{QX})(\mathbf{PX} - \mathbf{QX})'. \tag{3.6}$$

Hence if $\mathbf{PXX}' = \mathbf{QXX}'$, the left-hand side of (3.6) is null and therefore so is the right-hand side; hence by (3.3), $\mathbf{PX} - \mathbf{QX} = \mathbf{0}$; i.e., $\mathbf{PX} = \mathbf{QX}$, and (3.5) is established.

d.   Products of vectors

The inner product (Section 2.6a) of two vectors is a scalar; and through thinking of a scalar as a 1×1 matrix (Section 1.8h) we can say it is symmetric. For example,

$$\mathbf{x}'\mathbf{y} = \begin{bmatrix} 1 & 0 & 2 \end{bmatrix} \begin{bmatrix} 3 \\ 2 \\ 1 \end{bmatrix} = 5 = \begin{bmatrix} 3 & 2 & 1 \end{bmatrix} \begin{bmatrix} 1 \\ 0 \\ 2 \end{bmatrix} = \mathbf{y}'\mathbf{x} = (\mathbf{x}'\mathbf{y})'. \tag{3.7}$$

Moreover, with the scalar $\mathbf{x}'\mathbf{y}$ being deemed a 1×1 matrix we can consider the trace of $\mathbf{x}'\mathbf{y}$ which, by Section 2.6k, is

$$\text{tr}(\mathbf{x}'\mathbf{y}) = \text{tr}(\mathbf{yx}'),$$

where $\mathbf{yx}'$ is an outer product (Section 2.6e):

$$\text{tr}(\mathbf{x}'\mathbf{y}) = \text{tr}(5) = 5 = \text{tr}\left(\begin{bmatrix} 3 \\ 2 \\ 1 \end{bmatrix} \begin{bmatrix} 1 & 0 & 2 \end{bmatrix} \right) = \text{tr} \begin{bmatrix} 3 & 0 & 6 \\ 2 & 0 & 4 \\ 1 & 0 & 2 \end{bmatrix} = 3 + 2.$$

e. Sums of outer products

In the notation of (2.7) and (2.8) of Section 2.2d, where $\mathbf{a_j}$ is the jth column of $\mathbf{A}$ and $\boldsymbol{\beta}'_j$ is the jth row of $\mathbf{B}$, an application of multiplying partitioned matrices (Section 2.6m) is

$$\mathbf{AB} = [\mathbf{a}_1 \quad \mathbf{a}_2 \quad \cdots \quad \mathbf{a}_c] \begin{bmatrix} \boldsymbol{\beta}'_1 \\ \boldsymbol{\beta}'_2 \\ \vdots \\ \boldsymbol{\beta}'_c \end{bmatrix} = \sum_{j=1}^{c} \mathbf{a}_j \boldsymbol{\beta}'_j. \tag{3.8}$$

Thus $\mathbf{AB}$ is the sum of the outer products of each column of $\mathbf{A}$ with its corresponding row in $\mathbf{B}$.

Example

$$\begin{aligned} \mathbf{AB} &= \begin{bmatrix} 1 & 4 \\ 2 & 5 \\ 3 & 6 \end{bmatrix} \begin{bmatrix} 7 & 8 \\ 9 & 10 \end{bmatrix} = \begin{bmatrix} 43 & 48 \\ 59 & 66 \\ 75 & 84 \end{bmatrix} \\ &= \begin{bmatrix} 1 \\ 2 \\ 3 \end{bmatrix} [7 \quad 8] + \begin{bmatrix} 4 \\ 5 \\ 6 \end{bmatrix} [9 \quad 10] = \begin{bmatrix} 7 & 8 \\ 14 & 16 \\ 21 & 24 \end{bmatrix} + \begin{bmatrix} 36 & 40 \\ 45 & 50 \\ 54 & 60 \end{bmatrix}. \end{aligned}$$

A special case of (3.8) is when $\mathbf{B} = \mathbf{A}'$:

$$\mathbf{AA}' = \sum_{j=1}^{c} \mathbf{a}_j \mathbf{a}'_j. \tag{3.9}$$

$\mathbf{AA}'$ is thus the sum of outer products of each column of $\mathbf{A}$ with itself:

$$\begin{bmatrix} 1 & 4 \\ 2 & 5 \\ 3 & 6 \end{bmatrix} \begin{bmatrix} 1 & 2 & 3 \\ 4 & 5 & 6 \end{bmatrix} = \begin{bmatrix} 1 \\ 2 \\ 3 \end{bmatrix} [1 \quad 2 \quad 3] + \begin{bmatrix} 4 \\ 5 \\ 6 \end{bmatrix} [4 \quad 5 \quad 6].$$

It is left to the reader to confirm this simple arithmetic.

f. Elementary vectors

A special case of (3.9) is

$$\mathbf{I}_n = \sum_{i=1}^{n} \mathbf{e}_i \mathbf{e}'_i$$

for $\mathbf{e}_i$ being the ith column of $\mathbf{I}_n$, namely a vector with unity for its ith element and zeros elsewhere. $\mathbf{e}_i$ is called an *elementary vector*. When necessary, its order n can be identified by denoting $\mathbf{e}_i$ as $\mathbf{e}_i^{(n)}$.

The outer product of one elementary vector with another is a null matrix except for one element of unity; for example, with

$$\mathbf{e}_1 = \begin{bmatrix} 1 \\ 0 \\ 0 \end{bmatrix} \quad \text{and} \quad \mathbf{e}_2 = \begin{bmatrix} 0 \\ 1 \\ 0 \end{bmatrix}, \quad \mathbf{E}_{12} = \mathbf{e}_1\mathbf{e}_2' = \begin{bmatrix} 0 & 1 & 0 \\ 0 & 0 & 0 \\ 0 & 0 & 0 \end{bmatrix}.$$

In general, $\mathbf{E}_{ij} = \mathbf{e}_i\mathbf{e}_j'$ is null except for element (i,j) being unity; and, of course, $\mathbf{I} = \Sigma_i\mathbf{E}_{ii}$. These $\mathbf{E}_{ij}$-matrices are particularly useful in applications of calculus to matrix algebra (see Section 12.9f).

The e-vectors are also useful for delineating individual rows and columns of a matrix. Thus $\mathbf{e}_i'\mathbf{A} = \boldsymbol{\alpha}_i'$, the ith row of $\mathbf{A}$, and $\mathbf{A}\mathbf{e}_j = \mathbf{a}_j$, the jth column of $\mathbf{A}$; e.g., for

$$\mathbf{e}_2 = \begin{bmatrix} 0 \\ 1 \end{bmatrix} \quad \text{and} \quad \mathbf{A} = \begin{bmatrix} 2 & 3 \\ 5 & 7 \end{bmatrix}, \quad \mathbf{e}_2'\mathbf{A} = \begin{bmatrix} 5 & 7 \end{bmatrix} \quad \text{and} \quad \mathbf{A}\mathbf{e}_2 = \begin{bmatrix} 3 \\ 7 \end{bmatrix}.$$

g. Skew-symmetric matrices

A symmetric matrix $\mathbf{A}$ has the property $\mathbf{A} = \mathbf{A}'$. In contrast there are also (square) matrices $\mathbf{B}$ having the property $\mathbf{B}' = -\mathbf{B}$; they are said to be *skew-symmetric*. Their diagonal elements are zero and each off-diagonal element is minus its symmetric partner; i.e., $b_{ii} = 0$ and $b_{ij} = -b_{ji}$. An example is

$$\mathbf{B} = \begin{bmatrix} 0 & 1 & -3 \\ -1 & 0 & 2 \\ 3 & -2 & 0 \end{bmatrix}.$$

3.3 Summing Vectors and Their Products

Vectors having 1.0 for every element are called *summing vectors*. Their prime use is in expressing sums as inner products. For example,

$$\mathbf{1}'\mathbf{x} = \begin{bmatrix} 1 & 1 & 1 & 1 \end{bmatrix} \begin{bmatrix} x_1 \\ x_2 \\ x_3 \\ x_4 \end{bmatrix} = x_1 + x_2 + x_3 + x_4 = \mathbf{x}'\mathbf{1}.$$

For clarity, the order of a summing vector can be used as a subscript: $\mathbf{1}'_4 = \begin{bmatrix} 1 & 1 & 1 & 1 \end{bmatrix}$. For example,

$$\mathbf{1}'_3 \mathbf{X} = \begin{bmatrix} 1 & 1 & 1 \end{bmatrix} \begin{bmatrix} 2 & -1 \\ -5 & -3 \\ 4 & 5 \end{bmatrix} = \begin{bmatrix} 1 & 1 \end{bmatrix} = \mathbf{1}'_2.$$

Were this equation to be written as $\mathbf{1}'\mathbf{X} = \mathbf{1}'$, one might think that both $\mathbf{1}'$ vectors had the same order. That they do not is made clear by denoting $\mathbf{X}$ of order $r \times c$ as $\mathbf{X}_{r \times c}$ so that $\mathbf{1}'\mathbf{X}_{r \times c} = \mathbf{1}'$ is $\mathbf{1}'_r \mathbf{X}_{r \times c} = \mathbf{1}'_c$.

The inner product of a summing vector with itself is a scalar, the vector's order:

$$\mathbf{1}'_n \mathbf{1}_n = n. \tag{3.10}$$

And outer products are matrices with all elements unity:

$$\mathbf{1}_3 \mathbf{1}'_2 = \begin{bmatrix} 1 \\ 1 \\ 1 \end{bmatrix} \begin{bmatrix} 1 & 1 \end{bmatrix} = \begin{bmatrix} 1 & 1 \\ 1 & 1 \\ 1 & 1 \end{bmatrix} = \mathbf{J}_{3 \times 2}.$$

In general, $\mathbf{1}_r \mathbf{1}'_s$ has order $r \times s$ and is often denoted by the symbol $\mathbf{J}$ or $\mathbf{J}_{r \times s}$:

$$\mathbf{1}_r \mathbf{1}'_s = \mathbf{J}_{r \times s}, \text{ having all elements unity.} \tag{3.11}$$

Clearly, $\lambda \mathbf{J}_{r \times s}$ has all elements λ.

Products of $\mathbf{J}$s with each other and with $\mathbf{1}$s are, respectively, $\mathbf{J}$s and $\mathbf{1}$s (multiplied by scalars):

$$\mathbf{J}_{r \times s} \mathbf{J}_{s \times t} = s \mathbf{J}_{r \times t}, \quad \mathbf{1}'_r \mathbf{J}_{r \times s} = r \mathbf{1}'_s \quad \text{and} \quad \mathbf{J}_{r \times s} \mathbf{1}_s = s \mathbf{1}_r. \tag{3.12}$$

Particularly useful are square $\mathbf{J}$s and a variant thereof:

$$\mathbf{J}_n = \mathbf{1}_n \mathbf{1}'_n \quad \text{with} \quad \mathbf{J}_n^2 = n \mathbf{J}_n; \tag{3.13}$$

and $\bar{\mathbf{J}}_n$ is defined as square, of order n, with every element being $\frac{1}{n}$:

$$\bar{\mathbf{J}}_n = \frac{1}{n} \mathbf{J}_n \quad \text{with} \quad \bar{\mathbf{J}}_n^2 = \bar{\mathbf{J}}_n.$$

And for statistics

$$\mathbf{C}_n = \mathbf{I} - \bar{\mathbf{J}}_n = \mathbf{I} - \frac{1}{n} \mathbf{J}_n, \tag{3.14}$$

known as a *centering matrix*, is especially useful, with properties

$$\mathbf{C} = \mathbf{C}' = \mathbf{C}^2, \quad \mathbf{C1} = \mathbf{0} \quad \text{and} \quad \mathbf{CJ} = \mathbf{JC} = \mathbf{0}, \tag{3.15}$$

which the reader can easily verify.

Illustration   (Statistics)

The mean and sum of squares about the mean for data $x_1, x_2, \ldots, x_n$ can be expressed succinctly using 1-vectors and $\mathbf{J}$-matrices. Define

$$\mathbf{x}' = [x_1 \quad x_2 \quad \cdots \quad x_n].$$

Then the mean of the xs is

$$\bar{x} = (x_1 + x_2 + \cdots + x_n)/n = \sum_{i=1}^{n} x_i/n = \frac{1}{n}\mathbf{x}'\mathbf{1} = \frac{1}{n}\mathbf{1}'\mathbf{x},$$

the last equality arising from $\mathbf{x}'\mathbf{y} = \mathbf{y}'\mathbf{x}$ of (3.7). And using $\mathbf{C}$ of (3.14) and (3.15),

$$\mathbf{x}'\mathbf{C} = \mathbf{x}' - \mathbf{x}'\bar{\mathbf{J}} = \mathbf{x}' - \frac{1}{n}\mathbf{x}'\mathbf{1}\mathbf{1}' = \mathbf{x}' - \bar{x}\mathbf{1}' = \left\{_r \; x_i - \bar{x}\right\}_{i=1}^{n}$$

is the data vector with each observation expressed as a deviation from $\bar{x}$. (This is the origin of naming $\mathbf{C}$ as centering matrix.) Post-multiplying $\mathbf{x}'\mathbf{C}$ by $\mathbf{x}$ then gives

$$\mathbf{x}'\mathbf{C}\mathbf{x} = (\mathbf{x}' - \bar{x}\mathbf{1}')\mathbf{x} = \mathbf{x}'\mathbf{x} - \bar{x}(\mathbf{1}'\mathbf{x}) = \mathbf{x}'\mathbf{x} - n\bar{x}^2.$$

Hence, using a standard result in statistics, we get

$$\sum_{i=1}^{n}(x_i - \bar{x})^2 = \sum_{i=1}^{n} x_i^2 - n\bar{x}^2 = \mathbf{x}'\mathbf{x} - n\bar{x}^2 = \mathbf{x}'\mathbf{C}\mathbf{x}. \tag{3.16}$$

Thus for $\mathbf{x}'$ being a data vector, $\mathbf{x}'\mathbf{1}/n$ is the mean, $\mathbf{x}'\mathbf{C}$ is the vector of deviations about the mean and $\mathbf{x}'\mathbf{C}\mathbf{x}$ is the sum of squares about the mean.

Expression (3.16) is a special case of the form $\mathbf{x}'\mathbf{A}\mathbf{x}$, known as a quadratic form (see Section 3.6), which can be used for sums of squares generally. Expressed in this manner, and with the aid of other matrix concepts (notably idempotency; see Section 3.4), sums of squares of normally distributed xs are known to be distributed as χ^2 under very simply stated (in matrix notation) conditions.

Illustration   (Taxicab, *continued*)

It was noted in equation (2.26) of Section 2.6c that row sums of a transition probability matrix are always unity. For example, with the taxi illustration

$$\mathbf{P} = \begin{bmatrix} .2 & .8 \\ .4 & .6 \end{bmatrix} \text{ with } \begin{bmatrix} .2 + .8 \\ .4 + .6 \end{bmatrix} = \begin{bmatrix} 1 \\ 1 \end{bmatrix}; \text{ i.e., } \begin{bmatrix} .2 & .8 \\ .4 & .6 \end{bmatrix}\begin{bmatrix} 1 \\ 1 \end{bmatrix} = \begin{bmatrix} 1 \\ 1 \end{bmatrix}.$$

This last result is $\mathbf{P1} = \mathbf{1}$, which is true generally for any transition probability matrix. Furthermore, because $\mathbf{P^2 1} = \mathbf{P(P1)} = \mathbf{P1} = \mathbf{1}$, this in turn extends to $\mathbf{P^n 1} = \mathbf{1}$, showing that row sums of any power of a transition probability matrix are also unity. Furthermore, since the elements of a state vector (see Section 2.6e) add to unity, $\mathbf{x'_0 1} = 1$, so do those of any state vector derived from it because $\mathbf{x'_n 1} = \mathbf{x'_0 P^n 1} = \mathbf{x'_0 1} = 1$.

3.4 Idempotent Matrices

In the preceding section, two matrices ($\bar{\mathbf{J}}$ and $\mathbf{C}$) are each shown as having the property that its square equals itself. Many different matrices have this property: they are said to be *idempotent* (from Latin *idem* meaning "same" and *potent* "power"). In general,

$$\mathbf{K}^2 = \mathbf{K} \text{ defines } \mathbf{K} \text{ as idempotent.}$$

All idempotent matrices are square (so that $\mathbf{K}^2$ exists); and two of the simplest kind of idempotent matrices are identity matrices and square null matrices. If $\mathbf{K}$ is idempotent, then

$\mathbf{K}$ is square;

$\mathbf{K}^r = \mathbf{K}$ for r being a positive integer;

$\mathbf{I} - \mathbf{K}$ is idempotent;

$\mathbf{K} - \mathbf{I}$ is not idempotent; and

if $\mathbf{L}$ is idempotent so is $\mathbf{KL}$, provided $\mathbf{KL} = \mathbf{LK}$.

The suffix "potent" occurs in two other special matrices: $\mathbf{A}$ satisfying $\mathbf{A}^2 = \mathbf{0}$ is *nilpotent*; and $\mathbf{B}$ satisfying $\mathbf{B}^2 = \mathbf{I}$ is *unipotent*. These are nowhere nearly as important or useful as are idempotent matrices, which play an especially important role in properties of sums of squares in statistics.

Examples

$$\mathbf{A} = \begin{bmatrix} 1 & 2 & 5 \\ 2 & 4 & 10 \\ -1 & -2 & -5 \end{bmatrix} \text{ is nilpotent; } \quad \mathbf{B} = \begin{bmatrix} \mathbf{I} & \mathbf{X} \\ \mathbf{0} & -\mathbf{I} \end{bmatrix} \text{ is unipotent;}$$

$$\mathbf{C} = \frac{1}{3} \begin{bmatrix} 2 & -1 & -1 \\ -1 & 2 & -1 \\ -1 & -1 & 2 \end{bmatrix} \text{ is idempotent; } \quad \text{but } \mathbf{X} = \begin{bmatrix} 0 & 0 & 6 \\ \frac{1}{2} & 0 & 0 \\ 0 & \frac{1}{3} & 0 \end{bmatrix}$$

takes none of these names: $\mathbf{X}^3 = \mathbf{I}$, but $\mathbf{X}^2 \neq \mathbf{I}$. It arises from characteristics of Bernadelli's (1941) beetles!

3.5 Orthogonal Matrices

Another useful class of matrices is typified by having the property $\mathbf{AA'} = \mathbf{I} = \mathbf{A'A}$. Such matrices are called *orthogonal*. We lead up to them with the following intertwined definitions.

The *norm* of a real vector $\mathbf{x}' = \begin{bmatrix} x_1 & x_2 & \cdots & x_n \end{bmatrix}$ is defined as

$$(\text{norm of } \mathbf{x}) = \sqrt{\mathbf{x'x}} = \left(\sum_{i=1}^{n} x_i^2 \right)^{\frac{1}{2}}. \tag{3.17}$$

For example, the norm of $\mathbf{x}' = \begin{bmatrix} 1 & 2 & 2 & 4 \end{bmatrix}$ is $(1 + 4 + 4 + 16)^{\frac{1}{2}} = 5$. (The square root is taken as positive.) A vector is said to be a *unit vector* when its norm is unity; i.e., when $\mathbf{x'x} = 1$. An example is $\mathbf{x}' = \begin{bmatrix} .2 & .4 & .4 & .8 \end{bmatrix}$. Any non-null vector $\mathbf{x}$ can be changed into a unit vector by multiplying it by the scalar $1/\sqrt{\mathbf{x'x}}$; i.e.,

$$\mathbf{u} = \left(\frac{1}{\sqrt{\mathbf{x'x}}} \right) \mathbf{x}$$

is the *normalized* form of $\mathbf{x}$ (because $\mathbf{u'u} = 1$).

Non-null vectors $\mathbf{x}$ and $\mathbf{y}$ are described as being *orthogonal* when $\mathbf{x'y} = 0$ (equivalent, of course, to $\mathbf{y'x} = 0$); e.g., $\mathbf{x}' = \begin{bmatrix} 1 & 2 & 2 & 4 \end{bmatrix}$ and $\mathbf{y}' = \begin{bmatrix} 6 & 3 & -2 & -2 \end{bmatrix}$ are orthogonal vectors because

$$\mathbf{x'y} = \begin{bmatrix} 1 & 2 & 2 & 4 \end{bmatrix} \begin{bmatrix} 6 \\ 3 \\ -2 \\ -2 \end{bmatrix} = 6 + 6 - 4 - 8 = 0.$$

Two vectors are defined as *orthonormal vectors* when they are orthogonal *and* normal. Thus $\mathbf{u}$ and $\mathbf{v}$ are orthonormal when $\mathbf{u'u} = 1 = \mathbf{v'v}$ and $\mathbf{u'v} = 0$; for example, $\mathbf{u}' = \frac{1}{6}\begin{bmatrix} 1 & 1 & 3 & 3 & 4 \end{bmatrix}$ and $\mathbf{v}' = \begin{bmatrix} -.1 & -.9 & -.1 & -.1 & .4 \end{bmatrix}$ are orthonormal vectors.

A group, or collection, of vectors all of the same order is called a *set* of vectors. A set of vectors $\mathbf{x}_i$ for $i = 1, 2, \ldots, n$ is said to be an *orthonormal set* of vectors when every vector in the set is normal, $\mathbf{x}_i'\mathbf{x}_i = 1$ for all i, and when every pair of different vectors in the set is orthogonal, $\mathbf{x}_i'\mathbf{x}_j = 0$ for $i \neq j = 1, 2, \ldots, n$. We can say that the vectors of an orthonormal set are all normal, and pairwise orthogonal.

A matrix $\mathbf{P}_{r \times c}$ whose rows constitute an orthonormal set of vectors is said to have orthonormal rows, whereupon $\mathbf{PP}' = \mathbf{I}_r$. But then $\mathbf{P}'\mathbf{P}$ is not necessarily an identity matrix $\mathbf{I}_c$, as the following example shows:

$$\mathbf{P} = \begin{bmatrix} 1 & 0 & 0 \\ 0 & 1 & 0 \end{bmatrix}, \quad \mathbf{PP}' = \mathbf{I} \quad \text{but} \quad \mathbf{P}'\mathbf{P} = \begin{bmatrix} 1 & 0 & 0 \\ 0 & 1 & 0 \\ 0 & 0 & 0 \end{bmatrix} \neq \mathbf{I}_3.$$

Conversely, when $\mathbf{P}_{r \times c}$ has orthonormal columns $\mathbf{P}'\mathbf{P} = \mathbf{I}_c$, but $\mathbf{PP}'$ may not be an identity matrix. Square matrices having orthonormal rows are in a special class: their columns are also orthonormal. (This is proved in Section 5.7c.) The matrix $\mathbf{P}$, say, is then such that

$$\mathbf{PP}' = \mathbf{P}'\mathbf{P} = \mathbf{I}.$$

This equation defines $\mathbf{P}$ as being an *orthogonal matrix*; it implies that $\mathbf{P}$ has orthonormal rows and orthonormal columns. These are characteristics of any orthogonal matrix $\mathbf{P}$. Actually, as is shown in Chapter 5, any two of the conditions (i) $\mathbf{P}$ square, (ii) $\mathbf{P}'\mathbf{P} = \mathbf{I}$ and (iii) $\mathbf{PP}' = \mathbf{I}$ implies the third, i.e., implies that $\mathbf{P}$ is orthogonal.

A simple property of orthogonal matrices is that products of them are orthogonal. (See E 3.16.)

Example

$$\mathbf{A} = \frac{1}{\sqrt{6}} \begin{bmatrix} \sqrt{2} & \sqrt{2} & \sqrt{2} \\ \sqrt{3} & -\sqrt{3} & 0 \\ 1 & 1 & -2 \end{bmatrix}$$

is an orthogonal matrix, as the reader may easily verify.

The boundless variety of matrices that are orthogonal includes many that are carefully proscribed, three of which are known by their originators' names, Givens, Helmert and Householder, a few details of which are available in Searle (1982, Section 3.4b). The simplest is $\mathbf{H} = \mathbf{I} - 2\mathbf{hh}'$ for any unit vector $\mathbf{h}$.

3.6 Quadratic Forms

a. Definition

The simplest kind of production function in economics takes the form $y = \theta + \lambda x + ax^2$, where y is an output and x an input. The generalization of this when there are two inputs, x_1 and x_2, is

$$y = \theta + \lambda_1 x_1 + \lambda_2 x_2 + a_{11} x_1^2 + a_{22} x_2^2 + a_{12} x_1 x_2,$$

which can be written as

$$
\begin{aligned}
y &= \theta + [\lambda_1 \ \ \lambda_2] \begin{bmatrix} x_1 \\ x_2 \end{bmatrix} + [x_1 \ \ x_2] \begin{bmatrix} a_{11} & \frac{1}{2}a_{12} \\ \frac{1}{2}a_{12} & a_{22} \end{bmatrix} \begin{bmatrix} x_1 \\ x_2 \end{bmatrix} \\
&= \theta + \boldsymbol{\lambda}' \boldsymbol{x} + \boldsymbol{x}' \mathbf{A} \mathbf{x},
\end{aligned}
$$

so defining $\boldsymbol{\lambda}, \mathbf{A}$ and $\mathbf{x}$ in an obvious way. The feature of this which we here direct attention to is $\mathbf{x}'\mathbf{A}\mathbf{x}$: it is called a *quadratic form*. It is the product of a row vector $\mathbf{x}'$, a matrix $\mathbf{A}$ and the column vector $\mathbf{x}$. Expressions of this form have many uses, particularly in the general theory of analysis of variance in statistics wherein, with appropriate choice of $\mathbf{A}$, any sum of squares can be represented as $\mathbf{x}'\mathbf{A}\mathbf{x}$.

b.   The matrix A is always symmetric in x'Ax

Suppose we have a quadratic form $\mathbf{x}'\mathbf{B}\mathbf{x}$, where we do not know whether $\mathbf{B}$ is symmetric or not. Then the following equalities apply:

$$
\begin{aligned}
\mathbf{x}'\mathbf{B}\mathbf{x} &= (\mathbf{x}'\mathbf{B}\mathbf{x})', \quad \text{because } \mathbf{x}'\mathbf{B}\mathbf{x} \text{ is a scalar} \\
&= \mathbf{x}'\mathbf{B}'\mathbf{x} \\
&= \tfrac{1}{2}(\mathbf{x}'\mathbf{B}\mathbf{x} + \mathbf{x}'\mathbf{B}'\mathbf{x}), \quad \text{because } \quad t = \tfrac{1}{2}(t + t) \\
&= \mathbf{x}'\left[\tfrac{1}{2}(\mathbf{B} + \mathbf{B}')\right]\mathbf{x} \\
&= \mathbf{x}'\mathbf{A}\mathbf{x}
\end{aligned}
$$

for

$$\mathbf{A} = \tfrac{1}{2}(\mathbf{B} + \mathbf{B}'), \tag{3.18}$$

which is obviously symmetric. Therefore, no matter what the matrix $\mathbf{B}$ is, the quadratic form $\mathbf{x}'\mathbf{B}\mathbf{x}$ can *always* be written as $\mathbf{x}'\mathbf{A}\mathbf{x}$, where $\mathbf{A}$ is symmetric. Thus every quadratic form can be thought of as $\mathbf{x}'\mathbf{A}\mathbf{x}$ with symmetric $\mathbf{A}$. This is a very important result because symmetric matrices have a host of special properties many of which are useful in dealing with quadratic forms.

c. Numerical example

We start with a non-symmetric matrix **B** and consider

$$\mathbf{x'Bx} = [x_1 \quad x_2 \quad x_3] \begin{bmatrix} 2 & 3 & 8 \\ 2 & 7 & -5 \\ 4 & 11 & 9 \end{bmatrix} \begin{bmatrix} x_1 \\ x_2 \\ x_3 \end{bmatrix}. \tag{3.19}$$

Then from (3.18) the symmetric **A** is

$$\mathbf{A} = \tfrac{1}{2}(\mathbf{B} + \mathbf{B'}) = \tfrac{1}{2}\left[\begin{pmatrix} 2 & 3 & 8 \\ 2 & 7 & -5 \\ 4 & 11 & 9 \end{pmatrix} + \begin{pmatrix} 2 & 2 & 4 \\ 3 & 7 & 11 \\ 8 & -5 & 9 \end{pmatrix} \right] = \begin{bmatrix} 2 & 2.5 & 6 \\ 2.5 & 7 & 3 \\ 6 & 3 & 9 \end{bmatrix};$$

$$\tag{3.20}$$

and, aided by $x_i x_j = x_j x_i$ for every $i \neq j$, one finds

$$\begin{aligned} \mathbf{x'Ax} &= \mathbf{x'Bx} \\ &= 2x_1^2 + 7x_2^2 + 9x_3^2 + 5x_1x_2 + 12x_1x_3 + 6x_2x_3. \end{aligned} \tag{3.21}$$

A is not only symmetric but, no matter what a particular **B** is, the corresponding symmetric **A** is unique. For example, (3.21) can be written as

$$2x_1^2 + 7x_2^2 + 9x_3^2 + 2.5(x_1x_2 + x_2x_1) + 6(x_1x_3 + x_3x_1) + 3(x_2x_3 + x_3x_2), \tag{3.22}$$

which is exactly what one gets from multiplying out $\mathbf{x'Ax}$ using **A** of (3.20). And there is no other way of rewriting (3.21) so as to have the same coefficient for $x_i x_j$ as for $x_j x_i$. For example, $12x_1x_3$ in (3.21) can be written as $\lambda(x_1x_3 + x_3x_1)$ only with $\lambda = 6$ as in (3.22); and no value for λ other than 6 is possible. Moreover, although $\mathbf{A} = \mathbf{A'}$ is unique for the **B** in (3.19), there is an infinite number of non-symmetric matrices **B** that yield the same expression for $\mathbf{x'Bx}$ as in (3.21). For example,

$$\mathbf{B}_1 = \begin{bmatrix} 2 & 95 & 2.4 \\ -90 & 7 & 0 \\ 9.6 & 6 & 9 \end{bmatrix} \quad \text{and} \quad \mathbf{B}_2 = \begin{bmatrix} 2 & -111 & 32 \\ 116 & 7 & 982 \\ -20 & -976 & 9 \end{bmatrix}.$$

Every **B** for (3.21) will be like these, with diagonal elements 2, 7 and 9; and with $b_{12} + b_{21} = 5$, $b_{13} + b_{31} = 12$ and $b_{23} + b_{32} = 6$. And for each **B**, as in (3.20), $\mathbf{A} = \tfrac{1}{2}(\mathbf{B} + \mathbf{B'}) = \mathbf{A'}$. Therefore we always think of a quadratic form $\mathbf{x'Ax}$ as having symmetric **A**.

d.   Explicit expansion

On reference to $\mathbf{A}$ of (3.20), the general form of (3.21) is readily seen to be

$$\mathbf{x}'\mathbf{A}\mathbf{x} = \sum_i a_{ii}x_i^2 + \sum\sum_{j>i} x_ix_j(a_{ij} + a_{ji}). \qquad (3.23)$$

With $\mathbf{A}$ being symmetric, $a_{ji} = a_{ij}$, and so (3.23) becomes

$$\mathbf{x}'\mathbf{A}\mathbf{x} = \sum_i a_{ii}x_i^2 + 2\sum\sum_{j>i} x_ix_ja_{ij}. \qquad (3.24)$$

As an example, (3.23) for $\mathbf{A}$ of order 3 is

$$\mathbf{x}'\mathbf{A}\mathbf{x} = a_{11}x_1^2 + a_{22}x_2^2 + a_{33}x_3^2 + 2(a_{12}x_1x_2 + a_{13}x_1x_3 + a_{23}x_2x_3).$$

e.   Bilinear form

A slightly more general (but not so useful) function is the second-degree function in two sets of variables x and y, say. For example,

$$\begin{aligned}
\mathbf{x}'\mathbf{M}\mathbf{y} &= [x_1 \quad x_2]\begin{bmatrix} 2 & 4 & 3 \\ 7 & 6 & 5 \end{bmatrix}\begin{bmatrix} y_1 \\ y_2 \\ y_3 \end{bmatrix} \\
&= 2x_1y_1 + 4x_1y_2 + 3x_1y_3 + 7x_2y_1 + 6x_2y_2 + 5x_2y_3.
\end{aligned}$$

It is called a *bilinear form* and, as illustrated here, its matrix $\mathbf{M}$ does not have to be square as does the matrix in a quadratic form. Clearly, quadratic forms are special cases of bilinear forms — when $\mathbf{M}$ is square and $\mathbf{y} = \mathbf{x}$.

f.   Positive (and negative) definite matrices

It is obvious that $\mathbf{x} = \mathbf{0}$ gives $\mathbf{x}'\mathbf{A}\mathbf{x} = 0$ for every $\mathbf{A}$. However, there are some matrices $\mathbf{A}$ for which $\mathbf{x}'\mathbf{A}\mathbf{x} = 0$ *only* for $\mathbf{x} = \mathbf{0}$, whereas for other matrices $\mathbf{x}'\mathbf{A}\mathbf{x} = 0$ not only for $\mathbf{x} = \mathbf{0}$ but also for some $\mathbf{x} \neq \mathbf{0}$. In addition to this dichotomy among $\mathbf{A}$s, there are also $\mathbf{A}$s for which non-zero values of $\mathbf{x}'\mathbf{A}\mathbf{x}$ are always positive, and others for which non-zero values of $\mathbf{x}'\mathbf{A}\mathbf{x}$ are always negative. A series of names distinguishes these various cases.

– i. Positive definiteness

Consider the example

$$\mathbf{x}'\mathbf{A}\mathbf{x} = \begin{bmatrix} x_1 & x_2 & x_3 \end{bmatrix} \begin{bmatrix} 2 & 2 & 1 \\ 2 & 5 & 1 \\ 1 & 1 & 2 \end{bmatrix} \begin{bmatrix} x_1 \\ x_2 \\ x_3 \end{bmatrix}$$

$$= 2x_1^2 + 5x_2^2 + 2x_3^2 + 4x_1x_2 + 2x_1x_3 + 2x_2x_3$$

$$= (x_1 + 2x_2)^2 + (x_1 + x_3)^2 + (x_2 + x_3)^2,$$

which, by the nature of its last expression is positive (for real $\mathbf{x}$) unless all elements of $\mathbf{x}$ are zero, i.e., $\mathbf{x} = \mathbf{0}$. Such a quadratic form is described as being positive definite. More formally,

when $\mathbf{x}'\mathbf{A}\mathbf{x} > 0$ for all $\mathbf{x}$ other than $\mathbf{x} = \mathbf{0}$,

then $\mathbf{x}'\mathbf{A}\mathbf{x}$ is a *positive definite* quadratic form, and $\mathbf{A} = \mathbf{A}'$ is correspondingly a *positive definite* (p.d.) matrix.

There are also symmetric matrices $\mathbf{A}$ for which $\mathbf{x}'\mathbf{A}\mathbf{x}$ is zero and for some non-null $\mathbf{x}$ as well as for $\mathbf{x} = \mathbf{0}$; e.g.,

$$\mathbf{x}'\mathbf{A}\mathbf{x} = \begin{bmatrix} x_1 & x_2 & x_3 \end{bmatrix} \begin{bmatrix} 10 & -2 & -6 \\ -2 & 4 & 0 \\ -6 & 0 & 4 \end{bmatrix} \begin{bmatrix} x_1 \\ x_2 \\ x_3 \end{bmatrix}$$

$$= 10x_1^2 + 4x_2^2 + 4x_3^2 - 4x_1x_2 - 12x_1x_3$$

$$= (x_1 - 2x_2)^2 + (3x_1 - 2x_3)^2.$$

This is zero for $\mathbf{x}' = \begin{bmatrix} 2 & 1 & 3 \end{bmatrix}$, and for any scalar multiple thereof, as well as for $\mathbf{x} = \mathbf{0}$. This kind of quadratic form is called positive semi-definite and has for its formal definition:

when $\mathbf{x}'\mathbf{A}\mathbf{x} \geq 0$ for all $\mathbf{x}$ *and* $\mathbf{x}'\mathbf{A}\mathbf{x} = 0$ for some $\mathbf{x} \neq \mathbf{0}$

then $\mathbf{x}'\mathbf{A}\mathbf{x}$ is a *positive semi-definite* quadratic form and hence $\mathbf{A} = \mathbf{A}'$ is a *positive semi-definite* (p.s.d.) matrix. The two classes of (forms and) matrices taken together, positive definite and positive semi-definite, are called *non-negative definite* (n.n.d.).

Illustration (Statistics)

The sum of squares of (3.16),

$$\sum_{i=1}^{n} (x_i - \bar{x})^2 = \mathbf{x}'\mathbf{C}\mathbf{x},$$

is a positive semi-definite quadratic form because it is positive, except for being zero when all the x_is are equal. Its matrix, $\mathbf{I} - \bar{\mathbf{J}}$, which is idempotent, is also p.s.d., as are all symmetric idempotent matrices (except $\mathbf{I}$, which is the only p.d. idempotent matrix).

Unfortunately there is no universal agreement on the definition of positive semi-definite. Most writers use it with the meaning defined here, but some use it in the sense of meaning non-negative definite. But on one convention there is universal agreement: p.d., p.s.d., and n.n.d. matrices are always taken as being symmetric. This is so because the definitions of these matrices are in terms of quadratic forms which can always be expressed utilizing symmetric matrices.

– ii. Negative definiteness

If everywhere in the preceding subsection we replace $>$ with $<$ and also interchange the words positive and negative, we get definitions of negative definite (n.d.), negative semi-definite (n.s.d.) and non-positive definite (n.p.d.). Thus

$$\text{when } \mathbf{x}'\mathbf{A}\mathbf{x} < 0 \text{ for all } \mathbf{x} \text{ other than } \mathbf{x} = \mathbf{0}$$

then $\mathbf{A}$ is negative definite. And

$$\text{when } \mathbf{x}'\mathbf{A}\mathbf{x} \leq 0 \text{ for all } \mathbf{x} \text{ and } \mathbf{x}'\mathbf{A}\mathbf{x} = 0 \text{ for some } \mathbf{x} \neq \mathbf{0}$$

then $\mathbf{A}$ is negative semi-definite; and non-positive definiteness combines negative definite and negative semi-definite.

Illustration   (Production function)

Economic theory uses the notion of a negative definite quadratic form $\mathbf{x}'\mathbf{A}\mathbf{x}$ in its characterization of the production function $y = \theta + \boldsymbol{\lambda}'\mathbf{x} + \mathbf{x}'\mathbf{A}\mathbf{x}$ mentioned in Section 3.6a. In this function the elements of $\mathbf{x}$, being inputs, are always positive or zero (but never all of them zero, obviously), and elements of $\boldsymbol{\lambda}$ are all positive. But, in order to ensure diminishing marginal returns to inputs, $\mathbf{x}'\mathbf{A}\mathbf{x}$ is always negative definite.

3.7   Variance-Covariance Matrices

Suppose $x_1, x_2, \ldots, x_k$ are k random variables with means μ_1, μ_2, $\ldots, \mu_k$, variances $\sigma_1^2, \sigma_2^2, \ldots, \sigma_k^2$ and covariances $\sigma_{12}, \sigma_{13}, \ldots, \sigma_{k-1,k}$.

On representing the random variables and their means by the vectors

$$\mathbf{x} = \left\{ {}_c\, x_i \right\}_{i=1}^{k} \qquad \boldsymbol{\mu} = \left\{ {}_c\, \mu_i \right\}_{i=1}^{k}$$

we have

$$\mathrm{E}(\mathbf{x}) = \boldsymbol{\mu}.$$

And assembling the variances and covariances into a matrix gives

$$\mathrm{var}(\mathbf{x}) = \mathbf{V} = \left\{ {}_m\, \sigma_{ij} \right\}_{i=1\ j=1}^{k\quad k} \quad \text{with} \quad \sigma_{ii} = \sigma_i^2. \tag{3.25}$$

This is the *variance-covariance matrix* or *dispersion matrix*. It is symmetric, $\mathbf{V} = \mathbf{V}'$; its ith diagonal element is the variance σ_{ii} of x_i and its (i, j)th off-diagonal element is the covariance between x_i and x_j.

The familiar definitions of variance and covariance are $\sigma_i^2 = \sigma_{ii} = \mathrm{E}(x_i - \mu_i)^2$ and $\sigma_{ij} = \mathrm{E}(x_i - \mu_i)(x_j - \mu_j)$, respectively. They provide the wherewithal for expressing $\mathbf{V}$ as

$$\mathbf{V} = \mathrm{var}(\mathbf{x}) = \mathrm{E}[(\mathbf{x} - \boldsymbol{\mu})(\mathbf{x} - \boldsymbol{\mu})'] = \mathrm{E}(\mathbf{xx}') - \boldsymbol{\mu}\boldsymbol{\mu}'.$$

An important property of $\mathbf{V}$ is that it is n.n.d. in fact usually p.d. The proof of this is to consider the variance of $\mathbf{t}'\mathbf{x}$ for some vector $\mathbf{t}'$:

$$
\begin{aligned}
\mathrm{var}(\mathbf{t}'\mathbf{x}) \;&=\; \mathrm{E}\left\{[\mathbf{t}'\mathbf{x} - \mathrm{E}(\mathbf{t}'\mathbf{x})][\mathbf{t}'\mathbf{x} - \mathrm{E}(\mathbf{t}'\mathbf{x})]'\right\} \\
&=\; \mathrm{E}[\mathbf{t}'(\mathbf{x} - \boldsymbol{\mu})(\mathbf{x} - \boldsymbol{\mu})'\mathbf{t}] \\
&=\; \mathbf{t}'\mathrm{E}[(\mathbf{x} - \boldsymbol{\mu})(\mathbf{x} - \boldsymbol{\mu})']\mathbf{t} \\
&=\; \mathbf{t}'\mathbf{V}\mathbf{t} \\
&\geq\; 0, \quad \text{because every variance} \geq 0.
\end{aligned}
$$

3.8   Correlation Matrices

The end of Section 2.1c shows a correlation matrix for citrus production. Like all correlation matrices, it has diagonal elements 1.0 and off-diagonal elements which are correlations of the form

$$\rho_{ij} = \frac{\sigma_{ij}}{\sigma_i \sigma_j} = \frac{\sigma_{ij}}{\sqrt{\sigma_{ii}\sigma_{jj}}},$$

using elements from $\mathbf{V}$ of (3.25). Now define the correlation matrix as

$$\mathbf{R} = \left\{ {}_m\, \rho_{ij} \right\}_{i=1\ j=1}^{k\quad k} \quad \text{with} \quad \rho_{ii} = 1.$$

Then on using the diagonal matrix of elements $1/\sigma_i = 1/\sqrt{\sigma_{ii}}$,

$$\mathbf{D} = \left\{{}_d\, 1/\sqrt{\sigma_{ii}}\right\}_{i=1}^{k}.$$

$\mathbf{R}$ can be expressed in terms of $\mathbf{D}$ and $\mathbf{V}$ as $\mathbf{R} = \mathbf{DVD}$.

3.9   LDU Decomposition

Many forms of square matrices $\mathbf{A}$ can be expressed as a product $\mathbf{A} = \mathbf{LDU}$, where $\mathbf{L}$ and $\mathbf{U}$ are, respectively, unit lower and upper triangular matrices, and $\mathbf{D}$ is a diagonal matrix. Triangular matrices are described in Section 1.8d; they are unit triangular when their non-zero diagonal elements are 1.

Formal proof of being able to have $\mathbf{A} = \mathbf{LDU}$ is available in Harville (1997, Section 14.5). A clear description is given there of precisely what forms of $\mathbf{A}_{n \times n}$ can have the LDU decomposition. In particular, for example, every symmetric non-negative definite matrix can be expressed as $\mathbf{U'DU}$. We give no formal proof here, only a 3×3 example to illustrate that it can be possible.

Thus suppose we have the $\mathbf{LDU}$ as

$$\begin{bmatrix} 1 & \cdot & \cdot \\ a & 1 & \cdot \\ b & c & 1 \end{bmatrix} \begin{bmatrix} d_1 & \cdot & \cdot \\ \cdot & d_2 & \cdot \\ \cdot & \cdot & d_3 \end{bmatrix} \begin{bmatrix} 1 & x & y \\ \cdot & 1 & z \\ \cdot & \cdot & 1 \end{bmatrix} = \begin{bmatrix} 2 & 4 & 8 \\ 14 & 29 & 58 \\ 6 & 13 & 30 \end{bmatrix}.$$

First note on the left-hand side of this equation that there are nine unknowns and on the right are nine elements to be used for calculating the unknowns. That is, of course, no assurance that solutions for the unknowns can be found, although in many cases they can. That is the situation here. We just equate both sides of the equation, element by element. Thus from that equality

$$
\begin{array}{rclcrcll}
1(d_1) & = & 2 & \implies & d_1 & = & 2 & \\
d_1 x & = & 4 & & x & = & 4/2 & = 2 \\
d_1 y & = & 8 & & y & = & 8/2 & = 4 \\
a d_1 & = & 14 & & a & = & 14/2 & = 7 \\
a d_1 x + d_2 & = & 29 & & d_2 & = & 29 - 28 & = 1 \\
a d_1 y + d_2 z & = & 58 & & z & = & (58 - 56)/1 & = 2
\end{array}
$$

and so on, until we finish up with

$$\begin{bmatrix} 1 & \cdot & \cdot \\ 7 & 1 & \cdot \\ 3 & 1 & 1 \end{bmatrix} \begin{bmatrix} 2 & \cdot & \cdot \\ \cdot & 1 & \cdot \\ \cdot & \cdot & 4 \end{bmatrix} \begin{bmatrix} 1 & 2 & 4 \\ \cdot & 1 & 2 \\ \cdot & \cdot & 1 \end{bmatrix} = \begin{bmatrix} 2 & 4 & 8 \\ 14 & 29 & 58 \\ 6 & 13 & 30 \end{bmatrix}.$$

Clearly, any time that equations of the nature used here lead to a denominator of zero, the **LDU** method fails. An example of this would be where the above matrix had 28 in place of 29.

3.10 Exercises

E 3.1. Show that $\begin{bmatrix} 3 & 8 & 4 \\ 8 & 7 & -1 \\ 4 & -1 & 2 \end{bmatrix} + \begin{bmatrix} 1 & -1 & 3 \\ -1 & 2 & 4 \\ 3 & 4 & 6 \end{bmatrix}$ is symmetric.

E 3.2. If **x** and **y** are $n \times 1$ column vectors and **A** and **B** are $n \times n$ matrices, which of the following expressions are undefined? Of those that are defined, which are bilinear forms, quadratic forms or linear transformations?

(a) $\mathbf{y} = \mathbf{Ax}$ (b) $\mathbf{x'} = \mathbf{y'B'}$

(c) $\mathbf{xy} = \mathbf{A'B}$ (d) $\mathbf{x'Ay}$

(e) $\mathbf{x'Bx}$ (f) $\mathbf{y'A'By}$

(g) $\mathbf{yBx}$ (h) $\mathbf{xy'} = \mathbf{B'}$

(i) $\mathbf{y'B'Ax}$ (j) $\mathbf{y} + \begin{bmatrix} 30 \\ 40 \end{bmatrix} = \mathbf{AB'x}$ for $n = 2$

E 3.3. Using $\mathbf{A} = \begin{bmatrix} 1 & 1 \\ 2 & 1 \end{bmatrix}$, $\mathbf{B} = \begin{bmatrix} 3 & 6 \\ 4 & 8 \end{bmatrix}$, $\mathbf{x} = \begin{bmatrix} 1 \\ 2 \end{bmatrix}$ and $\mathbf{y} = \begin{bmatrix} 3 \\ 4 \end{bmatrix}$ in E 3.2:

(a) Check the validity of those equations that are defined.

(b) Calculate the value of the other defined expressions.

E 3.4. Use $\mathbf{x} = \begin{bmatrix} x_1 \\ x_2 \end{bmatrix}$, $\mathbf{y} = \begin{bmatrix} y_1 \\ y_2 \end{bmatrix}$ and **A** and **B** of E 3.3 to write down (after multiplication) expressions (a), (b), (d), (e), (f) and (i) of E 3.2.

E 3.5. Demonstrate that for

$$\mathbf{B} = \begin{bmatrix} 1/\sqrt{6} & 1/\sqrt{3} & 1/\sqrt{2} \\ -2/\sqrt{6} & 1/\sqrt{3} & 0 \\ 1/\sqrt{6} & 1/\sqrt{3} & -1/\sqrt{2} \end{bmatrix}, \ \mathbf{y} = \begin{bmatrix} y_1 \\ y_2 \\ y_3 \end{bmatrix} \text{ and } \mathbf{x} = \begin{bmatrix} x_1 \\ x_2 \\ x_3 \end{bmatrix} :$$

(a) $\mathbf{B'B} = \mathbf{BB'} = \mathbf{I}$.

(b) $\mathbf{x'B'Bx} = (\mathbf{Bx})'\mathbf{Bx} = \mathbf{x'}(\mathbf{B'B})\mathbf{x} = x_1^2 + x_2^2 + x_3^3$.

(c) $\mathbf{x'B'By} = x_1 y_1 + x_2 y_2 + x_3 y_3$.

E 3.6. For $\mathbf{A} = \begin{bmatrix} 3 & 0 & 3 & 2 \\ 0 & -1 & 5 & 0 \\ -1 & 6 & 0 & 8 \end{bmatrix}$ and $\mathbf{B} = \begin{bmatrix} -1 & 0 & 8 & 3 \\ 0 & -2 & -7 & -1 \\ -1 & 3 & 0 & 2 \end{bmatrix} :$

find $\mathbf{AA'}$, $\mathbf{BB'}$, $\mathbf{AB'}$ and $\mathbf{BA'}$ and verify that

$$\begin{aligned} (\mathbf{A+B})(\mathbf{A+B})' &= (\mathbf{A+B})(\mathbf{A'+B'}) = \mathbf{AA'} + \mathbf{BA'} \\ &\quad + \mathbf{AB'} + \mathbf{BB'} \\ &= \mathbf{AA'} + (\mathbf{AB'})' + \mathbf{AB'} + \mathbf{BB'} \\ &= \begin{bmatrix} 150 & -27 & 46 \\ -27 & 14 & -37 \\ 46 & -37 & 185 \end{bmatrix}. \end{aligned}$$

E 3.7. (Variance-covariance matrix) Suppose three random variables x_1, x_2 and x_3 have the following characteristics.

(i.) Their means are 5, 10 and 8, respectively.

(ii.) Their variances are 6, 14 and 1, respectively.

(iii.) The covariance between x_1 and x_2 is 3, that between x_1 and x_3 is 1 and that between x_2 and x_3 is 0.

(iv.) Three linear combinations of the xs are

$$\begin{aligned} y_1 &= x_1 + 3x_2 - 2x_3 \\ y_2 &= 7x_1 - 4x_2 + x_3 \ . \\ y_3 &= -2x_1 - x_2 + 4x_3 \end{aligned}$$

   (a) Write down the vector of means and the variance-covariance matrix of the xs.

   (b) Write down the matrix **T** which represents the linear transformation of the xs into ys.

   (c) Use **T** as obtained in (b) to derive the vector of means and the variance-covariance matrix of the ys.

   (d) Derive the correlation matrix of the xs using a linear transformation for standardizing the xs.

E 3.8. If **A** is skew-symmetric prove that

   (a) $a_{ii} = 0$ and $a_{ij} = -a_{ji}$ for $i \neq j$;

   (b) $\mathbf{I} + \mathbf{A}$ is positive definite.

E 3.9. (a) Write down $\mathbf{M} = \{i(j+1)\}_{i,j=1}^{3}$.

   (b) Demonstrate that **M** is the outer product of $[2 \quad 3 \quad 4]$ and some other vector; specify that vector.

   (c) Calculate $\mathbf{M} + \mathbf{M}'$.

   (d) What property must **T** have so that $\mathbf{T} + \mathbf{T}'$ exists?

   (e) Prove that $\mathbf{T} + \mathbf{T}'$ is symmetric but $\mathbf{T} - \mathbf{T}'$ is skew-symmetric.

   (f) For square **A** prove that it is the sum of a symmetric matrix and a skew-symmetric matrix.

E 3.10. Prove that the product of two symmetric matrices is symmetric if and only if the matrices commute in multiplication.

E 3.11. Show that if $\mathbf{X}'\mathbf{X} = \mathbf{X}$ then $\mathbf{X} = \mathbf{X}' = \mathbf{X}^2$.

E 3.12. (a) Show that the only real symmetric matrix whose square is null is itself null.

   (b) Explain why $\mathbf{X}'\mathbf{XGX}'\mathbf{X} = \mathbf{X}'\mathbf{X}$ implies $\mathbf{XGX}'\mathbf{X} = \mathbf{X}$.

E 3.13. For $\mathbf{X}_{q\times t} = \left\{ _r \; \mathbf{x}_j \right\}_{j=1}^{t}$ and for symmetric $\mathbf{A}$, prove that

$$\text{tr}(\mathbf{A}\mathbf{X}\mathbf{X}')^2 = \sum_{j=1}^{t} \sum_{k=1}^{t} \left(\mathbf{x}_j' \mathbf{A} \mathbf{x}_k \right)^2 .$$

E 3.14. For $\mathbf{I}$ and $\mathbf{J}$ of order n, and for scalars p, q, r and s:

(a) Write down $p\mathbf{I} + q\mathbf{J}$ of order 4.

(b) Show that $(p\mathbf{I} + q\mathbf{J})(r\mathbf{I} + s\mathbf{J}) = pr\mathbf{I} + (ps + qr + nqs)\mathbf{J}$.

(c) Simplify $(p\mathbf{I} + q\mathbf{J})\dfrac{1}{p}\left(\mathbf{I} - \dfrac{q}{p+qn}\mathbf{J} \right)$.

(d) In (c), what limitations are there on p and q?

(e) Explain why $\mathbf{J}\mathbf{X}\mathbf{J} = x_{..}\mathbf{J}$.

(f) Prove that $\mathbf{I} - \bar{\mathbf{J}}$ is idempotent, but $\mathbf{I} - \mathbf{J}$ is not.

E 3.15. (Transition probabilities) Suppose the probabilities of members of three political parties maintaining or changing their affiliation during a year are represented by the probability transition matrix

$$\mathbf{P} = \begin{bmatrix} .4 & .3 & .3 \\ .3 & .6 & .1 \\ .5 & .1 & .4 \end{bmatrix} .$$

The 1000 members of a country club are divided in their political leanings so that 600 support the first party, 300 the second and 100 the third. The vector $\mathbf{x}_0' = \begin{bmatrix} 600 & 300 & 100 \end{bmatrix}$ represents these political affiliations, and the expected political affiliations one year from now will be $\mathbf{x}_0'\mathbf{P}$.

(a) Calculate $\mathbf{x}_0'\mathbf{P}$.

(b) What will the affiliations be 2, 3 and 4 years from now?

(c) Express in matrix-vector form the fact that elements in each row of $\mathbf{P}$ add to unity.

(d) Prove that each row of $\mathbf{P}^k$ has elements summing to unity; k is a positive scalar.

(e) Prove that all elements of $\mathbf{x}_0'\mathbf{P}^k$ sum to 1000.

(f) Find the steady-state vector corresponding to **P**.

E 3.16. Prove that products of orthogonal matrices are orthogonal.

E 3.17. (a) For **A** being idempotent and symmetric prove that it is
             positive semi-definite.

   (b) Prove that a product of two idempotent matrices is idem-
        potent if the matrices commute in multiplication.

   (c) Prove that for **K** being real $\mathbf{I} + \mathbf{KK}'$ is positive definite.

E 3.18. For $\mathbf{x}' = \begin{bmatrix} 1 & 3 & 5 & 7 & 9 \end{bmatrix}$ ascertain the numerical value of **A**, **B**
         and **C** such that

   (a) $1^2 + 3^2 + 5^2 + 7^2 + 9^2 = \mathbf{x}'\mathbf{A}\mathbf{x}$,

   (b) $(1 + 3 + 5 + 7 + 9)^2/5 = \mathbf{x}'\mathbf{B}\mathbf{x}$,

   (c) $(1-5)^2 + (3-5)^2 + (5-5)^2 + (7-5)^2 + (9-5)^2 = \mathbf{x}'\mathbf{C}\mathbf{x}$.

E 3.19. Explain why

   (a) $\mathbf{x}'\mathbf{T}\mathbf{x} = \mathbf{x}'\mathbf{T}'\mathbf{x}$ even when **T** is not symmetric,

   (b) $\mathbf{x}'\mathbf{R}\mathbf{x} = \text{tr}(\mathbf{x}'\mathbf{R}\mathbf{x})$,

   (c) $\mathbf{x}'\mathbf{S}\mathbf{x} = \text{tr}(\mathbf{S}\mathbf{x}\mathbf{x}')$.

E 3.20. (a) For transition probability matrix **P** and $\mathbf{A} = \mathbf{P} - \bar{\mathbf{J}}$, prove
             that $\mathbf{A}^n = \mathbf{A}\mathbf{P}^{n-1}$ for n a positive integer.

   (b) Explain why $\text{tr}(\mathbf{T}\mathbf{Q}') = \mathbf{1}'(\mathbf{T} \bullet \mathbf{Q})\mathbf{1}$.

   (c) For symmetric, idempotent **X** and symmetric **LX**, prove
        that **LX=XLX**.

E 3.21. A transition probability matrix which has both its row sums
         and its column sums being 1.0 is called *doubly stochastic*.
         Prove that powers of such a matrix are doubly stochastic.

E 3.22. Suppose **D** is the diagonal matrix having as its jth diagonal
         element the largest element in the jth column of **P**, a pay-

off matrix as described in Section 2.5. Show that the regret matrix corresponding to $\mathbf{P}$ is $\mathbf{R} = \mathbf{JD} - \mathbf{P}$.

E 3.23. (Manufacturing) Suppose elements of the vectors $\mathbf{w}$, $\mathbf{x}$, $\mathbf{y}$ and $\mathbf{z}$ represent different types of raw materials, component parts, assemblies and finished products, respectively. Suppose the matrices $\mathbf{A}$, $\mathbf{B}$ and $\mathbf{C}$ represent the requirements from one level of product to the next lower level, such that $\mathbf{y} = \mathbf{Az}$, $\mathbf{x} = \mathbf{By}$, $\mathbf{w} = \mathbf{Cx}$; i.e., the equations relating the need for component parts to assemblies is $\mathbf{x} = \mathbf{By}$.

(a) Write down the equation describing $\mathbf{w}$ as a function of $\mathbf{z}$; i.e., derive the raw material requirements as a function of the finished products.

(b) If the amount of excess inventory at each level is represented by the vectors $\mathbf{w}^0$, $\mathbf{x}^0$, $\mathbf{y}^0$ and $\mathbf{z}^0$, write down the matrix equation describing net raw material requirements as a function of finished products.

E 3.24. (Correlation matrix) Section 2.1c shows a correlation matrix concerning citrus production. Suppose the variances of orange, lemon and tangelo production are 10,000, 62,500 and 1600, respectively; and that the covariances of production of oranges with lemons, of oranges with tangelos, and of lemons with tangelos are, respectively, 1250, -600 and 4800.

(a) What is the variance-covariance matrix?

(b) Use your result in (a) to calculate the correlation matrix.

Chapter 4

DETERMINANTS

We now discuss an operation applicable to a square matrix that leads to a scalar value known as the determinant of the matrix. Knowledge of this operation is necessary for understanding the counterpart of division in matrix algebra in Chapter 5; and it is also useful in succeeding chapters. The literature of determinants is extensive but the development in this book is brief and deals only with elementary methods of evaluation.

4.1 Introduction

A *determinant* is a scalar which is the sum of selected products of the elements of the matrix from which it is derived, each product being multiplied by +1 or −1 according to well-defined rules. We present a number of the customary procedures which implicitly follow these rules, and which can yield the value of the determinant of a square matrix of any order. Although a formal, mathematically rigorous definition is available, it is lengthy, convoluted and not needed for those who are interested primarily in practical applications of matrix algebra. We therefore avoid the definition: readers interested in it are referred to Searle (1966, pages 61-63, or 1982, pages 90-92).

Determinants are defined only for square matrices—the determinant of a non-square matrix is undefined and does not exist. The determinant of a square matrix of order n is referred to as an *n-order determinant*. The customary notation for the determinant of the matrix $\mathbf{A}$ is $|\mathbf{A}|$, where $\mathbf{A}$ is square. The notations $\|\mathbf{A}\|$, $[\mathbf{A}]$ and $\det(\mathbf{A})$ also exist, but $|\mathbf{A}|$ is more common. Obtaining the value of $|\mathbf{A}|$ by adding

adding the appropriate products of the elements of **A** (with the correct
+1 or -1 factor included in the product) is referred to as evaluating,
expanding or *reducing* the determinant. A procedure for doing this is
now illustrated by a series of numerical examples.

The determinant of a 1×1 matrix is the value of its sole element.
For example, the determinant of **B** = 7 is $|\mathbf{B}| = 7$. The value of a
second-order determinant is the product of its diagonal terms minus
the product of its off-diagonal terms. For example, the determinant of

$$\mathbf{A} = \begin{bmatrix} 7 & 3 \\ 4 & 6 \end{bmatrix} \qquad \text{is written} \qquad |\mathbf{A}| = \begin{vmatrix} 7 & 3 \\ 4 & 6 \end{vmatrix}$$

and is calculated as

$$|\mathbf{A}| = 7(6) - 3(4) = 30.$$

This example illustrates the general result for expanding a second-order
determinant: it consists of the product (multiplied by +1) of the diagonal terms plus the product (multiplied by -1) of the off-diagonal
terms. Hence in general

$$|\mathbf{A}| = \begin{vmatrix} a_{11} & a_{12} \\ a_{21} & a_{22} \end{vmatrix} = a_{11}a_{22} - a_{12}a_{21}.$$

For brevity we often use the word *determinant* (or the symbol $|\mathbf{A}|$)
to refer to a determinant in its arrayed form as well as to the scalar
value to which it reduces. Thus if

$$\mathbf{A} = \begin{bmatrix} 9 & 3 \\ 7 & -2 \end{bmatrix}$$

the symbol $|\mathbf{A}|$ might refer to the arrayed form

$$|\mathbf{A}| = \begin{vmatrix} 9 & 3 \\ 7 & -2 \end{vmatrix}$$

and to its expanded value

$$|\mathbf{A}| = 9(-2) - 7(3) = -39.$$

This dual usage is very customary, and is seldom confusing. It is used
regardless of the order of a determinant.

A third-order determinant can be expanded as a linear function of three second-order determinants derived from it. Their coefficients are elements of a row (or column) of the main determinant, each product being multiplied by $+1$ or -1. For example, the expansion of

$$|\mathbf{A}| = \begin{vmatrix} 1 & 2 & 3 \\ 4 & 5 & 6 \\ 7 & 8 & 10 \end{vmatrix}$$

based on the elements of the first row, 1, 2 and 3, is

$$
\begin{aligned}
|\mathbf{A}| &= 1(+1) \begin{vmatrix} 5 & 6 \\ 8 & 10 \end{vmatrix} + 2(-1) \begin{vmatrix} 4 & 6 \\ 7 & 10 \end{vmatrix} + 3(+1) \begin{vmatrix} 4 & 5 \\ 7 & 8 \end{vmatrix} \\
&= 1(50 - 48) - 2(40 - 42) + 3(32 - 35) = -3.
\end{aligned}
$$

The determinant is computed by adding the signed products of each element of the chosen row (in this case the first row) with the determinant derived from $|\mathbf{A}|$ by crossing out the row and column containing the element concerned. For example, the first element, 1, is multiplied by the determinant $\begin{vmatrix} 5 & 6 \\ 8 & 10 \end{vmatrix}$, which is obtained from $|\mathbf{A}|$ by crossing out the first row and first column, and the element 2 is multiplied (apart from the factor of -1) by the determinant derived from $|\mathbf{A}|$ by deleting the row and column containing that element—namely, the first row and second column, leaving $\begin{vmatrix} 4 & 6 \\ 7 & 10 \end{vmatrix}$. Determinants obtained in this way are called *minors* of $\mathbf{A}$; that is to say, $\begin{vmatrix} 5 & 6 \\ 8 & 10 \end{vmatrix}$ is the minor of the element 1 in $\mathbf{A}$; and $\begin{vmatrix} 4 & 6 \\ 7 & 10 \end{vmatrix}$ is the minor of the element 2.

The $(+1)$ and (-1) factors are determined by the following rule: if $\mathbf{A}$ is written in the form $\mathbf{A} = \{a_{ij}\}$, the product of a_{ij} and its minor in the expansion of the determinant $|\mathbf{A}|$ is multiplied by $(-1)^{i+j}$. The element 1 in the example is the element a_{11}; thus the product of a_{11} and its minor is multiplied by $(-1)^{1+1} = +1$. Similarly, the product of the element 2, a_{12}, with its minor is multiplied by $(-1)^{1+2} = -1$.

The minor of an element of a square matrix of order n is necessarily a determinant of order $n - 1$. However, minors are not all of order $n - 1$. Deleting any r rows and any r columns from a square matrix of order n leaves a sub-matrix of order $n - r$; and the determinant of this sub-matrix is a *minor of order $n - r$*, or an $(n - r)$-*order minor*.

4.2   Expansion by Minors

Suppose we denote the minor of the element a_{11} by $|M_{11}|$, where M_{11} is a sub-matrix of A obtained by deleting the first row and first column. Then, in the above example $|M_{11}| = \begin{vmatrix} 5 & 6 \\ 8 & 10 \end{vmatrix}$. Similarly, if $|M_{12}|$ is the minor of a_{12}, then $|M_{12}| = \begin{vmatrix} 4 & 6 \\ 7 & 10 \end{vmatrix}$; and if $|M_{13}|$ is the minor of a_{13}, then $|M_{13}| = \begin{vmatrix} 4 & 5 \\ 7 & 8 \end{vmatrix}$. With this notation, the expansion of $|A|$ given above is

$$|A| = a_{11}(-1)^{1+1}|M_{11}| + a_{12}(-1)^{1+2}|M_{12}| + a_{13}(-1)^{1+3}|M_{13}|.$$

This method of expanding a determinant is known as *expansion by the elements of a row (or column)* or as *expansion by minors*. It has been illustrated using elements of the first row, but it can also be applied to the elements of any row (or column). For example, the expansion of $|A|$ just considered, using elements of the second row, gives

$$|A| = 4(-1)\begin{vmatrix} 2 & 3 \\ 8 & 10 \end{vmatrix} + 5(+1)\begin{vmatrix} 1 & 3 \\ 7 & 10 \end{vmatrix} + 6(-1)\begin{vmatrix} 1 & 2 \\ 7 & 8 \end{vmatrix}$$

$$= -4(-4) + 5(-11) - 6(-6) = -3$$

as before, and using elements of the first column the expansion yields the same value:

$$|A| = 1(+1)\begin{vmatrix} 5 & 6 \\ 8 & 10 \end{vmatrix} + 4(-1)\begin{vmatrix} 2 & 3 \\ 8 & 10 \end{vmatrix} + 7(+1)\begin{vmatrix} 2 & 3 \\ 5 & 6 \end{vmatrix}$$

$$= 1(2) - 4(-4) + 7(-3) = -3.$$

The minors in these expansions are derived in exactly the same manner as above. For example, the minor of the element 4 is $|A|$ with the second row and first column deleted, and since 4 is a_{21} its product with its minor is multiplied by $(-1)^{2+1} = -1$. Other terms are obtained in a similar manner.

The foregoing example illustrates the expansion of the general third-order determinant

$$|\mathbf{A}| = \begin{vmatrix} a_{11} & a_{12} & a_{13} \\ a_{21} & a_{22} & a_{23} \\ a_{31} & a_{32} & a_{33} \end{vmatrix}.$$

Expanding this by elements of the first row gives

$$|\mathbf{A}| = a_{11}(+1)\begin{vmatrix} a_{22} & a_{23} \\ a_{32} & a_{33} \end{vmatrix} + a_{12}(-1)\begin{vmatrix} a_{21} & a_{23} \\ a_{31} & a_{33} \end{vmatrix} + a_{13}(+1)\begin{vmatrix} a_{21} & a_{22} \\ a_{31} & a_{32} \end{vmatrix}$$

$$= a_{11}a_{22}a_{33} - a_{11}a_{23}a_{32} - a_{12}a_{21}a_{33} + a_{12}a_{23}a_{31} + a_{13}a_{21}a_{32} - a_{13}a_{22}a_{31}.$$

The reader should be satisfied that expansion by the elements of any other row or column leads to the same result.

No matter by what row or column the expansion is made, the value of the determinant is the same. Note that once a row or column is decided on and the sign calculated for the product of the first element therein with its minor, the signs for the succeeding products alternate from plus to minus and minus to plus.

The expansion of an n-order determinant by this method is an extension of the expansion of a third-order determinant as just given. Thus the determinant of the $n \times n$ matrix $\mathbf{A} = \{a_{ij}\}$ for $i, j = 1, 2, \ldots, n$ is obtained as follows. Consider the elements of any one row (or column): multiply each element, a_{ij}, of this row (or column) by its minor, $|\mathbf{M}_{ij}|$, the determinant derived from $|\mathbf{A}|$ by crossing out the row and column containing a_{ij}; multiply the product by $(-1)^{i+j}$; add the signed products; and their sum is the determinant $|\mathbf{A}|$. This expansion is used recursively when n is large; i.e., each $|\mathbf{M}_{ij}|$ is in turn expanded by the same procedure.

Example

Expansion by elements of the first row of

$$|\mathbf{A}| = \begin{vmatrix} 1 & -2 & 3 & 0 \\ 0 & 4 & 5 & -6 \\ 1 & 0 & 2 & 3 \\ 4 & 0 & 2 & 3 \end{vmatrix}$$

gives

$$|\mathbf{A}| = 1(-1)^2|\mathbf{M}_{11}| + (-2)(-1)^3|\mathbf{M}_{12}| + 3(-1)^4|\mathbf{M}_{13}| + 0(-1)^5|\mathbf{M}_{14}|.$$

The needed minors are

$$|\mathbf{M}_{11}| = \begin{vmatrix} 4 & 5 & -6 \\ 0 & 2 & 3 \\ 0 & 2 & 3 \end{vmatrix} = 4(-1)^2 \begin{vmatrix} 2 & 3 \\ 2 & 3 \end{vmatrix} + 5(-1)^3 \begin{vmatrix} 0 & 3 \\ 0 & 3 \end{vmatrix}$$

$$+ (-6)(-1)^4 \begin{vmatrix} 0 & 2 \\ 0 & 2 \end{vmatrix} = 0,$$

$$|\mathbf{M}_{12}| = \begin{vmatrix} 0 & 5 & -6 \\ 1 & 2 & 3 \\ 4 & 2 & 3 \end{vmatrix} = 0(1) \begin{vmatrix} 2 & 3 \\ 2 & 3 \end{vmatrix} + 5(-1) \begin{vmatrix} 1 & 3 \\ 4 & 3 \end{vmatrix}$$

$$+ (-6)(1) \begin{vmatrix} 1 & 2 \\ 4 & 2 \end{vmatrix} = 81$$

and

$$|\mathbf{M}_{13}| = \begin{vmatrix} 0 & 4 & -6 \\ 1 & 0 & 3 \\ 4 & 0 & 3 \end{vmatrix} = 0(1) \begin{vmatrix} 0 & 3 \\ 0 & 3 \end{vmatrix} + 4(-1) \begin{vmatrix} 1 & 3 \\ 4 & 3 \end{vmatrix}$$

$$+ (-6)(1) \begin{vmatrix} 1 & 0 \\ 4 & 0 \end{vmatrix} = 36.$$

Hence

$$|\mathbf{A}| = 1(1)0 + (-2)(-1)81 + 3(1)36 + 0(-1)|\mathbf{M}_{14}| = 270.$$

Expansion in a similar manner by elements of some other row or column leads to the same result. Note that in this case $|\mathbf{M}_{14}|$ is not calculated because the element a_{14} is zero. By judicious choice of the row (or column) upon which a determinant will be expanded, arithmetic efforts can be minimized. Methods for reducing the arithmetic still further are discussed in the next section.

When expanding an n-order determinant by elements of a row we may write

$$|\mathbf{A}| = \sum_{j=1}^{n} a_{ij}(-1)^{i+j}|\mathbf{M}_{ij}| \qquad \text{for any } i, \tag{4.1}$$

and when expanding by elements of a column

$$|\mathbf{A}| = \sum_{i=1}^{n} a_{ij}(-1)^{i+j}|\mathbf{M}_{ij}| \qquad \text{for any } j. \tag{4.2}$$

Thus, as shown in the previous example, a fourth-order determinant is first expanded as four signed products each involving a third-order minor, and each of these is expanded as a sum of three signed products involving a second-order determinant. Consequently a fourth-order determinant ultimately involves $4 \times 3 \times 2 = 24$ products of its elements, each product containing four elements. This leads us to the general statement that the determinant of a square matrix of order n is a sum of $n!$ signed products.[1] The determinant is referred to as an *n-order* determinant. Utilizing methods given in Aiken (1948), it can be shown that each product involves n elements of the matrix, that each product has one and only one element from each row and column and that all such products are included and none occur more than once.

This method of evaluating a determinant involves tedious calculations for determinants of order greater than three: for order n there are $n!$ terms to be calculated, and for $n = 10$ this means 3,628,800 terms! Fortunately, easier methods exist, but because the method already discussed forms the basis of these easier methods it has been considered in detail. Furthermore, it is found useful in developing additional properties of determinants.

4.3 Elementary Expansions

a. Determinant of a transpose

The determinant of the transpose of a matrix is the same as the determinant of the matrix: $|\mathbf{A}'| = |\mathbf{A}|$.

This is so because expanding $|\mathbf{A}|$ using minors of elements in a row of $\mathbf{A}$ is the same as expanding $|\mathbf{A}'|$ using minors of elements in a column of $\mathbf{A}'$; except that the minors used for $|\mathbf{A}'|$ will be of matrices that are the transpose of those used for $|\mathbf{A}|$. This will be true right down to the minors of matrices of order 2×2 in each case. But these are equal: e.g.,

$$\begin{vmatrix} a & b \\ x & y \end{vmatrix} = ay - bx = \begin{vmatrix} a & x \\ b & y \end{vmatrix}.$$

Thus $|\mathbf{A}'| = |\mathbf{A}|$.

---

[1] $n!$, read as "factorial n," is the product of all integers 1 through n inclusive; e.g., $4! = 1(2)(3)(4) = 24$. Also, $0!$ is defined as unity.

Example

$$
\begin{vmatrix} 1 & -1 & 0 \\ 2 & 1 & 2 \\ 4 & 4 & 9 \end{vmatrix} = \begin{vmatrix} 1 & 2 \\ 4 & 9 \end{vmatrix} + \begin{vmatrix} 2 & 2 \\ 4 & 9 \end{vmatrix} = 1 + 10 = 11
$$

and

$$
\begin{vmatrix} 1 & 2 & 4 \\ -1 & 1 & 4 \\ 0 & 2 & 9 \end{vmatrix} = \begin{vmatrix} 1 & 4 \\ 2 & 9 \end{vmatrix} + \begin{vmatrix} 2 & 4 \\ 2 & 9 \end{vmatrix} = 1 + 10 = 11.
$$

An important consequence (or corollary, as mathematicians call it) of this is the following.

Corollary. All properties of $\mathbf{A}$ in terms of rows can be stated equivalently in terms of columns—because expanding $|\mathbf{A}'|$ by rows is identical to expanding $|\mathbf{A}|$ by columns. Thus although the following subsections b, c, and d are stated in terms of rows, they apply equally to columns.

b. Two rows the same

If two rows of a square matrix $\mathbf{A}$ are the same, its determinant is zero; i.e., $|\mathbf{A}| = 0$. This occurs because in expanding $|\mathbf{A}|$ by minors so that the 2×2 minors in the last step of the expansion are from two equal rows, all those minors are zero, and so $|\mathbf{A}| = 0$.

Example

$$
\begin{vmatrix} 1 & 4 & 3 \\ 7 & 5 & 2 \\ 7 & 5 & 2 \end{vmatrix} = \begin{vmatrix} 5 & 2 \\ 5 & 2 \end{vmatrix} - 4 \begin{vmatrix} 7 & 2 \\ 7 & 2 \end{vmatrix} + 3 \begin{vmatrix} 7 & 5 \\ 7 & 5 \end{vmatrix} = 0.
$$

c. Adding to a row a multiple of another row

Adding to one row of a determinant any multiple of another row does not affect the value of the determinant.

Example

$$
|\mathbf{A}| = \begin{vmatrix} 1 & 3 & 2 \\ 8 & 17 & 21 \\ 2 & 7 & 1 \end{vmatrix} \tag{4.3}
$$

$$
= 1(17 - 147) - 3(8 - 42) + 2(56 - 34) = 16.
$$

And adding four times row 1 to row 2 does not affect the value of $|\mathbf{A}|$:

$$|\mathbf{A}| = \begin{vmatrix} 1 & 3 & 2 \\ 8+4 & 17+12 & 21+8 \\ 2 & 7 & 1 \end{vmatrix} = \begin{vmatrix} 1 & 3 & 2 \\ 12 & 29 & 29 \\ 2 & 7 & 1 \end{vmatrix} \qquad (4.4)$$

$$= 1(29 - 203) - 3(12 - 58) + 2(84 - 58)$$

$$= -174 + 138 + 52 = 16.$$

This facility of adding to a row a multiple of some other row can be used repeatedly to simplify expansion of determinants. And although we describe these operations mostly in terms of rows they also apply for columns. Repeating the example, adding to the second row (-8) times the first row does not affect $|\mathbf{A}|$ and gives

$$|\mathbf{A}| = \begin{vmatrix} 1 & 3 & 2 \\ 0 & -7 & 5 \\ 2 & 7 & 1 \end{vmatrix}. \qquad (4.5)$$

Now, on adding to the third row (-2) times the first row, we get

$$|\mathbf{A}| = \begin{vmatrix} 1 & 3 & 2 \\ 0 & -7 & 5 \\ 0 & 1 & -3 \end{vmatrix}. \qquad (4.6)$$

Expansion by elements of the first column is now straightforward because two elements of the column are zero.

$$|\mathbf{A}| = 1 \begin{vmatrix} -7 & 5 \\ 1 & -3 \end{vmatrix} = 16,$$

the same result as before. By repeated use of the process of adding a multiple of one row to another we can reduce one column (not necessarily the first) to having all elements except one equal to zero; expansion by elements of that column then involves only one minor, which is a determinant of order one less than the original order. This in turn can be reduced also, and so on until the original determinant is a multiple of a single 2×2 determinant. At each stage the process of adding to one row multiples of other rows can be combined into one operation; e.g., in the previous example the steps that got us to (4.5) and then to (4.6) can be made simultaneously and then (4.6) can be written directly from (4.4).

The process can be started with any row; for example, consider bas-
ing the reduction of the foregoing example on its third row. If to row
2 we add -21 times row 3, and to row 1 is added -2 times row 3, we
get

$$|\mathbf{A}| = \begin{vmatrix} 1 & 3 & 2 \\ 8 & 17 & 21 \\ 2 & 7 & 1 \end{vmatrix} = \begin{vmatrix} -3 & -11 & 0 \\ -34 & -130 & 0 \\ 2 & 7 & 1 \end{vmatrix},$$

so that expansion by elements of the third column gives

$$|\mathbf{A}| = \begin{vmatrix} -3 & -11 \\ -34 & -130 \end{vmatrix} = 390 - 374 = 16$$

as before. The choice of which row is used and which column is simplified is dependent on observing which might lead to the simplest
arithmetic. For example, in expanding

$$\begin{vmatrix} 13 & 5 & 17 \\ 4 & 1 & 3 \\ 11 & 3 & 7 \end{vmatrix}$$

the arithmetic will be easier if multiples of the second row are added to
the other rows to put zero in the first and third elements of the second
column than if multiples of either the second or third row are added to
the others.

An explanation of the validity of adding to a row a multiple of an-
other row is as follows. Suppose $\mathbf{B}$ of order n has $[b_{11} \ b_{12} \ \cdots \ b_{1n}]$
and $[b_{21} \ b_{22} \ \cdots \ b_{2n}]$ as its first two rows. Let $\mathbf{A}$ be $\mathbf{B}$ with λ times
its second row added to its first. We show that $|\mathbf{A}| = |\mathbf{B}|$.

With $|\mathbf{M}_{ij}|$ being the minor of b_{ij} in $|\mathbf{B}|$, expansion of $|\mathbf{A}|$ by elements
of its first row, using (4.1), gives

$$|\mathbf{A}| \;=\; \sum_{j=1}^{n} (b_{1j} + \lambda b_{2j})(-1)^{1+j} |\mathbf{M}_{1j}|$$

$$=\; \sum_{j=1}^{n} b_{1j}(-1)^{1+j} |\mathbf{M}_{1j}| + \lambda \sum_{j=1}^{n} b_{2j}(-1)^{1+j} |\mathbf{M}_{1j}|.$$

The first sum here is $|\mathbf{B}|$, directly from (4.1). The second sum is an
example of equation (5.7) and is zero (see Section 5.4). Therefore
$|\mathbf{A}| = |\mathbf{B}|$.

The foregoing properties can be applied in endless variation in expanding determinants, with efficiency in perceiving a procedure that leads to minimal effort in any particular case being largely a matter of practice. The underlying method might be summarized as follows. Through adding multiples of one row to other rows of the determinant a column is reduced to having only one non-zero element. Expansion by elements of that column then involves only one non-zero product of an element with its minor, which is a determinant of order one less than the original determinant. Successive applications of this method reduce the determinant to a multiple of one of order 2×2. If at any stage these reductions lead to the elements of a row containing a common factor, this can be factored out as a factor of the determinant, and if they lead to a row of zeros or to two rows being identical, the determinant is zero. (See Section 4.3b.)

d. Adding a row to a multiple of a row

We have just seen how adding to a row a multiple of another row does not alter the value of a determinant. But beware, do not do the reverse: adding a row to a multiple of a row is not the same thing.

Example

$$\begin{vmatrix} 2 & 3 \\ 5 & 9 \end{vmatrix} = \begin{vmatrix} 2+5\lambda & 3+9\lambda \\ 5 & 9 \end{vmatrix} = 3, \qquad \text{but} \qquad \begin{vmatrix} 2\lambda+5 & 3\lambda+9 \\ 5 & 9 \end{vmatrix} = 3\lambda.$$

e. Products

When $\mathbf{A}$ and $\mathbf{B}$ are square and of the same order

$$|\mathbf{AB}| = |\mathbf{A}||\mathbf{B}|. \tag{4.7}$$

Example

With

$$\mathbf{AB} = \begin{bmatrix} 1 & 2 \\ 3 & 8 \end{bmatrix} \begin{bmatrix} 4 & 3 \\ 6 & 6 \end{bmatrix} = \begin{bmatrix} 16 & 15 \\ 60 & 57 \end{bmatrix}, \quad |\mathbf{AB}| = 912 - 900 = 12$$

and

$$|\mathbf{A}||\mathbf{B}| = \begin{vmatrix} 1 & 2 \\ 3 & 8 \end{vmatrix} \begin{vmatrix} 4 & 3 \\ 6 & 6 \end{vmatrix} = 2(6) = 12.$$

This result, (4.7), is known as the *product rule* of determinants. It is easy to state: the determinant of a matrix product is the product of the determinants. But unfortunately it is not as easy to verify. Nevertheless, its verification does include other useful results, so we proceed to give it. It requires as preliminaries: (i) the determinants of a triangular matrix; (ii) reducing a determinant to triangular form; and (iii) the determinants of two special partitioned matrices. Then comes the establishment of $|\mathbf{AB}| = |\mathbf{A}||\mathbf{B}|$.

− i.   Determinant of a triangular matrix

The determinant of a lower triangular matrix is the product of its diagonal elements. For example,

$$\begin{vmatrix} 6 & 0 & 0 \\ 3 & -1 & 0 \\ 7 & 3 & -5 \end{vmatrix} = 6(-1)(-5) = 30.$$

Verification is easy—through expansion by elements of successive rows.

− ii.   Reducing a determinant to triangular form

The determinant of any square matrix can, by a series of adding to rows multiples of other rows, be reduced to being a determinant of a lower triangular matrix.

Example

The operations

$$\text{to row 2} \qquad \text{add} \qquad (-2)\text{row 3}$$

and

$$\text{to row 1} \qquad \text{add} \qquad (-3\tfrac{1}{2})\text{row 3}$$

give

$$|\mathbf{P}| = \begin{vmatrix} 3 & 8 & 7 \\ 1 & 2 & 4 \\ -1 & 3 & 2 \end{vmatrix} = \begin{vmatrix} 6\tfrac{1}{2} & -2\tfrac{1}{2} & 0 \\ 3 & -4 & 0 \\ -1 & 3 & 2 \end{vmatrix}. \tag{4.8}$$

Then the operation

$$\text{to row 1} \quad \text{add} \quad -[(-2\tfrac{1}{2})/(-4)]\text{row 2} \tag{4.9}$$

gives

$$|\mathbf{P}| = \begin{vmatrix} 37/8 & 0 & 0 \\ 3 & -4 & 0 \\ -1 & 3 & 2 \end{vmatrix} = (37/8)(-4)2 = -37. \tag{4.10}$$

This procedure can be extended to a square matrix of any order.

– iii. Two partitioned determinants

For any $\mathbf{P}$ and $\mathbf{Q}$ which are square of order n

$$\begin{vmatrix} \mathbf{P} & \mathbf{0} \\ \mathbf{X} & \mathbf{Q} \end{vmatrix} = |\mathbf{P}||\mathbf{Q}|. \tag{4.11}$$

And for $\mathbf{R}$ and $\mathbf{S}$ square of order n

$$\begin{vmatrix} \mathbf{0} & \mathbf{R} \\ -\mathbf{I} & \mathbf{S} \end{vmatrix} = |\mathbf{R}|. \tag{4.12}$$

Result (4.11) comes from carrying out on the n rows of $[\mathbf{P} \ \mathbf{0}]$ the row operations necessary to reduce $\mathbf{P}$ to lower triangular form as demonstrated in (4.8) through (4.10). This does not alter the value of the determinant in (4.11) which, on then expanding by those first n rows, yields $|\mathbf{P}||\mathbf{Q}|$.

Result (4.12) comes from expanding the left-hand side by successive columns through the $-\mathbf{I}$. This gives

$$\begin{vmatrix} \mathbf{0} & \mathbf{R} \\ -\mathbf{I} & \mathbf{S} \end{vmatrix} = [(-1)^{(n+1)+1}(-1)]^n |\mathbf{R}| = (-1)^{n(n+3)} |\mathbf{R}| = |\mathbf{R}| \tag{4.13}$$

with n being the positive integer that it is. The reason for the term in square braces is as follows. The first diagonal element in $-\mathbf{I}$ of (4.12) is in row $n+1$ and column 1 of (4.12). Therefore the sign term is $(-1)^{(n+1)+1}$, and with that goes the (-1) from $-\mathbf{I}$. After deleting from (4.12) the row and column containing that first -1 in $-\mathbf{I}$, the next (-1) in $-\mathbf{I}$ is in row $n+1$ and column 1 of the now-reduced (4.12). This continues for all n diagonal elements of $-\mathbf{I}$, and so we get (4.13).

– iv. Verification of the product rule

The product

$$\begin{bmatrix} \mathbf{I} & \mathbf{A} \\ \mathbf{0} & \mathbf{I} \end{bmatrix} \begin{bmatrix} \mathbf{A} & \mathbf{0} \\ -\mathbf{I} & \mathbf{B} \end{bmatrix} = \begin{bmatrix} \mathbf{0} & \mathbf{AB} \\ -\mathbf{I} & \mathbf{B} \end{bmatrix} \tag{4.14}$$

$$= \begin{bmatrix} [\mathbf{A} \ \mathbf{0}] \ + \ \mathbf{A}[-\mathbf{I} \ \mathbf{B}] \\ -\mathbf{I} \qquad\qquad \mathbf{B} \end{bmatrix}. \tag{4.15}$$

(4.15) shows that the left-hand side of (4.14) is just $\begin{bmatrix} \mathbf{A} & \mathbf{0} \\ -\mathbf{I} & \mathbf{B} \end{bmatrix}$ with multiples of some of its rows (those of $[-\mathbf{I} \ \mathbf{B}]$) added to others of its rows (those of $[\mathbf{A} \ \mathbf{0}]$). This operation does not alter determinantal values. Therefore

$$\left| \begin{bmatrix} \mathbf{I} & \mathbf{A} \\ \mathbf{0} & \mathbf{I} \end{bmatrix} \begin{bmatrix} \mathbf{A} & \mathbf{0} \\ -\mathbf{I} & \mathbf{B} \end{bmatrix} \right| = \left| \begin{matrix} \mathbf{A} & \mathbf{0} \\ -\mathbf{I} & \mathbf{B} \end{matrix} \right|$$

$$= |\mathbf{A}||\mathbf{B}|, \text{ from (4.11)}. \tag{4.16}$$

But in taking the determinant of both sides of (4.14)

$$\left| \begin{bmatrix} \mathbf{I} & \mathbf{A} \\ \mathbf{0} & \mathbf{I} \end{bmatrix} \begin{bmatrix} \mathbf{A} & \mathbf{0} \\ -\mathbf{I} & \mathbf{B} \end{bmatrix} \right| = \left| \begin{matrix} \mathbf{0} & \mathbf{AB} \\ -\mathbf{I} & \mathbf{B} \end{matrix} \right|$$

$$= |\mathbf{AB}|, \text{ from (4.12)}. \tag{4.17}$$

Equating the right-hand sides of (4.16) and (4.17), because their left-hand sides are the same, gives the result we want:

$$|\mathbf{AB}| = |\mathbf{A}||\mathbf{B}|.$$

Example

For

$$\mathbf{A} = \begin{bmatrix} 1 & 3 \\ 2 & 5 \end{bmatrix} \text{ and } \mathbf{B} = \begin{bmatrix} -4 & 7 \\ 8 & 9 \end{bmatrix}, \ \mathbf{AB} = \begin{bmatrix} 20 & 34 \\ 32 & 59 \end{bmatrix}.$$

Then

$$|\mathbf{A}| = -1, \ |\mathbf{B}| = -92$$

and

$$|\mathbf{AB}| = 1180 - 1088 = 92 = (-1)(-92) = |\mathbf{A}||\mathbf{B}|.$$

– v.   Extensions to rectangular matrices

For $\mathbf{A}$ and $\mathbf{B}$ rectangular, $|\mathbf{AB}|$ exists for $\mathbf{A}$ being $r \times c$ only if $\mathbf{B}$ is $c \times r$. Then, if $r > c$ the determinant $|\mathbf{AB}| = 0$; and, as explained by Aitken (1948, p. 86), if $r < c$ the determinant $|\mathbf{AB}|$ is a sum of products of r-order minors from $\mathbf{A}$ and from $\mathbf{B}$.

– vi.   Useful corollaries

1. $|\mathbf{AB}| = |\mathbf{BA}|$ (because $|\mathbf{A}||\mathbf{B}| = |\mathbf{B}||\mathbf{A}|$), $\mathbf{A}$ and $\mathbf{B}$ square of the same order.

2. $|\mathbf{A}^2| = \mathbf{A}|^2$ (each equals $|\mathbf{A}|\,|\mathbf{A}|$); $|\mathbf{A}^k| = |\mathbf{A}|^k$ is the extension.

3. For orthogonal $\mathbf{A}$, $|\mathbf{A}| = \pm 1$ (because $\mathbf{AA}' = \mathbf{I}$ implies $|\mathbf{A}|^2 = 1$).

4. For idempotent $\mathbf{A}$, $|\mathbf{A}| = 0$ or 1 (because $\mathbf{A}^2 = \mathbf{A}$ implies $|\mathbf{A}|^2 = |\mathbf{A}|$).

f.   Cramer's rule for linear equations

An easy-to-understand and long-established method for solving linear equations is what is known as *Cramer's rule* (Cramer, 1704-1752). It is easy to understand because it involves only determinants and in that regard is quite practical for a small number (2, 3 or 4, say) of equations. Although it works perfectly well for any number of equations, n say, it involves calculating $n + 1$ determinants of order n, and for $n > 4$ this can be computationally tedious if done by hand. If done with the aid of a computer, there is always the possibility of wrong answers arising from the inherent difficulty of computers to calculate large-order determinants correctly. In practice, large sets of linear equations are customarily solved using computing software that involves a matrix inverse (see Chapter 5) or some kind of iterative process. Nevertheless, Cramer's rule is a useful start for first learning about solving systems of linear equations.

Linear equations are equations wherein the unknowns, x_1, x_2, x_3 and x_4, say, occur only in that form and not as functions such as powers, products or whatever. One such equation is

$$a_{11}x_1 + a_{12}x_2 + a_{13}x_3 + a_{14}x_4 = y_1, \tag{4.18}$$

where the as and the y_1 are known numbers. A common occurrence is to have the same number of equations as unknowns and then write them as

$$\mathbf{Ax} = \mathbf{y}$$

where, for (4.18) $\mathbf{A}$ is 4×4 and $\mathbf{x}$ and $\mathbf{y}$ are each 4×1. Let us write $\mathbf{A}$ in its column partitioned form $\mathbf{A} = [\mathbf{a}_1\ \mathbf{a}_2\ \mathbf{a}_3\ \mathbf{a}_4]$. Then the equations are

$$[\mathbf{a}_1\ \mathbf{a}_2\ \mathbf{a}_3\ \mathbf{a}_4]\mathbf{x} = \mathbf{y}.$$

Cramer's rule is now easily stated: e.g.,

$$x_2 = |\mathbf{a}_1 \ \mathbf{y} \ \mathbf{a}_3 \ \mathbf{a}_4|/|\mathbf{A}|.$$

The numerator for x_2 is simply the determinant of the matrix which is $\mathbf{A}$ with column 2 (because we want x_2) replaced by $\mathbf{y}$. In general for the jth unknown,

$$x_j = |\mathbf{A} \text{ with column } j \text{ replaced by } \mathbf{y}|/|\mathbf{A}|.$$

– i.   Illustration: Input-output equations

Suppose a manufacturing firm needs three items of input x_1 and two of x_2 to make 32 items of output 1; and it needs five of x_1 and three of x_2 to make 52 items of output 2. This means the manufacturing process is represented by the equations

$$\begin{array}{rcl} 3x_1 + 2x_2 & = & 32, \\ 5x_1 + 3x_2 & = & 52, \end{array} \quad \text{i.e.,} \quad \begin{bmatrix} 3 & 2 \\ 5 & 3 \end{bmatrix} \begin{bmatrix} x_1 \\ x_2 \end{bmatrix} = \begin{bmatrix} 32 \\ 52 \end{bmatrix}.$$

The Cramer rule gives

$$x_1 = \frac{\begin{vmatrix} 32 & 2 \\ 52 & 3 \end{vmatrix}}{\begin{vmatrix} 3 & 2 \\ 5 & 3 \end{vmatrix}} = \frac{-8}{-1} = 8 \quad \text{and} \quad x_2 = \frac{\begin{vmatrix} 3 & 32 \\ 5 & 52 \end{vmatrix}}{\begin{vmatrix} 3 & 2 \\ 5 & 3 \end{vmatrix}} = \frac{-4}{-1} = 4.$$

Therefore the firm needs a production schedule based on eight units of input x_1 and four of x_2.

– ii.   Illustration: Supply and demand

Consider a market involving just two commodities, Q_1 and Q_2 with prices P_1 and P_2. The demand and supply curves (planes) are as follows:

Commodity	Demand	Supply
Q_1	$a_0 + a_1 P_1 + a_2 P_2$	$b_0 + b_1 P_1 + b_2 P_2$
Q_2	$c_0 + c_1 P_1 + c_2 P_2$	$d_0 + d_1 P_1 + d_2 P_2$

Setting the demand and supply equal to each other for each commodity gives equations

$$(a_1 - b_1)P_1 + (a_2 - b_2)P_2 = b_0 - a_0$$

and

$$(c_1 - d_1)P_1 + (c_2 - d_2)P_2 = d_0 - c_0.$$

Writing these as

$$e_1 P_1 + e_2 P_2 = e_0$$

$$f_1 P_1 + f_2 P_2 = f_0$$

defines e_1, e_2, e_0 and f_1, f_2 and f_0 in an obvious way. And then Cramer's rule gives

$$\bar{P}_1 = \frac{\begin{vmatrix} e_0 & e_2 \\ f_0 & f_2 \end{vmatrix}}{\begin{vmatrix} e_1 & e_2 \\ f_1 & f_2 \end{vmatrix}} = \frac{e_0 f_2 - e_2 f_0}{e_1 f_2 - e_2 f_1} \text{ and } \bar{P}_2 = \frac{\begin{vmatrix} e_1 & e_0 \\ f_1 & f_0 \end{vmatrix}}{\begin{vmatrix} e_1 & e_2 \\ f_1 & f_2 \end{vmatrix}} = \frac{e_1 f_0 - e_0 f_1}{e_1 f_2 - e_2 f_1}.$$

Replacing P_1 and P_2 in the demand and supply functions by $\bar{P}_1$ and $\bar{P}_2$ gives the quantities needed of the two commodities.

4.4 Elementary Row Operations

a. Definitions

In Section 4.3d we introduced the operation of adding a multiple of a row (column) of a determinant to another row (column). It did not alter the value of the determinant. That same operation on a matrix can be represented by a matrix product. For example, the determinant in (4.4) came from that in (4.3) by the operation

$$\text{to row 2 add 4(row 1).} \tag{4.19}$$

This is achieved in the matrix product

$$\begin{bmatrix} 1 & 0 & 0 \\ 4 & 1 & 0 \\ 0 & 0 & 1 \end{bmatrix} \begin{bmatrix} 1 & 3 & 2 \\ 8 & 17 & 21 \\ 2 & 7 & 1 \end{bmatrix} = \begin{bmatrix} 1 & 3 & 2 \\ 12 & 29 & 29 \\ 2 & 7 & 1 \end{bmatrix}, \tag{4.20}$$

where the left-most matrix specifically achieves (4.19).

That matrix is one of three kinds of matrices known as *elementary operator matrices*. Its general form of $\mathbf{P}_{ij}(\lambda)$ is $\mathbf{I}$ with λ replacing 0 for the (i,j)th element, with

$$[\mathbf{P}_{ij}(\lambda)]\,\mathbf{A} = \mathbf{A} \text{ with row } i \text{ becoming } [\text{row } i + \lambda(\text{row } j)].$$

The other two elementary operator matrices are

$$\mathbf{E}_{ij} \;=\; \mathbf{I} \text{ with its } i\text{th and } j\text{th rows interchanged}$$

with

$$\mathbf{E}_{ij}\mathbf{A} \;=\; \mathbf{A} \text{ with its } i\text{th and } j\text{th rows interchanged;}$$

and

$$\mathbf{R}_{ii}(\lambda) \;=\; \mathbf{I} \text{ with its } i\text{th diagonal element replaced by } \lambda$$

with

$$[\mathbf{R}_{ii}(\lambda)]\,\mathbf{A} \;=\; \mathbf{A} \text{ with its } i\text{th row multiplied by } \lambda.$$

Examples

$\mathbf{P}_{12}(4)\mathbf{A}$ is shown in (4.20).

$$\mathbf{E}_{12}\mathbf{A} \;=\; \begin{bmatrix} 0 & 1 & 0 \\ 1 & 0 & 0 \\ 0 & 0 & 1 \end{bmatrix} \begin{bmatrix} 1 & 3 & 2 \\ 8 & 17 & 21 \\ 2 & 7 & 1 \end{bmatrix} = \begin{bmatrix} 8 & 17 & 21 \\ 1 & 3 & 2 \\ 2 & 7 & 1 \end{bmatrix}$$

and

$$\mathbf{R}_{33}(5)\mathbf{A} \;=\; \begin{bmatrix} 1 & 0 & 0 \\ 0 & 1 & 0 \\ 0 & 0 & 5 \end{bmatrix} \begin{bmatrix} 1 & 3 & 2 \\ 8 & 17 & 21 \\ 2 & 7 & 1 \end{bmatrix} = \begin{bmatrix} 1 & 3 & 2 \\ 8 & 17 & 21 \\ 10 & 35 & 5 \end{bmatrix}.$$

All three elementary operator matrices have simple determinants:

$$|\mathbf{P}_{ij}(\lambda)| = 1, \qquad |\mathbf{E}_{ij}| = -1 \quad \text{and} \quad |\mathbf{R}_{ii}(\lambda)| = \lambda. \qquad (4.21)$$

These elementary operators used in combination with $|\mathbf{AB}| = |\mathbf{A}||\mathbf{B}|$ provide easy establishment of methods for simplifying the expansion of determinants. This we now show. But before we do, we strongly emphasize that these elementary operator matrices play very important roles in later chapters: in Chapter 7, dealing with rank, and in Chapter 8 on canonical forms.

b. Factorization

If $\mathbf{A}$ has λ as a factor of a row (column) then λ is also a factor of $|\mathbf{A}|$; i.e.,

$$|\mathbf{A}| = \lambda \left(|\mathbf{A} \text{ with } \lambda \text{ factored out of a row or column}| \right). \qquad (4.22)$$

Example

$$\begin{vmatrix} 4 & 6 \\ 1 & 7 \end{vmatrix} = 2 \begin{vmatrix} 2 & 3 \\ 1 & 7 \end{vmatrix}, \qquad \text{i.e., } 4(7) - 1(6) = 22 = 2[2(7) - 1(3)].$$

The reason for (4.22) being true is that in expanding $\mathbf{A}$ by elements of the row (column) containing λ, every term will have λ as a factor. In point of fact,

$$\mathbf{A} = \mathbf{R}_{ii}(\lambda) [\mathbf{A} \text{ with } \lambda \text{ factored out of its } i\text{th row}] \qquad (4.23)$$

so that $|\mathbf{R}_{ii}(\lambda)| = \lambda$ of (4.21), and applying the product rule to (4.23) yields (4.22).

There is also the useful result that

$$|\lambda \mathbf{A}_{n \times n}| = \lambda^n |\mathbf{A}_{n \times n}|,$$

because λ is a factor of every row of $\lambda \mathbf{A}$.

Example

$$\begin{vmatrix} 3 & 0 & 27 \\ -9 & 3 & 0 \\ 15 & 6 & -3 \end{vmatrix} = 3^3 \begin{vmatrix} 1 & 0 & 9 \\ -3 & 1 & 0 \\ 5 & 2 & -1 \end{vmatrix} = -2700.$$

c. One row being a multiple of another

When a determinant has one row (column) which is a multiple of another, the determinant is zero. (Factoring out the multiple reduces the determinant to having two rows the same. Hence the determinant is zero.)

Example

$$\begin{vmatrix} -3 & 6 & 12 \\ 2 & -4 & -8 \\ 7 & 5 & 9 \end{vmatrix} = -(1.5) \begin{vmatrix} 2 & -4 & -8 \\ 2 & -4 & -8 \\ 7 & 5 & 9 \end{vmatrix} = 0.$$

d. Row (column) of zeros

When a determinant has zero for every element of a row (or column), the determinant is zero—because (4.22) then has $\lambda = 0$.

Example

$$\begin{vmatrix} 0 & 0 & 0 \\ 3 & 6 & 5 \\ 2 & 9 & 7 \end{vmatrix} = 0 \begin{vmatrix} 6 & 5 \\ 9 & 7 \end{vmatrix} - 0 \begin{vmatrix} 3 & 5 \\ 2 & 7 \end{vmatrix} + 0 \begin{vmatrix} 3 & 6 \\ 2 & 9 \end{vmatrix} = 0.$$

e. Interchanging rows (columns)

Interchanging two rows (columns) of a determinant changes its sign, because $|\mathbf{E}_{ij}\mathbf{A}| = |\mathbf{E}_{ij}||\mathbf{A}| = -|\mathbf{A}|$.

Example

$$\begin{vmatrix} 1 & 0 & 0 \\ 7 & 2 & 3 \\ 9 & 4 & 8 \end{vmatrix} = 4 \qquad \text{but} \qquad \begin{vmatrix} 9 & 4 & 8 \\ 7 & 2 & 3 \\ 1 & 0 & 0 \end{vmatrix} = -4.$$

4.5 Diagonal Expansion

A square matrix can always be expressed as a sum $\mathbf{A} + \mathbf{D}$, where $\mathbf{D} = \left\{_d \, x_i \right\}$ is a diagonal matrix (see Section 1.8b). The determinant $|\mathbf{A} + \mathbf{D}|$ can then be expressed as a polynomial in the x_is.

a. Notation for minors

To describe this we use an abbreviated notation for minors of $\mathbf{A} = \{a_{ij}\}$, whereby they are denoted by just their diagonal elements; for example, $\begin{vmatrix} a_{11} & a_{12} \\ a_{21} & a_{22} \end{vmatrix}$ is written as $|a_{11} \quad a_{22}|$ and $\begin{vmatrix} a_{12} & a_{13} \\ a_{22} & a_{23} \end{vmatrix}$ as $|a_{12} \quad a_{23}|$. Combined with the notation $\mathbf{A} = \{a_{ij}\}$ no confusion can arise because, for example, $|a_{21} \quad a_{33} \quad a_{44}|$ has a_{21}, a_{33} and a_{44} as diagonal elements with the 3×3 being filled with elements in the same rows and columns from $\mathbf{A}$. Thus

$$|a_{21} \quad a_{33} \quad a_{44}| = \begin{vmatrix} a_{21} & a_{23} & a_{24} \\ a_{31} & a_{33} & a_{34} \\ a_{41} & a_{43} & a_{44} \end{vmatrix}.$$

b. Determinant of $\mathbf{A} + \mathbf{D}$

We begin with $|\mathbf{A} + \mathbf{D}|$ of order 2. It is

$$
|\mathbf{A} + \mathbf{D}| = \begin{vmatrix} a_{11} + x_1 & a_{12} \\ a_{21} & a_{22} + x_2 \end{vmatrix}
$$

$$
= (a_{11} + x_1)(a_{22} + x_2) - a_{12}a_{21}
$$

$$
= x_1 x_2 + x_1 a_{22} + x_2 a_{11} + \begin{vmatrix} a_{11} & a_{12} \\ a_{21} & a_{22} \end{vmatrix}. \qquad (4.24)
$$

Similarly, it can be shown that

$$
\begin{vmatrix} a_{11} + x_1 & a_{12} & a_{13} \\ a_{21} & a_{22} + x_2 & a_{23} \\ a_{31} & a_{32} & a_{33} + x_3 \end{vmatrix} = x_1 x_2 x_3 + x_1 x_2 a_{23} + x_1 x_3 a_{22}
$$

$$
+ x_1 |a_{22}\ a_{33}| + x_2 |a_{11}\ a_{33}| + x_3 |a_{11}\ a_{22}|
$$

$$
\qquad (4.25)
$$

$$
+ |a_{11}\ a_{22}\ a_{33}|.
$$

A direct extension of the second-order and third-order polynomials (in the xs) of (4.24) and (4.25) applies to $|\mathbf{A} + \mathbf{D}|$ of order n. It consists of the sum of all possible products of the x_i taken r at a time for $r = 0, 1, \ldots, n$, each product being multiplied by its complementary principal minor of order $n - r$ in $|\mathbf{A}|$. (See the definitions of principal minors in subsection c which follows.) Interest in these results is likely to be stronger among mathematicians than among economists, so we have described them but briefly and pursue them no further, save for the useful special case in subsection c.

c. Principal minors

A *principal minor* of $|\mathbf{A}|$ is a minor that has diagonal elements which are in the diagonal of $|\mathbf{A}|$; an example is $|a_{22}\ a_{55}|$. A *complementary principal minor* is the minor obtained from $|\mathbf{A}|$ after deleting the rows and columns of $|\mathbf{A}|$ that contain some other principal minor; e.g., in $|\mathbf{A}|$ of order 6 the complementary principal minor for $|a_{22}\ a_{55}|$ is $|a_{11}\ a_{33}\ a_{44}\ a_{66}|$. And connected with principal minors is the *trace-of-order* k of $\mathbf{A}$. It is

$$\text{tr}_k(\mathbf{A}) = \text{ sum of all } \binom{n}{k} \text{ principal minors of order } k \text{ in } \mathbf{A},$$

where $\binom{n}{k}$ is the number of ways of choosing k things out of n.[2]

More broadly, a general *complementary minor* is the minor remaining after deleting the rows and columns of $|\mathbf{A}|$ that contain some other (not necessarily principal) minor; e.g., for $|\mathbf{A}_{6\times 6}|$, the complementary minor for $|a_{12}\ a_{54}|$ is $|a_{21}\ a_{33}\ a_{45}\ a_{66}|$.

d.   Special case: $\mathbf{A} + \lambda\mathbf{I}$

When in $|\mathbf{A}+\mathbf{D}|$ every x_i is the same, λ say, we find that for order n

$$|\mathbf{A} + \lambda\mathbf{I}| = \sum_{k=0}^{n} \lambda^k \text{tr}_{n-k}(\mathbf{A})$$

where, as usual, $\lambda^0 = 1$, and $\text{tr}(\mathbf{A}) = |\mathbf{A}|$ and $\text{tr}_0(\mathbf{A}) = 1$. This is particularly useful whenever needing to algebraically (as distinct from numerically, using a computer) solve the equation $|\mathbf{A} - \lambda\mathbf{I}| = 0$ for λ, which one does for deriving eigenvalues of $\mathbf{A}$, discussed in Chapter 11.

Example   For $\mathbf{A} = b\mathbf{J}_n$ and $\lambda = a$,

$$
\begin{aligned}
|\mathbf{A} + \lambda\mathbf{I}| &= |b\mathbf{J}_n + a\mathbf{I}_n| \\
&= a^n \text{tr}_0(b\mathbf{J}_n) + a^{n-1}\text{tr}_1(b\mathbf{J}_n) = a^{n-1}(a + nb).
\end{aligned}
$$

4.6   Laplace Expansion

In the expansion of

$$|\mathbf{A}| = \begin{vmatrix} a_{11} & a_{12} & a_{13} & a_{14} \\ a_{21} & a_{22} & a_{23} & a_{24} \\ a_{31} & a_{32} & a_{33} & a_{34} \\ a_{41} & a_{42} & a_{43} & a_{44} \end{vmatrix}$$

the minor of a_{11} is $|a_{22}\ a_{33}\ a_{44}|$. An easily verified extension of this is that $|a_{11}\ a_{22}|$ is multiplied by its complementary minor $|a_{33}\ a_{44}|$. Likewise, $|a_{11}\ a_{24}|$ is multiplied by $|a_{32}\ a_{43}|$. Each minor multiplying

---

[2] $\binom{n}{k} = n!/k!(n-k)!$ where $n! = 1(2)3(4)\cdots(n-1)n$.

a particular minor of $|\mathbf{A}|$ is the complementary minor in $|\mathbf{A}|$ of that particular minor. This is simply an extension of the procedure for finding the coefficient of an individual element in $|\mathbf{A}|$ as derived in the expansion by elements of a row or column discussed earlier. In that case the particular minor is a single element and its coefficient in $|\mathbf{A}|$ is $|\mathbf{A}|$ amended by deletion of the one row and column containing the element concerned. A sign factor is also involved, namely $(-1)^{i+j}$ for the coefficient of a_{ij} in $|\mathbf{A}|$. In the extension to coefficients of minors the sign factor is minus one raised to the power of the sum of the subscripts of the diagonal elements of the chosen minor: for example, the sign factor for the coefficient of $|a_{32}\ a_{43}|$ is $(-1)^{3+2+4+3} = -1$. The complementary minor multiplied by this sign factor is the coefficient of the particular minor concerned. Furthermore, just as the expansion of a determinant is the sum of products of elements of a row (or column) with their coefficients, so also is it the sum of products of all minors of order m that can be derived from any set of m rows, each multiplied by its complementary minor and sign. This method of expanding a determinant by minors of a set of rows was first established by Laplace and so bears his name. Aitken (1948) and Ferrar (1941) give proof of the procedure; we shall be satisfied here with a general statement of the method and an example illustrating its use.

The Laplace expansion of a determinant $|\mathbf{A}|$ of order n is as follows. (i) Consider any m rows of $|\mathbf{A}|$. They contain $n!/[m!(n-m)!]$ minors of order m (see footnote, Section 4.2). (ii) Multiply each of these minors, $\mathbf{M}$ say, by its complementary minor and by a sign factor, $(-1)^{\mu}$, where μ is the sum of the subscripts of the diagonal elements of $\mathbf{M}$. (iii) The sum of all such products is $|\mathbf{A}|$.

Example

Interchanging the second and fourth rows of

$$|\mathbf{A}| = \begin{vmatrix} 1 & 2 & 3 & 0 & 0 \\ 1 & 0 & 4 & 2 & 3 \\ 2 & 0 & 1 & 4 & 5 \\ -1 & 2 & -1 & 0 & 0 \\ 0 & 2 & 1 & 2 & 3 \end{vmatrix} \quad \text{gives } |\mathbf{A}| = - \begin{vmatrix} 1 & 2 & 3 & 0 & 0 \\ -1 & 2 & -1 & 0 & 0 \\ 2 & 0 & 1 & 4 & 5 \\ 1 & 0 & 4 & 2 & 3 \\ 0 & 2 & 1 & 2 & 3 \end{vmatrix}.$$

In this form we expand $|\mathbf{A}|$ using the Laplace expansion based on the first two rows ($m = 2$). There are ten minors of order 2 in these two rows; seven of them are zero because they involve a column of zeros. Hence $|\mathbf{A}|$ can be expanded as the sum of three products involving the three 2×2 nonzero minors in the first two rows, namely as

$$-|\mathbf{A}| = (-1)^{1+1+2+2} \begin{vmatrix} 1 & 2 \\ -1 & 2 \end{vmatrix} \begin{vmatrix} 1 & 4 & 5 \\ 4 & 2 & 3 \\ 1 & 2 & 3 \end{vmatrix}$$

$$+ (-1)^{1+1+2+3} \begin{vmatrix} 1 & 3 \\ -1 & -1 \end{vmatrix} \begin{vmatrix} 0 & 4 & 5 \\ 0 & 2 & 3 \\ 2 & 2 & 3 \end{vmatrix}$$

$$+ (-1)^{1+2+2+3} \begin{vmatrix} 2 & 3 \\ 2 & -1 \end{vmatrix} \begin{vmatrix} 2 & 4 & 5 \\ 1 & 2 & 3 \\ 0 & 2 & 3 \end{vmatrix}$$

which reduces to $|\mathbf{A}| = 16$. It will be found that expansion by a more direct method leads to the same result.

Numerous other methods of expanding determinants are based on extensions of the Laplace expansion, using it recurrently to expand a determinant not only by minors and their complementary minors but also to expand these minors themselves. Many of these expansions are identified by the names of their originators, for example, Cauchy, Binet-Cauchy and Jacoby. A good account of some of them is to be found in Aitken (1948) and Ferrar (1941).

4.7   Sums and Differences of Determinants

The sum of the determinants of each of two (or more) matrices generally does not equal the determinant of the sum. The simplest demonstration of this is

$$|\mathbf{A}| + |\mathbf{B}| = \begin{vmatrix} a_{11} & a_{12} \\ a_{21} & a_{22} \end{vmatrix} + \begin{vmatrix} b_{11} & b_{12} \\ b_{21} & b_{22} \end{vmatrix}$$

$$= a_{11}a_{22} - a_{12}a_{21} + b_{11}b_{22} - b_{12}b_{21} \neq |\mathbf{A} + \mathbf{B}|.$$

The same applies to the difference, $|\mathbf{A}| - |\mathbf{B}| \neq |\mathbf{A} - \mathbf{B}|$.

Note that both $|\mathbf{A}| + |\mathbf{B}|$ and $|\mathbf{A}| - |\mathbf{B}|$ have meaning even when $\mathbf{A}$ and $\mathbf{B}$ are square matrices of different orders, because a determinant is a scalar. This contrasts with $\mathbf{A} + \mathbf{B}$ and $\mathbf{A} - \mathbf{B}$, which have meaning only when the matrices are conformable for addition (have the same order).

Another point of interest is that although $|\mathbf{A}| + |\mathbf{B}|$ does not generally equal $|\mathbf{A} + \mathbf{B}|$, the latter can be written as the sum of certain other determinants. For example,

$$
|\mathbf{A} + \mathbf{B}| = \left| \begin{bmatrix} a_{11} & a_{12} \\ a_{21} & a_{22} \end{bmatrix} + \begin{bmatrix} b_{11} & b_{12} \\ b_{21} & b_{22} \end{bmatrix} \right| = \begin{vmatrix} a_{11} + b_{11} & a_{12} + b_{12} \\ a_{21} + b_{21} & a_{22} + b_{22} \end{vmatrix}
$$

$$
= \begin{vmatrix} a_{11} & a_{12} \\ a_{21} + b_{21} & a_{22} + b_{22} \end{vmatrix} + \begin{vmatrix} b_{11} & b_{12} \\ a_{21} + b_{21} & a_{22} + b_{22} \end{vmatrix}
$$

$$
= \begin{vmatrix} a_{11} & a_{12} \\ a_{21} & a_{22} \end{vmatrix} + \begin{vmatrix} a_{11} & a_{12} \\ b_{21} & b_{22} \end{vmatrix} + \begin{vmatrix} b_{11} & b_{12} \\ a_{21} & a_{22} \end{vmatrix} + \begin{vmatrix} b_{11} & b_{12} \\ b_{21} & b_{22} \end{vmatrix}
$$

In general if $\mathbf{A}$ and $\mathbf{B}$ are $n \times n$, $|\mathbf{A} + \mathbf{B}|$ can be expanded as the sum of 2^n n-order determinants.

4.8 Exercises

E 4.1. Show that

(a) $\begin{vmatrix} 1 & 5 & -5 \\ 3 & 2 & -5 \\ 6 & -2 & -5 \end{vmatrix}$ and $\begin{vmatrix} -3 & 2 & -6 \\ -3 & 5 & -7 \\ -2 & 3 & -4 \end{vmatrix}$ equal -5;

(b) $\begin{vmatrix} 2 & 6 & 5 \\ -2 & 7 & -5 \\ 2 & -7 & 9 \end{vmatrix}$ and $\begin{vmatrix} 2 & -1 & 9 \\ -1 & 7 & 2 \\ 3 & -21 & 2 \end{vmatrix}$ equal 104;

E 4.2. Explain why $\begin{vmatrix} 1 & 0 & 0 \\ 2 & 3 & 0 \\ 4 & 5 & 6 \end{vmatrix} = 18$, but $\begin{vmatrix} 0 & 0 & 1 \\ 0 & 3 & 2 \\ 6 & 5 & 4 \end{vmatrix} = -18$,

whereas $\begin{vmatrix} 1 & 0 & 0 & 0 \\ 2 & 3 & 0 & 0 \\ 4 & 5 & 6 & 0 \\ 7 & 8 & 9 & 10 \end{vmatrix} = \begin{vmatrix} 0 & 0 & 0 & 1 \\ 0 & 0 & 3 & 2 \\ 0 & 6 & 5 & 4 \\ 10 & 9 & 8 & 7 \end{vmatrix} = 180.$

E 4.3. Show that

$$\begin{vmatrix} 1 & 1 & 1 & 1 \\ 2 & 3 & -1 & 5 \\ 4 & 9 & 1 & 25 \\ 8 & 27 & -1 & 125 \end{vmatrix} = 432.$$

E 4.4. (Laplace expansion) Reduce the following determinants using expansions by minors and/or Laplace expansions, or other methods of expansion given in this chapter. Wherever possible utilize one result to obtain succeeding ones (e.g., d is the transpose of b). When expansion by minors is unnecessary—as it is in most cases—state why, and give the value of the determinant.

(a) $\begin{vmatrix} 1 & 4 & 1 & 8 \\ 2 & 8 & 2 & 16 \\ 1 & 0 & 0 & 12 \\ 7 & 8 & 1 & 2 \end{vmatrix}$

(b) $\begin{vmatrix} 2 & 3 & 7 \\ 4 & 1 & 8 \\ 1 & 0 & 2 \end{vmatrix}$

(c) $\begin{vmatrix} 1 & 4 & 1 \\ 2 & 6 & 3 \\ -1 & -2 & 7 \end{vmatrix}$

(d) $\begin{vmatrix} 2 & 4 & 1 \\ 3 & 1 & 0 \\ 7 & 8 & 2 \end{vmatrix}$

(e) $\begin{vmatrix} 0 & 0 & -1 \\ 0 & 2 & 0 \\ 1 & 0 & 0 \end{vmatrix}$

(f) $\begin{vmatrix} 3 & 20 & 62 \\ 2 & 22 & 67 \\ 0 & 4 & 16 \end{vmatrix} = \left| \begin{bmatrix} 2 & 3 & 7 \\ 4 & 1 & 8 \\ 1 & 0 & 2 \end{bmatrix} \begin{bmatrix} 2 & 8 & 2 \\ 2 & 6 & 3 \\ -1 & -2 & 7 \end{bmatrix} \right|$

(g) $\begin{vmatrix} 7 & 1 & 3 & 0 \\ 3 & 1 & 4 & 1 \\ 2 & 7 & 8 & 4 \\ 1 & 4 & 1 & 4 \end{vmatrix}.$

E 4.5. Show that:

(a) The determinant of $\mathbf{x}\mathbf{1}'$ is zero.

(b) If $\mathbf{A}\mathbf{1} = \mathbf{0}$, then $\mathbf{A}$ has a zero determinant.

E 4.6. Calculate the determinant of

$$\mathbf{A} = \begin{bmatrix} 0 & -a & b \\ a & 0 & -d \\ -b & d & 0 \end{bmatrix}.$$

E 4.7. Without expanding the determinants explain why

$$\begin{vmatrix} 1 & 1 & 1 \\ x & y & z \\ yz & xz & xy \end{vmatrix} = \begin{vmatrix} 1 & 1 & 1 \\ x & y & z \\ x^2 & y^2 & z^2 \end{vmatrix}.$$

E 4.8. Evaluate

$$\begin{vmatrix} 7 & 6 & 6 & 6 & 6 & 6 \\ 6 & 7 & 6 & 6 & 6 & 6 \\ 6 & 6 & 7 & 6 & 6 & 6 \\ 6 & 6 & 6 & 7 & 6 & 6 \\ 6 & 6 & 6 & 6 & 7 & 6 \\ 6 & 6 & 6 & 6 & 6 & 7 \end{vmatrix}.$$

E 4.9. Explain why the determinant of a diagonal matrix is the product of its diagonal elements. Is the same true for a lower triangular matrix? For an upper triangular matrix? Give examples.

E 4.10. Answer the following.

(a) When $\mathbf{K}$ has 12 rows and $\mathbf{K}^5 = 3\mathbf{K}$, what is a numerical value for $|\mathbf{K}|$? Is there more than one possibility? If so, give all possible values.

(b) When all row sums of $\mathbf{A}$ are s, explain why s is a factor of $|\mathbf{A}|$.

(c) For the partitioned square matrix $\mathbf{X} = [\mathbf{A} \ \mathbf{B}]$ show that

$$|\mathbf{X}|^2 = |\mathbf{A}\mathbf{A}' + \mathbf{B}\mathbf{B}'| = \begin{vmatrix} \mathbf{A}'\mathbf{A} & \mathbf{A}'\mathbf{B} \\ \mathbf{B}'\mathbf{A} & \mathbf{B}'\mathbf{B} \end{vmatrix}.$$

Construct a simple numerical example of order 4 or more.

E 4.11. Explain why $\begin{vmatrix} \mathbf{P'} & \mathbf{X'} \\ \mathbf{0} & \mathbf{Q'} \end{vmatrix} = |\mathbf{P}||\mathbf{Q}|.$

E 4.12. Do the following.

(a) Explain why $\begin{vmatrix} \mathbf{I} & -\mathbf{A} \\ \mathbf{0} & \mathbf{I} \end{vmatrix} = 1.$

(b) Using $\begin{bmatrix} \mathbf{I} & -\mathbf{A} \\ \mathbf{0} & \mathbf{I} \end{bmatrix}$ in the two products with $\begin{bmatrix} \mathbf{I} & \mathbf{A} \\ -\mathbf{B} & \mathbf{I} \end{bmatrix}$
show that $|\mathbf{I} + \mathbf{AB}| = |\mathbf{I} + \mathbf{BA}|.$

E 4.13. Denote the equations

$$\left. \begin{array}{rcrcrcr} x_1 & + & 2x_2 & + & 3x_3 & = & 26 \\ 3x_1 & + & 7x_2 & + & 10x_3 & = & 87 \\ 2x_1 & + & 11x_2 & + & 7x_3 & = & 73 \end{array} \right\} \text{ by } \mathbf{Ax} = \mathbf{b}.$$

(a) Solve the equation using successive elimination.

(b) Solve the equation using Cramer's rule.

E 4.14. (Supply and demand) Suppose in the market for books there
are four commodities: adventure books, textbooks, science
fiction books and mystery books. These can be designated by
A, T, S and M, respectively. The market supply and demand
functions for each of these commodities using $\mathbf{P}$ for price and
$\mathbf{Q}$ for quantity are:

Adventure Books   *Demand*: Q_A $=$ $10 - 5P_A + 3P_T - 2P_S - 9P_M$

Supply: Q_A $=$ $33 + 15P_A - 4P_T + 6P_S + 3P_M$

Textbooks   *Demand*: Q_T $=$ $3 + 2P_A - 1P_T + 9P_S - 4P_M$

Supply: Q_T $=$ $-10P_A + 2P_T$

Science Fiction   *Demand*: Q_S $=$ $8 + 4P_A + 1P_T - 3P_S + 2P_M$

Supply: Q_S $=$ $8 + 11P_A - 2P_T + 5P_S + 4P_M$

Mystery   *Demand*: Q_M $=$ $3 + 2P_A + 1P_T + 6P_S - 6P_M$

Supply: Q_M $=$ $1 + 6P_A - 1P_T + 4P_S + 6P_M$

(a) Simplify the supply and demand equations for each commodity assuming market equilibrium.

(b) Using Cramer's rule, solve for the equilibrium price of each book type.

(c) Calculate the equilibrium quantities for each type of book.

E 4.15. (Input-output) Sunshine, Inc. produces three types of juice: orange, grapefruit and citrus blend. Its processes are reflected in the following equations, where x_o, x_g and x_t refer to the inputs oranges, grapefruits and tangelos, respectively.

Orange juice	$10x_o$	$+$	$0x_g$	$+$	$2x_t$	$=$	1150
Grapefruit juice			$2x_g$			$=$	240
Citrus blend	$3x_o$	$+$	$2x_g$	$+$	$10x_t$	$=$	1290

Solve for Sunshine's optimal level of each input given the production processes.

Chapter 5

INVERSE MATRICES

The operations of addition, subtraction and multiplication of matrices have already been described; but division has not. Just as multiplying matrices is more complicated than multiplying scalars, so is the reverse operation, the counterpart of division for scalars. This counterpart is *not* division. There is no division in matrix algebra. One never "divides by a matrix". Instead, under certain circumstances one can often multiply by what is an inverse matrix, thus effecting for matrices the counterpart of division. This chapter describes the nature of inverse matrices, the conditions for their existence, and the manner of using them.

5.1 Introduction

In scalar arithmetic the operation of division is represented, for example, as $17 \div 6$, or $17/6$ or $17(6^{-1})$ or $6^{-1}(17)$. And in scalar algebra as $x \div a$ or x/a or $x(a^{-1})$ or $a^{-1}(x)$. In using the 6^{-1} or a^{-1} notation we acknowledge that

$$6^{-1}(6) = 1 = 6(6^{-1}) \qquad \text{and} \qquad a^{-1}(a) = 1 = a(a^{-1}). \qquad (5.1)$$

These equalities are virtually definitional. Therefore, for matrices, given some matrix $\mathbf{A}$ it seems reasonable to contemplate having a matrix to be denoted as $\mathbf{A}^{-1}$, such that, in concert with (5.1)

$$\mathbf{A}^{-1}\mathbf{A} = \mathbf{I} = \mathbf{A}\mathbf{A}^{-1}, \qquad (5.2)$$

$\mathbf{I}$ being the "one" of matrix algebra. The matrix $\mathbf{A}^{-1}$ satisfying (5.1) is called the inverse of $\mathbf{A}$, often spoken of as "$\mathbf{A}$ inverse", or "$\mathbf{A}$ to the minus one", or as just "$\mathbf{A}$ minus one".

There are a number of things to learn about $\mathbf{A}^{-1}$. First and foremost, even before knowing the conditions for its existence, it is essential to understand that the use of $\mathbf{A}^{-1}$ can never be represented by the symbols $\div$ or $/$, as can scalar division. There is no such thing as division by a matrix. In its place, we can often (but not always) multiply by $\mathbf{A}^{-1}$.

5.2   Existence and Uniqueness of an Inverse

Statement (5.2) tells us the major property that we want $\mathbf{A}^{-1}$ to have, that its products with $\mathbf{A}$ should be the (same) identity matrix. But (5.2) tells us nothing about how the elements of $\mathbf{A}^{-1}$ are related to those of $\mathbf{A}$. (See Section 5.5.) Despite this, (5.2) does imply conditions on $\mathbf{A}$ that must be satisfied for $\mathbf{A}^{-1}$ to exist. And provided those conditions are met, so that $\mathbf{A}^{-1}$ exists, it is then easily shown that $\mathbf{A}^{-1}$ is unique for given $\mathbf{A}$.

a.   Existence

Suppose in (5.2) that the $\mathbf{I}$-matrix has order r. Since $\mathbf{I}$ is square, we can easily show from (5.2) that both $\mathbf{A}$ and $\mathbf{A}^{-1}$ must be square, of order r. This is not a condition which ensures that $\mathbf{A}^{-1}$ *does* exist, only that it *can* exist. Contrariwise, if $\mathbf{A}$ is not square it cannot have an inverse. So it is only square matrices which can have an inverse.

Given that $\mathbf{A}$ must be square for $\mathbf{A}^{-1}$ to possibly exist, we use (5.2) again to derive the condition under which $\mathbf{A}^{-1}$ does exist. It is easy. Take the determinant of both sides of the first (or second) equality in (5.2):

$$|\mathbf{A}^{-1}\mathbf{A}| = |\mathbf{I}|.$$

Therefore, by the product rule for determinants (Section 4.3e),

$$|\mathbf{A}^{-1}||\mathbf{A}| = 1. \tag{5.3}$$

The important consequence of this is that $|\mathbf{A}| \neq 0$. This means that only if the determinant of $\mathbf{A}$ is non-zero will $\mathbf{A}^{-1}$ exist. Clearly, if $|\mathbf{A}| = 0$ the equality (5.3) cannot be satisfied.

So from (5.2) we have derived:

(i) $\mathbf{A}^{-1}$ *can* exist only if $\mathbf{A}$ is square, and

(ii) $\mathbf{A}^{-1}$ *does* exist only if $|\mathbf{A}| \neq 0$.

And when $|\mathbf{A}| \neq 0$ and $\mathbf{A}^{-1}$ exists, $\mathbf{A}$ is said to be *non-singular*; and when square $\mathbf{A}$ has $|\mathbf{A}| = 0$ then $\mathbf{A}$ is *singular*.

b. Uniqueness

Provided $\mathbf{A}^{-1}$ exists there is still one more important consequence to be derived from (5.2), namely that $\mathbf{A}^{-1}$ is unique. Suppose $\mathbf{S}$ is some matrix other than $\mathbf{A}^{-1}$ such that [like the left-hand part of (5.2)] $\mathbf{SA} = \mathbf{I}$. Post-multiplying both sides of this equation gives

$$\mathbf{SAA}^{-1} = \mathbf{A}^{-1}.$$

But, from (5.2)

$$\mathbf{AA}^{-1} = \mathbf{I}.$$

Therefore

$$\mathbf{SI} = \mathbf{A}^{-1}, \text{ i.e., } \mathbf{S} = \mathbf{A}^{-1}.$$

The same kind of argument applied to $\mathbf{T}$ if $\mathbf{AT} = \mathbf{I}$ yields $\mathbf{T} = \mathbf{A}^{-1}$, thus showing that if $\mathbf{A}$ is non-singular there is only one $\mathbf{A}^{-1}$ satisfying (5.2).

5.3 Rectangular Matrices

We have shown $\mathbf{A}^{-1}$ cannot exist if $\mathbf{A}$ is other than square, i.e., if it is rectangular. Nevertheless, for rectangular $\mathbf{A}$ there can be matrices which in products with $\mathbf{A}$ yield $\mathbf{I}$; but not both products with $\mathbf{A}$, as is the case when $\mathbf{A}^{-1}$ exists. The following examples illustrate possibilities. For

$$\mathbf{A} = \begin{bmatrix} 1 & 1 \\ -1 & 0 \\ 3 & -1 \end{bmatrix} \text{ each of } \begin{bmatrix} 1 & 3 & 1 \\ 2 & 5 & 1 \end{bmatrix} \text{ and } \begin{bmatrix} 4 & 15 & 4 \\ 7 & 25 & 6 \end{bmatrix}$$

used as $\mathbf{L}$ yield $\mathbf{LA} = \mathbf{I}_2$; but $\mathbf{AL} \neq \mathbf{I}_3$. Similarly for

$$\mathbf{B} = \begin{bmatrix} 0 & 3 & 7 & 6 \\ 0 & 2 & 5 & 7 \end{bmatrix} \text{ using each of } \begin{bmatrix} 3 & 8 \\ 24 & 12 \\ -11 & -6 \\ 1 & 1 \end{bmatrix} \text{ and } \begin{bmatrix} 1 & -2 \\ 24 & 31 \\ -11 & -15 \\ 1 & 2 \end{bmatrix}$$

as $\mathbf{R}$ yields $\mathbf{BR} = \mathbf{I}$; but there is no matrix $\mathbf{L}$ for which $\mathbf{LB} = \mathbf{I}$. Also for

$$\mathbf{C} = \begin{bmatrix} 0 & 0 & 0 \\ 0 & 2 & -1 \end{bmatrix}$$

there are no matrices $\mathbf{L}$ or $\mathbf{R}$ for which $\mathbf{LC}$ or $\mathbf{CR}$ are identity matrices.

The sole importance of these examples is to emphasize that not only does a rectangular matrix $\mathbf{A}$ have no inverse but it may have some matrices $\mathbf{L}$ such that $\mathbf{LA} = \mathbf{I}$, or some matrices $\mathbf{R}$ where $\mathbf{AR} = \mathbf{I}$, but not both $\mathbf{L}$ and $\mathbf{R}$; or it may have no $\mathbf{L}$ or $\mathbf{R}$ matrices.

5.4   Cofactors

Chapter 4 shows the equation for expanding the determinant of

$$\mathbf{A} = \left\{ {}_m\, a_{ij} \right\}_{i,j=1}^{n}$$

in terms of elements of its ith row:

$$|\mathbf{A}| = \sum_{j=1}^{n} a_{ij}(-1)^{i+j}|\mathbf{M}_{ij}| \quad \text{for any } i. \tag{5.4}$$

$|\mathbf{M}_{ij}|$ in (5.4) is the minor of a_{ij}. We now define the *cofactor* of a_{ij}:

$$c_{ij} = (-1)^{i+j}|\mathbf{M}_{ij}|. \tag{5.5}$$

It is the signed minor. Then (5.4) is

$$\sum_{j=1}^{n} a_{ij}c_{ij} = |\mathbf{A}| \quad \text{for any } i. \tag{5.6}$$

Cofactors are essential to understanding the relationship of elements of $\mathbf{A}^{-1}$ to those of $\mathbf{A}$. That relationship stems from (5.6) and from a second property of cofactors:

$$\sum_{j=1}^{n} a_{ij}c_{hj} = 0 \quad \text{for any } i \neq h. \tag{5.7}$$

Result (5.6) is the sum along a row of its elements multiplying their own cofactors equaling $|\mathbf{A}|$. But (5.7) is the sum along a row of its elements multiplying the cofactors of elements in some other row: and that sum is zero.

We have (5.4) as a definition of a determinant which can, as mentioned in Chapter 4, be shown equivalent to the rigorously formal definition. But (5.7) is true because $\sum_{j=1}^{n} a_{ij}c_{hj}$ is, like (5.6), the expansion of $\mathbf{A}$ by elements of its ith row *except* that the cofactors in (5.7) are those of the elements of the hth ($h \neq i$) row. Such cofactors come from all rows of $\mathbf{A}$ except its hth row, and so those rows contain the ith row. Therefore $\sum_{j=1}^{n} a_{ij}c_{hj}$ is effectively the expansion of a determinant having two rows the same. Hence $\sum_{j=1}^{n} a_{ij}c_{hj}$ is zero, as in (5.7).

5.5 Deriving the Inverse

a. For a matrix of order 3

We describe deriving an inverse matrix in terms of a 3×3 example,

$$\mathbf{A} = \begin{bmatrix} a_{11} & a_{12} & a_{13} \\ a_{21} & a_{22} & a_{23} \\ a_{31} & a_{32} & a_{33} \end{bmatrix}. \tag{5.8}$$

First, form a new matrix by replacing each element of $\mathbf{A}$ by its cofactor:

$$\mathbf{C} = \begin{bmatrix} c_{11} & c_{12} & c_{13} \\ c_{21} & c_{22} & c_{23} \\ c_{31} & c_{32} & c_{33} \end{bmatrix}. \tag{5.9}$$

Second, transpose that matrix:

$$\mathbf{C}' = \begin{bmatrix} c_{11} & c_{21} & c_{31} \\ c_{12} & c_{22} & c_{32} \\ c_{13} & c_{23} & c_{33} \end{bmatrix}. \tag{5.10}$$

Third, multiply $\mathbf{C}'$ by the scalar $(1/|\mathbf{A}|)$ to yield $\mathbf{A}^{-1}$:

$$\mathbf{A}^{-1} = \frac{1}{|\mathbf{A}|}\mathbf{C}' = \frac{1}{|\mathbf{A}|} \begin{bmatrix} c_{11} & c_{21} & c_{31} \\ c_{12} & c_{22} & c_{32} \\ c_{13} & c_{23} & c_{33} \end{bmatrix}. \tag{5.11}$$

The reason for (5.11) being the inverse of $\mathbf{A}$ is that in taking the product $\mathbf{A}\mathbf{A}^{-1}$ it will be found that diagonal elements of $\mathbf{A}\mathbf{A}^{-1}$ are of the form (5.6) and so equal $|\mathbf{A}|/|\mathbf{A}| = 1$; and off-diagonal elements of $\mathbf{A}\mathbf{A}^{-1}$ are of the form (5.7) and so are zero. For example, in using (5.8) and (5.11) the (1,1) element of $\mathbf{A}\mathbf{A}^{-1}$ is

$$(a_{11}c_{11} + a_{12}c_{12} + a_{13}c_{13})/|\mathbf{A}| = \sum_{j=1}^{3} a_{1j}c_{1j}/|\mathbf{A}| = |\mathbf{A}|/|\mathbf{A}| = 1$$

and the (1,2) element of $\mathbf{A}\mathbf{A}^{-1}$ is

$$\begin{aligned} (a_{11}c_{21} + a_{12}c_{22} + a_{13}c_{23})/|\mathbf{A}| &= \sum_{j=1}^{3} a_{1j}c_{2j}/|\mathbf{A}| \\ &= 0, \text{ from (5.7)}. \end{aligned}$$

Results of this nature occur for all elements of $\mathbf{A}\mathbf{A}^{-1}$ (diagonal elements are 1.0 and off-diagonal elements are zero). Hence $\mathbf{A}\mathbf{A}^{-1} = \mathbf{I}$.

Example

For

$$\mathbf{A} = \begin{bmatrix} 1 & 2 & 3 \\ 4 & 5 & 6 \\ 7 & 8 & 10 \end{bmatrix},$$

the cofactors of elements of the first row are

$$(-1)^{1+1} \begin{vmatrix} 5 & 6 \\ 8 & 10 \end{vmatrix} = 2, \ (-1)^{1+2} \begin{vmatrix} 4 & 6 \\ 7 & 10 \end{vmatrix} = 2 \text{ and } (-1)^{1+3} \begin{vmatrix} 4 & 5 \\ 7 & 8 \end{vmatrix} = -3.$$

Therefore, as an example of (5.6),

$$1(2) + 2(2) + 3(-3) = -3 = |\mathbf{A}|;$$

and an example of (5.7) for $i = 2$ and $h = 1$ is

$$4(2) + 5(2) + 6(-3) = 0.$$

Calculating the cofactors of each element it will then be found that the matrix of cofactors, $\mathbf{C}$ of (5.9), is

$$\mathbf{C} = \begin{bmatrix} 2 & 2 & -3 \\ 4 & -11 & 6 \\ -3 & 6 & -3 \end{bmatrix}; \text{ and so } \mathbf{A}^{-1} \text{ from (5.11) is}$$

$$\mathbf{A}^{-1} = \frac{1}{-3} \begin{bmatrix} 2 & 4 & -3 \\ 2 & -11 & 6 \\ -3 & 6 & -3 \end{bmatrix}.$$

And easy arithmetic shows that $\mathbf{A}\mathbf{A}^{-1} = \mathbf{I}$.

Up to this point we have described $\mathbf{A}^{-1}$ in a manner which yields $\mathbf{A}\mathbf{A}^{-1} = \mathbf{I}$, the second equality in the defining equation, (5.2). But we still need to show that the first part of (5.2), namely $\mathbf{A}^{-1}\mathbf{A} = \mathbf{I}$, holds true. This is easy. Assume that $\mathbf{P}$ is such that

$$\mathbf{P}\mathbf{A} = \mathbf{I}.$$

Post-multiply both sides of this equation by $\mathbf{A}^{-1}$:

$$\mathbf{P}\mathbf{A}\mathbf{A}^{-1} = \mathbf{I}\mathbf{A}^{-1},$$

i.e.,

$$\mathbf{PI} \;=\; \mathbf{A}^{-1}.$$

Therefore $\mathbf{A}^{-1}\mathbf{A} = \mathbf{I}$.

Note the procedure here: multiply by $\mathbf{A}^{-1}$ the two sides of an equation, just as was done in Section 5.1b. It is the standard way of using inverse matrices: not division by a matrix, but multiplication by its inverse (providing it exists).

b. General case

The sentences which precede equations (5.9), (5.10) and (5.11) describe the three steps required for calculating $\mathbf{A}^{-1}$ from $\mathbf{A}$. They apply, just as they stand, equally as well to any square $\mathbf{A}$ having $|\mathbf{A}| \neq 0$. The resulting $\mathbf{A}^{-1}$ satisfies $\mathbf{A}\mathbf{A}^{-1} = \mathbf{I} = \mathbf{A}^{-1}\mathbf{A}$. The derivation of $\mathbf{A}^{-1}$ is summarized as follows:

$$\mathbf{A}^{-1} = \frac{1}{|\mathbf{A}|}\begin{bmatrix}\mathbf{A} \text{ with every element} \\ \text{replaced by its cofactor}\end{bmatrix}^{\text{transposed}}. \tag{5.12}$$

Calculating $\mathbf{A}^{-1}$ from (5.12) is not irksome for $\mathbf{A}$ of small order, as in the preceding 3×3 example. But for $\mathbf{A}$ of order n it involves one determinant of order n and n^2 determinants of order $n - 1$. That is totally impractical, even for some computer software. This is a book about algebra, not computing, so we do not discuss software, sufficient to say that there are many software packages which include matrix inversion. Our only nod to computing is a brief discussion of rounding error in Section 5.12. Even with today's large and fast computers, rounding error can still be troublesome, as discussed by McCullough (1998) and McCullough and Vinod (1999).

c. Adjugate matrix

Part of one's initial acquaintance with any discipline is to learn its vocabulary. In this context we have the *adjugate* or *adjoint* of a matrix. It is the transpose of the matrix of cofactors; i.e., it is $\mathbf{C}'$ of (5.10):

$$\text{adj } \mathbf{A} = \{c_{ij}\}' = \{c_{ji}\}.$$

It exists whether $|\mathbf{A}| = 0$ or not, and always satisfies

$$(\text{adj } \mathbf{A})\mathbf{A} = \mathbf{A}(\text{adj } \mathbf{A}) = |\mathbf{A}|\mathbf{I}.$$

When $|\mathbf{A}| \neq 0$ and $\mathbf{A}^{-1}$ exists then adj $\mathbf{A} = |\mathbf{A}|\mathbf{A}^{-1}$.

5.6   Properties of the Inverse

If $\mathbf{A}$ is a square, non-singular matrix its inverse, $\mathbf{A}^{-1}$, has the following properties, the first two of which have already been established.

(i) The inverse commutes with $\mathbf{A}$, both products being the identity matrix: $\mathbf{A}^{-1}\mathbf{A} = \mathbf{A}\mathbf{A}^{-1} = \mathbf{I}$.

(ii) The inverse of $\mathbf{A}$ is unique.

(iii) The determinant of the inverse of $\mathbf{A}$ is the reciprocal of the determinant of $\mathbf{A}$: $|\mathbf{A}^{-1}| = 1/|\mathbf{A}|$.

(iv) The inverse matrix is non-singular.

(v) The inverse of $\mathbf{A}^{-1}$ is $\mathbf{A}$: $(\mathbf{A}^{-1})^{-1} = \mathbf{A}$.

(vi) The inverted transpose is the transposed inverse: $(\mathbf{A}')^{-1} = (\mathbf{A}^{-1})'$.

(vii) If $\mathbf{A}$ is symmetric, so is its inverse: if $\mathbf{A}' = \mathbf{A}$, then $(\mathbf{A}^{-1})' = \mathbf{A}^{-1}$.

(viii) The inverse of a product is the product of the inverses taken in reverse order: $(\mathbf{AB})^{-1} = \mathbf{B}^{-1}\mathbf{A}^{-1}$, provided $\mathbf{A}^{-1}$ and $\mathbf{B}^{-1}$ exist.

Evidence for these properties is as follows: (i) and (ii) are from Sections 5.1 and 5.2, (iii) comes from (5.3) which, with $|\mathbf{A}| \neq 0$ yields $|\mathbf{A}^{-1}| \neq 0$ and hence (iv). For (v), (vi) and (vii) consider the identity $\mathbf{I} = \mathbf{A}^{-1}\mathbf{A}$. Pre-multiplying it by $(\mathbf{A}^{-1})^{-1}$ leads to $(\mathbf{A}^{-1})^{-1} = \mathbf{A}$; transposing it leads to $(\mathbf{A}')^{-1} = (\mathbf{A}^{-1})'$. For $\mathbf{A}' = \mathbf{A}$ the latter becomes $\mathbf{A}^{-1} = (\mathbf{A}^{-1})'$. And for (viii)

$$\mathbf{B}^{-1}\mathbf{A}^{-1}\mathbf{AB} = \mathbf{B}^{-1}(\mathbf{A}^{-1}\mathbf{A})\mathbf{B} = \mathbf{B}^{-1}\mathbf{IB} = \mathbf{B}^{-1}\mathbf{B} = \mathbf{I},$$

and on post-multiplying by $(\mathbf{AB})^{-1}$, we have $(\mathbf{AB})^{-1} = \mathbf{B}^{-1}\mathbf{A}^{-1}$. The reversal rule that applies to transposing a product therefore applies also to inverting it.

5.7   Some Simple Special Cases

Certain matrices have inverses that are related to the matrices themselves in a simple manner. The following examples are particularly useful.

a. Inverses of order 2

The inverse of a general 2×2 matrix is easy:

$$\mathbf{A} = \begin{bmatrix} a & x \\ y & b \end{bmatrix} \text{ has } \mathbf{A}^{-1} = \frac{1}{ab - xy} \begin{bmatrix} b & -x \\ -y & a \end{bmatrix} \text{ for } ab - xy \neq 0.$$

$$(5.13)$$

One simply interchanges the diagonal elements and changes the sign of the off-diagonal elements, and divides by the determinant $|\mathbf{A}| = ab - xy$, which must not be zero.

b. Diagonal matrices

The inverse of a diagonal matrix is also easy (providing all diagonal elements are nonzero):

$$\left\{ {}_d x_i \right\}^{-1} = \left\{ {}_d 1/x_i \right\} \quad \text{when} \quad x_i \neq 0.$$

For example,

$$\begin{bmatrix} 2 & 0 & 0 \\ 0 & 4 & 0 \\ 0 & 0 & 3 \end{bmatrix}^{-1} = \begin{bmatrix} \frac{1}{2} & 0 & 0 \\ 0 & \frac{1}{4} & 0 \\ 0 & 0 & \frac{1}{3} \end{bmatrix}.$$

c. Orthogonal matrices

Section 3.5 shows that $\mathbf{P}$ having orthonormal rows means that $\mathbf{PP}' = \mathbf{I}$ which, for square $\mathbf{P}$, means that (as in Section 4.3e-vi) $|\mathbf{P}| = \pm 1 \neq 0$. Therefore $\mathbf{P}^{-1}$ exists and so pre-multiplying $\mathbf{PP}' = \mathbf{I}$ by $\mathbf{P}^{-1}$ gives $\mathbf{P}^{-1} = \mathbf{P}'$. And post-multiplying $\mathbf{P}^{-1} = \mathbf{P}'$ by $\mathbf{P}$ gives $\mathbf{P}'\mathbf{P} = \mathbf{I}$, thus showing that $\mathbf{P}$ also has orthonormal columns. Therefore orthogonal $\mathbf{P}$ (i) is square, (ii) has $|\mathbf{P}| = \pm 1$, (iii) has orthonormal rows, $\mathbf{PP}' = \mathbf{I}$, and (iv) has orthonormal columns, $\mathbf{P}'\mathbf{P} = \mathbf{I}$. Any combination of (i) or (ii) with (iii) or (iv) is necessary and sufficient for the other two of those four to hold, and for $\mathbf{PP}' = \mathbf{P}'\mathbf{P} = \mathbf{I}$ to hold also.

5.8   Solving Linear Equations

The simple equations of E 4.15,

$$10x_1 \qquad + \qquad 2x_3 \;=\; 1150$$

$$2x_2 \qquad\qquad =\; 240 \;,$$

$$3x_1 \;+\; 2x_2 \;+\; 10x_3 \;=\; 1290$$

are easily solved by direct elimination or, as asked for in Chapter 4, by Cramer's rule. They can also be solved by first rewriting them in a matrix-vector format.

$$\begin{bmatrix} 10 & 0 & 2 \\ 0 & 2 & 0 \\ 3 & 2 & 10 \end{bmatrix} \begin{bmatrix} x_1 \\ x_2 \\ x_3 \end{bmatrix} = \begin{bmatrix} 1150 \\ 240 \\ 1290 \end{bmatrix}. \tag{5.14}$$

The generalization of this, for n unknowns is

$$\mathbf{Ax} = \mathbf{b} \tag{5.15}$$

where $\mathbf{x}_{n \times 1}$ is the vector of unknown and $\mathbf{A}_{n \times n}$ and $\mathbf{b}_{n \times 1}$ are known values. Then, providing $\mathbf{A}^{-1}$ exists, pre-multiplying both sides of (5.15) gives

$$\mathbf{A}^{-1}\mathbf{Ax} = \mathbf{A}^{-1}\mathbf{b}$$

so that the solution to $\mathbf{Ax} = \mathbf{b}$ is

$$\mathbf{x} = \mathbf{A}^{-1}\mathbf{b}. \tag{5.16}$$

In the case of (5.14) this gives

$$\begin{bmatrix} x_1 \\ x_2 \\ x_3 \end{bmatrix} = \begin{bmatrix} 10 & 0 & 2 \\ 0 & 2 & 0 \\ 3 & 2 & 10 \end{bmatrix}^{-1} \begin{bmatrix} 1150 \\ 240 \\ 1290 \end{bmatrix}$$

$$= \frac{1}{188} \begin{bmatrix} 20 & 4 & -4 \\ 0 & 94 & 0 \\ -6 & -20 & 20 \end{bmatrix} \begin{bmatrix} 1150 \\ 240 \\ 1290 \end{bmatrix} = \begin{bmatrix} 100 \\ 120 \\ 75 \end{bmatrix}.$$

This is a very standard way of solving linear equations. It is used extensively in a wide variety of applications and with today's computing power it is often being used where there are as many as 1000 or more unknowns. The algebraic conditions remain and are immutable: $\mathbf{A}$ must be square with $|\mathbf{A}| \neq 0$, so that $\mathbf{A}^{-1}$ exists. However, even if $\mathbf{A}$ is not square, or is square with $|\mathbf{A}| = 0$, there are methods of solving $\mathbf{Ax} = \mathbf{b}$. They are dealt with in Chapter 10.

5.9 Algebraic Simplifications

Division by a matrix never occurs. Multiplication by the corresponding inverse (when it exists) is used in its place. But, as usual, distinction must always be made between pre-multiplication and post-multiplication. For example, suppose $\mathbf{A}^{-1}$ exists. Then $\mathbf{AX} = \mathbf{B}$ can be solved for $\mathbf{X}$ by pre-multiplying by $\mathbf{A}^{-1}$:

$$\mathbf{A}^{-1}\mathbf{AX} = \mathbf{A}^{-1}\mathbf{B}, \text{ i.e., } \mathbf{IX} = \mathbf{A}^{-1}\mathbf{B}, \text{ or } \mathbf{X} = \mathbf{A}^{-1}\mathbf{B}.$$

But post-multiplying $\mathbf{AX} = \mathbf{B}$ by $\mathbf{A}^{-1}$ (provided $\mathbf{X}$ is square) gives just $\mathbf{AXA}^{-1} = \mathbf{BA}^{-1}$, which cannot be simplified (other than back to $\mathbf{AX} = \mathbf{B}$). Unlike $axa^{-1} = x$, when a and x are scalars and so a and a^{-1} cancel out, in $\mathbf{AXA}^{-1}$ the $\mathbf{A}$ and $\mathbf{A}^{-1}$ do not cancel and $\mathbf{AXA}^{-1}$ is simply that, just $\mathbf{AXA}^{-1}$.

The non-existence of division, and the existence of inverse matrices only in special cases, also means that in matrix algebra an equation such as $\mathbf{PK} = \mathbf{QK}$ does not always imply that $\mathbf{P}$ equals $\mathbf{Q}$. An example of this is

$$\begin{bmatrix} 7 & 5 & 4 \\ 3 & 3 & 10 \end{bmatrix} \begin{bmatrix} 1 & 2 & 2 \\ -1 & 0 & 1 \\ 0 & 2 & 3 \end{bmatrix} = \begin{bmatrix} 3 & 1 & 8 \\ 2 & 2 & 11 \end{bmatrix} \begin{bmatrix} 1 & 2 & 2 \\ -1 & 0 & 1 \\ 0 & 2 & 3 \end{bmatrix}. \quad (5.17)$$

The reason that (5.17) does not imply equality of the two 2×3 matrices (which would clearly be false) is that the 3×3 matrix has no inverse. It is *only* the existence of $\mathbf{K}^{-1}$ that permits the equation $\mathbf{PK} = \mathbf{QK}$ to imply $\mathbf{P} = \mathbf{Q}$. This is so because when $\mathbf{K}^{-1}$ exists, the equation $\mathbf{PK} = \mathbf{QK}$ can be post-multiplied on both sides by $\mathbf{K}^{-1}$ to yield

$$\mathbf{PKK}^{-1} = \mathbf{QKK}^{-1}, \text{ i.e., } \mathbf{PI} = \mathbf{QI}, \text{ or } \mathbf{P} = \mathbf{Q}. \quad (5.18)$$

But when $\mathbf{K}^{-1}$ does not exist, as in (5.17), then (5.18) cannot be used.

The existence of matrix inverses often leads to algebraic equalities that might not otherwise be apparent. For example, it is easy to see that

$$\mathbf{R} + \mathbf{RST} = \mathbf{R(I + ST)},$$

but if $\mathbf{T}^{-1}$ exists this can further be written as

$$\mathbf{R} + \mathbf{RST} = \mathbf{R(T}^{-1} + \mathbf{S)T}. \quad (5.19)$$

On multiplying out the right-hand side of (5.19) one sees at once that (5.19) is correct. The useful thing to realize is that, in starting from the left-hand side of (5.19) one can, if $\mathbf{T}^{-1}$ is known to exist, develop the right-hand side.

A further example of the algebra of inverses is a matrix analogue of the scalar result that for scalar $x \neq 1$

$$1 + x + x^2 + \cdots + x^{n-1} = (x^n - 1)/(x - 1).$$

A matrix counterpart is that, provided $(\mathbf{X} - \mathbf{I})^{-1}$ exists,

$$\mathbf{I} + \mathbf{X} + \mathbf{X}^2 + \cdots + \mathbf{X}^{n-1} = (\mathbf{X}^n - \mathbf{I})(\mathbf{X} - \mathbf{I})^{-1}. \tag{5.20}$$

This is established by noting that the product

$$\begin{aligned}
(\mathbf{I} + \mathbf{X} &+ \mathbf{X}^2 + \cdots + \mathbf{X}^{n-1})(\mathbf{X} - \mathbf{I}) \\
&= \mathbf{X} + \mathbf{X}^2 + \mathbf{X}^3 + \cdots + \mathbf{X}^{n-1} + \mathbf{X}^n - (\mathbf{I} + \mathbf{X} + \mathbf{X}^2 + \cdots + \mathbf{X}^{n-1}) \\
&= \mathbf{X}^n - \mathbf{I}
\end{aligned} \tag{5.21}$$

and therefore if $(\mathbf{X} - \mathbf{I})^{-1}$ exists, post-multiplication of both sides of (5.23) yields (5.22). The similar procedure for establishing

$$\mathbf{I} + \mathbf{X} + \mathbf{X}^2 + \cdots + \mathbf{X}^{n-1} = (\mathbf{X} - \mathbf{I})^{-1}(\mathbf{X}^n - \mathbf{I})$$

is left as an exercise for the reader.

Illustration   (Transition probabilities)

Suppose a 6×6 transition probability matrix can be written as a partitioned matrix

$$\mathbf{P} = \begin{bmatrix} \mathbf{I} & \mathbf{0} \\ \mathbf{C} & \mathbf{B} \end{bmatrix}.$$

On making use of (5.21) for $\mathbf{B}$ in place of $\mathbf{X}$, it can be shown (using induction) that after n transitions

$$\mathbf{P}^n = \begin{bmatrix} \mathbf{I} & \mathbf{0} \\ (\mathbf{B} - \mathbf{I})^{-1}(\mathbf{B}^n - \mathbf{I})\mathbf{C} & \mathbf{B}^n \end{bmatrix}.$$

For large n this form of $\mathbf{P}^n$ is more readily calculated than is $\mathbf{P}^n$ directly. Also, it shows the relationship of $\mathbf{P}^n$ to $\mathbf{B}$ and $\mathbf{C}$ and provides easy consideration of the limiting form of $\mathbf{P}^n$ for infinitely large n.

The technique of establishing (5.20) through first deriving (5.21) and then post-multiplying by $(\mathbf{X} - \mathbf{I})^{-1}$ is very powerful. Many results using inverses are established in this manner, particularly those of showing that some matrix inverse equals some other matrix. For example, suppose we want to show that $\mathbf{A}^{-1}$ is some given matrix $\mathbf{Q}$. Provided we are told or can assume that $\mathbf{A}^{-1}$ exists (i.e., that $\mathbf{A}$ is square with nonzero determinant), we therefore anticipate $\mathbf{AQ}$ to be an identity matrix. Hence on post-multiplying $\mathbf{A}$ by $\mathbf{Q}$, we endeavor (hope) to simplify the product until it is an identity matrix. We will then have $\mathbf{AQ} = \mathbf{I}$, and so $\mathbf{A}^{-1}\mathbf{AQ} = \mathbf{A}^{-1}$ and hence $\mathbf{Q} = \mathbf{A}^{-1}$.

Example

To prove that $(\mathbf{I}+\mathbf{M}^{-1})^{-1}$ is $\mathbf{M}(\mathbf{M}+\mathbf{I})^{-1}$ we hope to show that $(\mathbf{I}+\mathbf{M}^{-1})[\mathbf{M}(\mathbf{M}+\mathbf{I})^{-1}]$ is an identity matrix; indeed it is, because

$$
\begin{aligned}
(\mathbf{I}+\mathbf{M}^{-1})[\mathbf{M}(\mathbf{M}+\mathbf{I})^{-1}] &= (\mathbf{I}+\mathbf{M}^{-1})\mathbf{M}(\mathbf{M}+\mathbf{I})^{-1} \\
&= (\mathbf{M}+\mathbf{M}^{-1}\mathbf{M})(\mathbf{M}+\mathbf{I})^{-1} \\
&= (\mathbf{M}+\mathbf{I})(\mathbf{M}+\mathbf{I})^{-1} = \mathbf{I}.
\end{aligned}
$$

Now, since we set out to prove something about the inverse of $\mathbf{I}+\mathbf{M}^{-1}$, we need to either prove or assume that it has a non-zero determinant. In the absence of evidence to prove this we assume it, and then make the step that is crucial to the logic of developing our desired result, of pre-multiplying the preceding equation by $(\mathbf{I}+\mathbf{M}^{-1})^{-1}$ to get

$$(\mathbf{I}+\mathbf{M}^{-1})^{-1}(\mathbf{I}+\mathbf{M}^{-1})[\mathbf{M}(\mathbf{M}+\mathbf{I})^{-1}] = (\mathbf{I}+\mathbf{M}^{-1})^{-1}\mathbf{I}.$$

This simplifies to what we want:

$$\mathbf{M}(\mathbf{M}+\mathbf{I})^{-1} = (\mathbf{I}+\mathbf{M}^{-1})^{-1}.$$

5.10 Partitioned Matrices

For reference purposes we give here, with but brief indication of their derivation, results concerning determinants and inverses of partitioned matrices that are often useful in applied problems. Searle (1982) has a chapter on partitioned matrices which contains details.

A determinantal result which can be established through applying arguments like those of Section 4.3e-iii and -iv to

$$
\begin{bmatrix} \mathbf{A} & \mathbf{B} \\ \mathbf{C} & \mathbf{D} \end{bmatrix}
\begin{bmatrix} \mathbf{I} & \mathbf{0} \\ -\mathbf{D}^{-1}\mathbf{C} & \mathbf{I} \end{bmatrix}
= \begin{bmatrix} \mathbf{A}-\mathbf{BD}^{-1}\mathbf{C} & \mathbf{B} \\ \mathbf{0} & \mathbf{D} \end{bmatrix}
$$

is

$$
\begin{vmatrix} \mathbf{A} & \mathbf{B} \\ \mathbf{C} & \mathbf{D} \end{vmatrix} = |\mathbf{D}||\mathbf{A}-\mathbf{BD}^{-1}\mathbf{C}|,
$$

providing $\mathbf{D}^{-1}$ exists. One can also derive, for $\mathbf{A}^{-1}$ existing,

$$\begin{vmatrix} \mathbf{A} & \mathbf{B} \\ \mathbf{C} & \mathbf{D} \end{vmatrix} = |\mathbf{A}||\mathbf{D} - \mathbf{C}\mathbf{A}^{-1}\mathbf{B}|.$$

Useful inversion formulae for a partitioned matrix, obtainable from solving

$$\begin{bmatrix} \mathbf{A} & \mathbf{B} \\ \mathbf{C} & \mathbf{D} \end{bmatrix}\begin{bmatrix} \mathbf{X} & \mathbf{Y} \\ \mathbf{Z} & \mathbf{W} \end{bmatrix} = \mathbf{I} = \begin{bmatrix} \mathbf{I} & \mathbf{0} \\ \mathbf{0} & \mathbf{I} \end{bmatrix}$$

for $\mathbf{X}$, $\mathbf{Y}$, $\mathbf{Z}$ and $\mathbf{W}$, are

$$\begin{bmatrix} \mathbf{A} & \mathbf{B} \\ \mathbf{C} & \mathbf{D} \end{bmatrix}^{-1} = \begin{bmatrix} \mathbf{A}^{-1} & \mathbf{0} \\ \mathbf{0} & \mathbf{0} \end{bmatrix} + \begin{bmatrix} -\mathbf{A}^{-1}\mathbf{B} \\ \mathbf{I} \end{bmatrix}$$

$$\times [\mathbf{D} - \mathbf{C}\mathbf{A}^{-1}\mathbf{B}]^{-1}[-\mathbf{C}\mathbf{A}^{-1} \quad \mathbf{I}]$$

$$= \begin{bmatrix} \mathbf{0} & \mathbf{0} \\ \mathbf{0} & \mathbf{D}^{-1} \end{bmatrix} + \begin{bmatrix} \mathbf{I} \\ -\mathbf{D}^{-1}\mathbf{C} \end{bmatrix}$$

$$\times [\mathbf{A} - \mathbf{B}\mathbf{D}^{-1}\mathbf{C}]^{-1}[\mathbf{I} \quad -\mathbf{B}\mathbf{D}^{-1}].$$

All of these results are, of course, valid only when the inverse matrices therein are of non-singular matrices.

5.11   Left and Right Inverses

Early in this chapter we showed a rectangular $\mathbf{A}$ for which we had matrices $\mathbf{R}$ such that $\mathbf{A}\mathbf{R} = \mathbf{I}$, and another rectangular $\mathbf{A}$ where matrices $\mathbf{L}$ gave $\mathbf{L}\mathbf{A} = \mathbf{I}$. In these examples and quite generally, $\mathbf{R}$ is called a *right inverse* and $\mathbf{L}$ is a *left inverse*. No rectangular matrix can have both a left and a right inverse. The only matrix which has both is a non-singular matrix, and its inverse is then also both a left and a right inverse.

Furthermore, for $\mathbf{A}_{r \times c}$, (with $r \neq c$), $\mathbf{A}$ may have left inverses only if $r > c$, and it may have right inverses only if $r < c$. Notice here that the statements are only that $\mathbf{A}$ *may* have left (or right) inverses; it is not that $\mathbf{A}$ *does* have them.

Verification can be found in Searle (1982, Section 5.10).

5.12 Using Computing Software

a. Problems of accuracy

Given that $|\mathbf{A}| \neq 0$, we have described how the elements of $\mathbf{A}^{-1}$ are related to those of $\mathbf{A}$, based on determinants including $|\mathbf{A}|$, and cofactors. That provides a clear understanding of the relationship between elements of $\mathbf{A}$ and $\mathbf{A}^{-1}$. But it is a totally impractical way of calculating $\mathbf{A}^{-1}$, unless $\mathbf{A}$ is of order 2 or 3. Until the advent of modern computers, there were methods of calculating $\mathbf{A}^{-1}$ based on solving linear equations by successive elimination. An example illustrating this is given in Searle (1982, Section 5.9a). By and large these methods are nothing more than different schemes for setting out the necessary arithmetic in a logical and economical manner which included a certain amount of self-checking of the arithmetic. These schemes were known by their originators names, beginning with Doolittle (1878), followed by Crout and Fox, and various combinations thereof.

The advent of modern computers and software, along with extensive research in computer science and numerical analysis over the last thirty years and more, has made these Doolittle-Crout-Fox "methods" effectively *passé*. We now make use of computers and software, in most cases assuming without question that the results they yield are numerically correct. In a broad sense this is a reasonable assumption, except for two aspects of computing packages which many users know little about: algorithmic error and rounding error. A good discussion of both sources of error is given by McCullough (1998) and McCullough and Vinod (1999). We treat each rather briefly.

b. Algorithmic error

Deciding what are good and bad computing algorithms so far as accuracy is concerned is far beyond the scope of this book. We give but one illustration, based on McCullough (1998): the well-known formula $\Sigma x_i^2 - n\bar{x}^2$ is, in general, not as accurate for calculating $\Sigma(x_i - \bar{x}_i)^2$ as that is itself. For example, suppose $n = 3$ and the three x-values are 10,001, 10,002 and 10,003. With a computer of 6-digit accuracy the first of the two preceding formulae yields 0; the second yields the correct answer, 2. One might argue that data of this nature never (don't usually) arise. That may be true. But good software has to cater to *all* kinds of data, including what might be deemed unusual.

c.   Rounding error

Rounding error is due to the hardware limitations of any computer: recurring decimals, requiring an infinite number of digits after the decimal point, cannot be represented with *complete* accuracy. No matter what finite number of 6s follow the decimal point in 4.666..., that value is not $4\frac{2}{3}$. Nor is this inexactitude alleviated if the last 6 is rounded to 7: 4.667 is not exactly $4\frac{2}{3}$ any more than is 4.666. This rounding error may appear inconsequential, but a series of such errors can be insidious. To quote from McCullough (1998), successive "rounding errors in a series of calculations do not cancel but, rather, accumulate, and the bound on the total error is proportional to the number of calculations. Sometimes this total error affects only the right-most digits of the final answer. Sometimes the total error can be so large as to completely swamp the answer, resulting in no accurate digits."

Critical analysis of rounding error generation and accumulation is difficult and not within the scope of this book. Instead, we give three illustrations of its effect, to alert the reader to the kinds of things that can occur with numbers. For numerical simplification, the illustrations are based on hypothetical computers of limited capacity. At best one can usually say that trouble as extreme as we demonstrate does not occur very often. At worst one must be cognizant of the possibility of trouble whenever rounding is involved, which is almost always.

− i.   Addition

Suppose a computer using only 2-digit numbers is used for adding 6.8, 7.4 and 8.5, rounding as it goes. Because computer addition is performed by adding each number to a preceding total, the sequence in which numbers are added can influence the computer total. Thus it gets $6.8 + 7.4 = 14$ and then $14 + 8.5 = 23$ (rounding up). But in a different sequence it gets $6.8 + 8.5 = 15$ and then $15 + 7.4 = 22$.

− ii.   Inverting a matrix

Suppose we seek the inverse of

$$\mathbf{A} = \begin{bmatrix} 2 & 1 \\ 6 & 3 \end{bmatrix}. \tag{5.22}$$

Clearly $|\mathbf{A}| = 0$ and $\mathbf{A}^{-1}$ does not exist. But suppose $\mathbf{A}$ comes not in the form of (5.22) but is generated within a computer as

$$\mathbf{A} = \begin{bmatrix} 6 \times (\frac{1}{3}) & 9 \times (\frac{1}{9}) \\ 66 \times (\frac{1}{11}) & 21 \times (\frac{1}{7}) \end{bmatrix}.$$

Furthermore, suppose that the computer carries only four decimal digits, so that numbers are rounded off in the fourth decimal place, a 5 in the fifth place being rounded up. Then the matrix computed as being $\mathbf{A}$, which we shall call $\mathbf{A}_c$, is

$$\mathbf{A}_c = \begin{bmatrix} 6 \times (.3333) & 9 \times (.1111) \\ 66 \times (.0909) & 21 \times (.1429) \end{bmatrix} = \begin{bmatrix} 1.9998 & .9999 \\ 5.9994 & 3.0009 \end{bmatrix}.$$

Its determinant is

$$|\mathbf{A}_c| = 1.9998(3.0009) - 5.9994(.9999) = 6.0012 - 5.9988 = .0024.$$

We note at once that although $|\mathbf{A}| = 0$ with $\mathbf{A}$ having no inverse, $|\mathbf{A}_c| \neq 0$ and $\mathbf{A}_c$ does have an inverse:

$$\mathbf{A}_c^{-1} = \frac{1}{.0024} \begin{bmatrix} 3.0009 & -.9999 \\ -5.9994 & 1.9998 \end{bmatrix} = \begin{bmatrix} 1250.3750 & -416.6250 \\ -2499.7500 & 833.2500 \end{bmatrix}.$$

Clearly $\mathbf{A}_c^{-1}$ exists but it cannot be $\mathbf{A}^{-1}$ because $\mathbf{A}^{-1}$ does not exist. This contrariness occurs because $\mathbf{A}_c$ is not $\mathbf{A}$, and whereas $|\mathbf{A}|$ is zero, $|\mathbf{A}_c|$ is nonzero, namely 0.0024. This is a common consequence of rounding error, a zero value being computed as nonzero, and in this instance it leads to the existence of $\mathbf{A}_c^{-1}$.

Knowing an inverse does not exist is almost self-evident for a matrix as simple as a 2×2, but this is seldom the case for matrices of large order encountered in real life. Fortunately, erroneous inverses do not occur frequently; nevertheless, even the best of programming safeguards against their occasional appearance may not be universally inviolable. It is therefore important to have an appreciation for the possible consequences of rounding error.

The important question is when should computer output (designated $\mathbf{A}_c^{-1}$) be accepted as $\mathbf{A}^{-1}$ and when should it not? Acceptance should occur only if some check calculations yield satisfactory results. These should include calculating at least some elements of $\mathbf{A}_c^{-1}\mathbf{A}$ and of

$\mathbf{AA}_c^{-1}$ to see that they are satisfactorily close to the ones and zeros of $\mathbf{I}$ if indeed $\mathbf{A}_c^{-1}$ is $\mathbf{A}^{-1}$. For our example this is found *not* to be so:

$$\mathbf{A}_c^{-1}\mathbf{A} = \begin{bmatrix} 1 & \frac{1}{2} \\ 0 & 0 \end{bmatrix} \neq \mathbf{I} \neq \mathbf{AA}_c^{-1} = \begin{bmatrix} 1 & 0 \\ 3 & 0 \end{bmatrix}.$$

Calculations like these immediately suggest that $\mathbf{A}_c^{-1}$ is suspect as a value of $\mathbf{A}^{-1}$; and, of course, the existence of $\mathbf{A}^{-1}$ is to be questioned, too. Although calculation of $\mathbf{A}_c^{-1}\mathbf{A}$ and $\mathbf{AA}_c^{-1}$ on a desk calculator is too onerous when $\mathbf{A}$ is of larger order, some elements of these products should always be so checked. In addition, the consequences of using $\mathbf{A}_c^{-1}$ should always be carefully scrutinized for reasonableness within the context of its use.

– iii. Solving linear equations

A computer's capacity for significant digits can materially affect the usefulness of its output, as has already been illustrated. It can also affect the input of a desired calculation to such an extent as to materially alter the whole problem. We illustrate this in terms of solving two linear equations.

The equations

$$\begin{bmatrix} 2.04 & 2.49 \\ 2.49 & 3.04 \end{bmatrix} \begin{bmatrix} x_1 \\ x_2 \end{bmatrix} = \begin{bmatrix} .45 \\ .55 \end{bmatrix} \quad \text{with solution} \quad \begin{bmatrix} x_1 \\ x_2 \end{bmatrix} = \begin{bmatrix} -1 \\ 1 \end{bmatrix} \tag{5.23}$$

will be called the real problem and the real solution, respectively. In contrast, for a computer that handles only 2-digit numbers, rounded in the right-most position, the problem has to be formulated as the computer problem

$$\begin{bmatrix} 2.0 & 2.5 \\ 2.5 & 3.0 \end{bmatrix} \begin{bmatrix} x_1 \\ x_2 \end{bmatrix} = \begin{bmatrix} .45 \\ .55 \end{bmatrix} \quad \text{with real solution} \quad \begin{bmatrix} x_1 \\ x_2 \end{bmatrix} = \begin{bmatrix} .1 \\ .1 \end{bmatrix}. \tag{5.24}$$

The alteration in these real solutions, between (5.23) and (5.24), is startling: a change in sign, and a change of an order of magnitude. And these changes, be it noted, have arisen solely as a result of having to alter the real problem so as to computerize it. Small changes in the problem have yielded considerable changes in the real solution.

Now consider the computer solution to the computer problem: taking $2.0(3.0) - (2.5)^2 = 6.0 - 6.25$ as -0.3, it is

$$\frac{1}{-.3}\begin{bmatrix} 3.0 & -2.5 \\ -2.5 & 2.0 \end{bmatrix}\begin{bmatrix} .45 \\ .55 \end{bmatrix} = \begin{bmatrix} -10 & 8.3 \\ 8.3 & -6.7 \end{bmatrix}\begin{bmatrix} .45 \\ .55 \end{bmatrix} = \begin{bmatrix} .1 \\ 0 \end{bmatrix}.$$

This differs from the real solution (5.24) of the same problem, which in turn is different from (5.23), the real solution to the real problem.

The reason for this apparent fickleness of numbers is that the 2×2 matrix in (5.23) is what is called *ill-conditioned*. Its determinant, relative to its elements, is quite small, namely 0.0015. As a result, the small changes made in the matrix in deriving (5.24) bring about a relatively large change in the determinant, which is now -0.25; and this in turn leads to even larger differences between the inverses:

$$\begin{bmatrix} 2.04 & 2.49 \\ 2.49 & 3.04 \end{bmatrix}^{-1} = \begin{bmatrix} 2026\frac{2}{3} & -1660 \\ -1660 & 1360 \end{bmatrix}$$

whereas

$$\begin{bmatrix} 2.0 & 2.5 \\ 2.5 & 3.0 \end{bmatrix}^{-1} = \begin{bmatrix} -12 & 10 \\ 10 & -8 \end{bmatrix}.$$

Further demonstration of the effect of the relatively large numbers in the first of these inverses is to change the real problem of (5.23) to be

$$\begin{bmatrix} 2.04 & 2.49 \\ 2.49 & 3.04 \end{bmatrix}\begin{bmatrix} x_1 \\ x_2 \end{bmatrix} = \begin{bmatrix} .451 \\ .55 \end{bmatrix}$$

with real solution

$$\begin{bmatrix} x_1 \\ x_2 \end{bmatrix} = \begin{bmatrix} 1.02\ddot{6} \\ -.66 \end{bmatrix}.$$

Simply adding .001 to one element of the right-hand side of the equations brings the substantive change of altering the sign of both elements of the solution.

These simple examples of inaccurate results arising from rounding might prompt the idea that this kind of material inaccuracy occurs frequently. Usually it does not (although if one is always using the same software, there is no basis for comparison). There is always rounding error in right-most digits, but the important question is to ascertain when they lead to inaccuracy of practical importance. Salient features of the examples here that delineate them from real life are that they are based on hypothetical computers, and that real-life data are not often as pathological as those of the examples which were designed to illustrate what *can* sometimes happen rather than what usually *does* happen. Consequently, computer routines designed for inverting matrices deliver results that are seldom spoiled by rounding error. But this can happen, and computer users need to be aware of such an eventuality, its possible causes and the consequences.

– iv.   Rounding at data input

Note that throughout the presentation of all the preceding examples, the origin of rounding error has always been attributed to hardware limitation—the inability to accommodate certain decimal numbers with complete accuracy. There is, however, another source of rounding: it can occur at the time of data input. Whoever is doing data-input work may, for whatever reason, do some rounding at that time—possibly under the misapprehension, for example, that the rounding from 2.04, 2.49 and 3.04 to 2.0, 2.5 and 3.0 in the matrix going from (5.23) to (5.24) would not affect anything. How mistaken that would be! Let people who do data input, and those who supervise it, be warned: danger lurks.

d.   Speed

The effort (time, amount of arithmetic, or cost) of inverting a matrix is considered to be proportional to n^3 for an n-order, non-singular matrix. Prior to the electronic age, inverting a 4×4 using determinants took approximately 15 minutes. And around 1951, prior to computers, a case is known of inverting a 40×40! Row echelon methods were used, with one person doing a row and then handing it to a co-worker who redid it, it being necessary to check the arithmetic and the transcription from (electric) desk calculator to paper. The whole process took six weeks, which at 40 hours a week is some 240 hours, almost exactly the $10^3(\frac{1}{4}$ hour) one would expect from the n^3 proportionality rule. By the late 1950s we thought 7 minutes to invert a 10×10 matrix was sheer magic. Nowadays inverting a matrix of order 1000 or so can take but a few seconds. That *is* magic.

5.13 Exercises

E 5.1. Show that if

(a) $\mathbf{A} = \begin{bmatrix} 6 & 13 \\ 5 & 12 \end{bmatrix}$, then $|\mathbf{A}| = 7$, $\mathbf{A}^{-1} = \frac{1}{7} \begin{bmatrix} 12 & -13 \\ -5 & 6 \end{bmatrix}$

and $\mathbf{A}\mathbf{A}^{-1} = \mathbf{A}^{-1}\mathbf{A} = \mathbf{I}$;

(b) $\mathbf{B} = \begin{bmatrix} 3 & -4 \\ 7 & 14 \end{bmatrix}$, then $|\mathbf{B}| = 70$, $\mathbf{B}^{-1} = \frac{1}{70} \begin{bmatrix} 14 & 4 \\ -7 & 3 \end{bmatrix}$

and $\mathbf{B}\mathbf{B}^{-1} = \mathbf{B}^{-1}\mathbf{B} = \mathbf{I}$.

E 5.2. Demonstrate the reversal rule for the inverse of a product of two matrices, using $\mathbf{A}$ and $\mathbf{B}$ in Exercise 5.1.

E 5.3. Show that if

(a) $\mathbf{A} = \begin{bmatrix} -1 & 3 & 0 \\ 0 & 2 & 1 \\ 1 & 0 & 4 \end{bmatrix}$, $|\mathbf{A}| = -5$, $\mathbf{A}^{-1} = -\frac{1}{5} \begin{bmatrix} 8 & -12 & 3 \\ 1 & -4 & 1 \\ -2 & 3 & -2 \end{bmatrix}$

and $\mathbf{A}^{-1}\mathbf{A} = \mathbf{A}\mathbf{A}^{-1} = \mathbf{I}$;

(b) $\mathbf{B} = \begin{bmatrix} 10 & 6 & -1 \\ 6 & 5 & 4 \\ -1 & 4 & 17 \end{bmatrix}$, $|\mathbf{B}| = 25$,

$$\mathbf{B}^{-1} = \begin{bmatrix} 2.76 & -4.24 & 1.16 \\ -4.24 & 6.76 & -1.84 \\ 1.16 & -1.84 & 0.56 \end{bmatrix}$$

and $\mathbf{B}^{-1}\mathbf{B} = \mathbf{B}\mathbf{B}^{-1} = \mathbf{I}$;

(c) $\mathbf{C} = \frac{1}{10} \begin{bmatrix} 0 & -6 & 8 \\ -10 & 0 & 0 \\ 0 & -8 & -6 \end{bmatrix}$, then $|\mathbf{C}| = 1$, $\mathbf{C}^{-1} = \mathbf{C}'$

and $\mathbf{C}\mathbf{C}' = \mathbf{C}'\mathbf{C} = \mathbf{I}$.

E 5.4. For each of the following matrices derive the determinant and
inverse. Check each inverse by multiplication.

(a) $\begin{bmatrix} 7 & 3 \\ 8 & 9 \end{bmatrix}$ (b) $\begin{bmatrix} 6 & 31 \\ 8 & 29 \end{bmatrix}$ (c) $\begin{bmatrix} -7 & -4 \\ 3 & 1 \end{bmatrix}$

(d) $\begin{bmatrix} 1 & 5 & -5 \\ 3 & 2 & -5 \\ 6 & -2 & -5 \end{bmatrix}$ (e) $\begin{bmatrix} -3 & 2 & -6 \\ -3 & 5 & -7 \\ -2 & 3 & -4 \end{bmatrix}$ (f) $\begin{bmatrix} 2 & 1 & 3 \\ -5 & 1 & 0 \\ 1 & 4 & -2 \end{bmatrix}$

(g) $\begin{bmatrix} -1 & -2 & 1 \\ -2 & 2 & -2 \\ 1 & -2 & -1 \end{bmatrix}$ (h) $\begin{bmatrix} 1 & -2 & 1 \\ -2 & 4 & -2 \\ 1 & -2 & 1 \end{bmatrix}$ (i) $\begin{bmatrix} 7 & 4 & -1 \\ 4 & 7 & -1 \\ -4 & -4 & 4 \end{bmatrix}$

E 5.5. Calculate the inverse of $\begin{bmatrix} 1 & 0 & 6 & 8 \\ 0 & 1 & 5 & 4 \\ 0 & 0 & -1 & 0 \\ 0 & 0 & 0 & -1 \end{bmatrix}$ and of $\begin{bmatrix} a & b & c \\ 0 & d & e \\ 0 & 0 & f \end{bmatrix}$.

E 5.6. Let $\mathbf{A} = \begin{bmatrix} 6 & -1 & 4 \\ 2 & 5 & -3 \\ 2 & 5 & 2 \end{bmatrix}$.

(a) Calculate the transpose of $\mathbf{A}^{-1}$ and the inverse of $\mathbf{A}'$.

(b) Calculate the inverse of $\mathbf{A}^{-1}$.

E 5.7. Find the inverse of $\begin{bmatrix} 0 & 0 & 0 & 3 \\ 0 & 0 & 7 & 0 \\ 0 & 4 & 0 & 0 \\ 5 & 0 & 0 & 0 \end{bmatrix}$.

E 5.8. For $\mathbf{P} = \frac{1}{15} \begin{bmatrix} 5 & -14 & 2 \\ -10 & -5 & -10 \\ 10 & 2 & -11 \end{bmatrix}$ demonstrate

(a) $\mathbf{P}' = \mathbf{P}^{-1}$, (b) $\mathbf{PP}' = \mathbf{P}'\mathbf{P} = \mathbf{I}$ and   (c) $|\mathbf{P}| = \pm 1$.

E 5.9. Write each of the following sets of equations in matrix and vector form. Solve the equations, after obtaining the necessary inverse matrices (if they exist). If an inverse does not exist, explain why.

(a)
$$3x + 4y - 2z = 4,$$
$$-x - y + 3z = 6,$$
$$x - 7y + z = -2.$$

(b)
$$-3x + 4y = 2,$$
$$x + 7z = -8,$$
$$-2x - 3y - 8z = -11.$$

(c)
$$2w + 8y - z = 0,$$
$$8x - 3z = -4,$$
$$4y + \tfrac{3}{2}z = 42,$$
$$\tfrac{3}{8}z = 7\tfrac{1}{2}.$$

(d)
$$x + 3y - z = 7,$$
$$4x - y + 2z = 8\tfrac{1}{2},$$
$$5x + 2y + z = 10.$$

E 5.10. Calculate the inverse of the following matrices. Check each inverse by calculating a product that is an identity matrix.

(a) $\begin{bmatrix} 1 & -4 & 1 \\ 2 & 7 & 0 \\ 0 & 1 & 0 \end{bmatrix}$ (b) $\begin{bmatrix} \tfrac{1}{2} & \tfrac{1}{4} & \tfrac{1}{8} \\ \tfrac{1}{4} & \tfrac{1}{8} & 0 \\ 4 & 0 & 1 \end{bmatrix}$ (c) $\begin{bmatrix} 1 & 1 & 0 \\ 0 & 1 & 1 \\ 1 & 0 & 1 \end{bmatrix}$

E 5.11. In Exercise 5.10 denote each matrix in turn by **A** and solve the equations **Ax = 1**.

E 5.12. (Advertising) Suppose a marketing department for a large automobile manufacturer wanted to reach three target groups consisting of 20,000 working mothers, 32,000 households with income greater than $50,000, and 35,000 people age 30 to 45. (It is of no concern that these groups are not mutually exclusive.) The department's advertising budget is split among three media: magazines, direct mail and television. It is known that twice the dollars spent on magazines (m) plus eight times the dollars spent on direct mail (d) plus twice the money spent on television (t) equals the number of working mothers (x) who will be reached on average by the advertising. Households with income greater than $50,000 ($y$) and people age 30 to 45 (z) are reached according to the following equations.

$$y = 7m + d$$

$$z = 6m + 6d + 4t$$

(a) If the market department's requirements are met exactly, how many dollars will be spent on magazines, direct mail and television?

(b) How many additional dollars must be spent to reach 5% more working mothers, 25% more households with income greater than \$50,000 and 33% more people age 30 to 45?

E 5.13. (Manufacturing) Suppose a manufacturer produces candy canes, chocolates and chewing gum using three production processes: mixing, shaping and wrapping. The capacity (in hours) of the three processes are 50, 50 and 120, respectively, while the hours required in each process to produce one unit of finished product are

$$\mathbf{A} = \begin{bmatrix} 2 & 2 & 1 \\ 1 & 4 & 1 \\ 1 & 6 & 4 \end{bmatrix} = \{a_{ij}\},$$

where a_{ij} represents the number of hours from process i needed for producing one unit of product j.

(a) Using matrix notation, write down the three equations representing the equality of capacity used to available capacity for each production process.

(b) What volumes of candy canes, chocolates and chewing gum are produced, assuming each production process is used to capacity?

(c) Suppose candy canes are more profitable than chocolates or chewing gum. Find the maximum number of candy canes that can be produced. Use the remaining production capacity to maximize chocolate output; and use any further remaining capacity to produce chewing gum. Are all three processes used to capacity? How many different products are produced?

E 5.14. (National income) Consider the following model of income determination, where X = exports and G = government expenditures, and all symbols are scalars:

$$Y = C + I + G + X - M = \text{net national product (NNP)}$$
$$C = a + bY = \text{consumption expenditures}$$
$$I = c + dY = \text{net investment expenditures}$$
$$M = e + fC = \text{imports}$$

Assume that G and X are determined exogenously and that $a, b, \ldots, f$ are known constants.

(a) Given values for G and X, is it possible to derive values for Y, C, I and M? Give reasons for your answer.

(b) With $\mathbf{x} = \begin{bmatrix} Y \\ C \\ I \\ M \end{bmatrix}$ and $\mathbf{y} = \begin{bmatrix} G + X \\ a \\ c \\ e \end{bmatrix}$ write down the

equation that relates $\mathbf{x}$ to $\mathbf{y}$, and solve for $\mathbf{x}$. [Hint: See equation (5.18) of the text.]

(c) Suppose government spending were to increase by a dollar. What is the algebraic form of the resultant increase in NNP? What economic term is used to describe this effect? By how much would consumption, investment and imports change?

E 5.15. Prove that the only nonsingular idempotent matrices are identity matrices.

E 5.16. Without inverting the matrix, explain why

$$[a_1 \ b_1 \ c_1 \ d_1] \begin{bmatrix} a_1 & b_1 & c_1 & d_1 \\ b_1 & a_2 & b_2 & c_2 \\ c_1 & b_2 & a_3 & b_3 \\ d_1 & c_2 & b_3 & a_4 \end{bmatrix}^{-1} \begin{bmatrix} a_1 \\ b_1 \\ c_1 \\ d_1 \end{bmatrix} = a_1.$$

E 5.17. When $\mathbf{A}$ is skew-symmetric, prove the following.

(a) When $\mathbf{A}^{-1}$ exists, it is skew-symmetric.

(b) $(\mathbf{I} - \mathbf{A})(\mathbf{I} + \mathbf{A})^{-1}$ is orthogonal.
(Hint: $\mathbf{I} - \mathbf{A}$ and $\mathbf{I} + \mathbf{A}$ commute in multiplication.)

E 5.18. For $\mathbf{A} = \{a_{ij}\}$, explain why

(a) $\mathbf{B} = \mathbf{A} - \mathbf{A}1(1'\mathbf{A}1)^{-1}1'\mathbf{A} = \{a_{ij} - a_i.a_{.j}/a_{..}\}$;

(b) $\mathbf{B}$ has row sums that are zero.

E 5.19. Prove the following statements.

(a) For $\mathbf{A}$ and $\mathbf{B}$ symmetric, $[(\mathbf{AB})']^{-1} = \mathbf{A}^{-1}\mathbf{B}^{-1}$.

(b) $\mathbf{C} = \mathbf{X}(\mathbf{X}'\mathbf{X})^{-1}\mathbf{X}'$ is symmetric and idempotent.

(c) The transpose and inverse of an orthogonal matrix are equal.

(d) All powers of a symmetric orthogonal matrix are the matrix itself, or an identity matrix.

(e) When $\mathbf{A} = \lambda\mathbf{B}$, of order n and for scalar λ, then $|\mathbf{A}| = \lambda^n|\mathbf{B}|$ and $\mathbf{A}^{-1} = (1/\lambda)\mathbf{B}^{-1}$. When is the latter not true?

(f) Prove that there is only one form of matrix that is orthogonal and idempotent, and find that form.

(g) If $\mathbf{A}$ is a square matrix of order n, $|\mathrm{adj}(\mathbf{A})| = |\mathbf{A}|^{n-1}$.

(h) The product of two matrices being a null matrix does not necessarily imply that one of them is null.

(i) When $\mathbf{A}$ is positive definite, so is $\mathbf{A}^{-1}$.

E 5.20. Assuming that each of the stated inverses exists, prove the following statements.

(a) $(\mathbf{I} + \mathbf{A}^{-1})^{-1} = \mathbf{A}(\mathbf{A} + \mathbf{I})^{-1}$.

(b) $(\mathbf{A} + \mathbf{BB}')^{-1}\mathbf{B} = \mathbf{A}^{-1}\mathbf{B}(\mathbf{I} + \mathbf{B}'\mathbf{A}^{-1}\mathbf{B})^{-1}$.

(c) $(\mathbf{A}^{-1} + \mathbf{B}^{-1})^{-1} = \mathbf{A}(\mathbf{A} + \mathbf{B})^{-1}\mathbf{B} = \mathbf{B}(\mathbf{A} + \mathbf{B})^{-1}\mathbf{A}$.

(d) $\mathbf{A} - \mathbf{A}(\mathbf{A} + \mathbf{B})^{-1}\mathbf{A} = \mathbf{B} - \mathbf{B}(\mathbf{A} + \mathbf{B})^{-1}\mathbf{B}$.

(e) $\mathbf{A}^{-1} + \mathbf{B}^{-1} = \mathbf{A}^{-1}(\mathbf{A} + \mathbf{B})\mathbf{B}^{-1}$.

(f) $(\mathbf{A} + \mathbf{B})^{-1} = \mathbf{A}^{-1} + \mathbf{B}^{-1}$ implies $\mathbf{AB}^{-1}\mathbf{A} = \mathbf{BA}^{-1}\mathbf{B}$.

(g) $(\mathbf{I} + \mathbf{AB})^{-1} = \mathbf{I} - \mathbf{A}(\mathbf{I} + \mathbf{BA})^{-1}\mathbf{B}$.

(h) $(\mathbf{I} + \mathbf{AB})^{-1}\mathbf{A} = \mathbf{A}(\mathbf{I} + \mathbf{BA})^{-1}$.

E 5.21. For $\mathbf{H} = \mathbf{I} - 2\mathbf{ww}'$ with $\mathbf{w}'\mathbf{w} = 1$, prove that $\mathbf{H} = \mathbf{H}^{-1}$.

E 5.22. By partitioning $M = \begin{bmatrix} -5 & 3 & 5 & 1 & 0 \\ -4 & 8 & 10 & 0 & -1 \\ 2 & 13 & 11 & 2 & 1 \\ 0 & 1 & 1 & 3 & 2 \\ 3 & 1 & 0 & 7 & 5 \end{bmatrix}$, find $\mathbf{M}^{-1}$.

E 5.23. Show that

(a) $\begin{bmatrix} \mathbf{I} & \mathbf{P} \\ \mathbf{Q} & \mathbf{I} \end{bmatrix}^{-1} = \begin{bmatrix} (\mathbf{I}-\mathbf{PQ})^{-1} & -(\mathbf{I}-\mathbf{PQ})^{-1}\mathbf{P} \\ -\mathbf{Q}(\mathbf{I}-\mathbf{PQ})^{-1} & \mathbf{I}+\mathbf{Q}(\mathbf{I}-\mathbf{PQ})^{-1}\mathbf{P} \end{bmatrix}$.

(b) $\begin{bmatrix} \mathbf{0} & \mathbf{P} \\ \mathbf{Q} & \mathbf{I} \end{bmatrix}^{-1} = \begin{bmatrix} -(\mathbf{PQ})^{-1} & (\mathbf{PQ})^{-1}\mathbf{P} \\ \mathbf{Q}(\mathbf{PQ})^{-1} & \mathbf{I}-\mathbf{Q}(\mathbf{PQ})^{-1}\mathbf{P} \end{bmatrix}$.

(c) on writing $\mathbf{T} = (\mathbf{PR}^{-1}\mathbf{Q})^{-1}$

$\begin{bmatrix} \mathbf{0} & \mathbf{P} \\ \mathbf{Q} & \mathbf{R} \end{bmatrix}^{-1} = \begin{bmatrix} -\mathbf{T} & \mathbf{TPR}^{-1} \\ \mathbf{R}^{-1}\mathbf{QT} & \mathbf{R}^{-1}-\mathbf{R}^{-1}\mathbf{QTPR}^{-1} \end{bmatrix}$.

(d) $\begin{bmatrix} \mathbf{S} & \mathbf{P} \\ \mathbf{0} & \mathbf{R} \end{bmatrix}^{-1} = \begin{bmatrix} \mathbf{S}^{-1} & -\mathbf{S}^{-1}\mathbf{PR}^{-1} \\ \mathbf{0} & \mathbf{R}^{-1} \end{bmatrix}$.

E 5.24. For $\mathbf{I}$ and $\mathbf{J}$ of order n, prove the following.

(a) For $p \neq 0$ and $p + nq \neq 0$,

$$(p\mathbf{I} + q\mathbf{J})^{-1} = \frac{1}{p}\left(\mathbf{I} - \frac{q}{p+nq}\mathbf{J}\right). \quad \text{(See E 3.14c.)}$$

(b) For $\mathbf{C}$ such that $\mathbf{JC} = \mathbf{0} = \mathbf{CJ}$,

$$\mathbf{J}(\mathbf{C}+a\mathbf{J})^{-1} = (\mathbf{C}+a\mathbf{J})^{-1}\mathbf{J} = \frac{1}{an}\mathbf{J}$$

and hence

$$(\mathbf{C}+a\mathbf{J})^{-1} - (\mathbf{C}+b\mathbf{J})^{-1} = \frac{b-a}{abn^2}\mathbf{J}.$$

E 5.25. For orthogonal $\mathbf{T}$, and $\mathbf{L}$ being a left inverse of $\mathbf{A}$, what is a left inverse of $\mathbf{TA}$ and of $\mathbf{AT}$?

E 5.26. When $\mathbf{G}$ represents either a left or a right inverse (whichever exists) of $\mathbf{A}$, show that $\mathbf{AGA} = \mathbf{A}$, and that $\mathbf{GAG} = \mathbf{G}$ and that $\mathbf{GA}$ is symmetric in one case and $\mathbf{AG}$ is in the other.

E 5.27. Prove

(a) For square $\mathbf{L}$, that $\mathbf{L}$ is non-singular if $\mathbf{LL}'$ is.

(b) For symmetric $\mathbf{V}$ that $\mathbf{K} = \mathbf{X}'(\mathbf{XV}^{-1}\mathbf{X}')^{-1}\mathbf{X}$ is symmetric and that the transpose of $\mathbf{D} = \mathbf{KV}^{-1}$ is $\mathbf{V}^{-1}\mathbf{DV}$. Assume $\mathbf{X}$ is rectangular.

E 5.28. For $\mathbf{w}' = [\mathbf{y}'\ \ \mathbf{0}]$ where $\mathbf{y}$ is a vector of order N, and with $\mathbf{Z}' = [\mathbf{0}_{k \times N}\ \ \mathbf{I}_{k \times k}]$, prove that

$$\mathbf{w}' \left(\mathbf{Z} - \frac{1}{N+k} \mathbf{JZ} \right) \left(\mathbf{Z}'\mathbf{Z} - \frac{1}{N+k} \mathbf{Z}'\mathbf{JZ} \right)^{-1} \left(\mathbf{Z} - \frac{1}{N+k} \mathbf{JZ} \right)' \mathbf{w}$$

$$= \frac{Nk\bar{y}^2}{N+k} \text{ for } \bar{y} \text{ being the average of the elements of } \mathbf{y}.$$

E 5.29. Show that

$$\begin{bmatrix} 10 & 3 & 2 & 4 \\ 3 & 3 & 0 & 0 \\ 2 & 0 & 2 & 0 \\ 4 & 0 & 0 & 4 \end{bmatrix}^{-1} = \begin{bmatrix} 1 & -1 & -1 & -1 \\ -1 & 1\frac{1}{3} & 1 & 1 \\ -1 & 1 & 1\frac{1}{2} & 1 \\ -1 & 1 & 1 & 1\frac{1}{4} \end{bmatrix}.$$

Show that the generalization of the above is also true, namely that

$$\begin{bmatrix} \mathbf{1}'\mathbf{u}+1 & \mathbf{u}' \\ \mathbf{u} & \left\{_d u_i \right\} \end{bmatrix}^{-1} = \begin{bmatrix} 1 & -\mathbf{1}' \\ -\mathbf{1} & \mathbf{J} + \left\{_d 1/u_i \right\} \end{bmatrix}.$$

Part II

NECESSARY THEORY

Chapter 6

LINEARLY (IN)DEPENDENT VECTORS

In the preceding chapters we have dealt with the arithmetic of matrices and described a number of commonly encountered special matrices such as those which are idempotent, orthogonal and so on. We now have three short chapters which revolve around certain mathematical concepts and relationships concerning individual rows (and/or columns) of a matrix. The three interrelated topics of these chapters are linear (in)dependence of vectors, rank and canonical forms. The first two of these provide methods for ascertaining if square **A** has a zero determinant, methods which are far easier to use than the laborious calculating of a determinant. And the third (Chapter 8) provides mechanisms for establishing many matrix results which are useful in applying matrices to practical problems. Illustrations in the next three chapters are limited since one needs to know some of the mathematics of matrices to deal with interesting complicated problems in economics.

6.1 Linear Combinations of Vectors

When x and y represent the axes in two-dimensional Cartesian coordinates, the equation of a straight line can be represented as $ax + by = c$ for constants a, b and c. The idea of linearity extends to vectors by calling $a\mathbf{x} + b\mathbf{y}$ a *linear combination* of the vectors $\mathbf{x}$ and $\mathbf{y}$. More generally, if we have n vectors all of the same order, which we call a

147

set of n vectors, $\mathbf{x}_1, \mathbf{x}_2, \ldots, \mathbf{x}_n$, then for scalars $a_1, a_2, \ldots, a_n$,

$$a_1\mathbf{x}_1 + a_2\mathbf{x}_2 + \cdots + a_n\mathbf{x}_n \tag{6.1}$$

is called a *linear combination of the set of n vectors.* For example, for the three vectors

$$\mathbf{x}_1 = \begin{bmatrix} 1 \\ 0 \\ -1 \\ 6 \end{bmatrix}, \quad \mathbf{x}_2 = \begin{bmatrix} -2 \\ 11 \\ 3 \\ 5 \end{bmatrix} \quad \text{and} \quad \mathbf{x}_3 = \begin{bmatrix} 0 \\ 1 \\ 2 \\ 7 \end{bmatrix}, \tag{6.2}$$

a linear combination of them (which is always a vector) is

$$a_1\mathbf{x}_1 + a_2\mathbf{x}_2 + a_3\mathbf{x}_3 = a_1 \begin{bmatrix} 1 \\ 0 \\ -1 \\ 6 \end{bmatrix} + a_2 \begin{bmatrix} -2 \\ 11 \\ 3 \\ 5 \end{bmatrix} + a_3 \begin{bmatrix} 0 \\ 1 \\ 2 \\ 7 \end{bmatrix}$$

$$= \begin{bmatrix} a_1 \\ 0 \\ -a_1 \\ 6a_1 \end{bmatrix} + \begin{bmatrix} -2a_2 \\ 11a_2 \\ 3a_2 \\ 5a_2 \end{bmatrix} + \begin{bmatrix} 0 \\ a_3 \\ 2a_3 \\ 7a_3 \end{bmatrix} = \begin{bmatrix} a_1 - 2a_2 + 0a_3 \\ 0a_1 + 11a_2 + a_3 \\ -a_1 + 3a_2 + 2a_3 \\ 6a_1 + 5a_2 + 7a_3 \end{bmatrix}. \tag{6.3}$$

This is a vector for any scalar values a_1, a_2 and a_3. These coefficients (the as) must be scalar, either positive or negative (real) numbers or even mathematical functions, so long as they are scalars.

Notice that the final result in (6.3) can also be written as

$$\begin{bmatrix} a_1 - 2a_2 + 0a_3 \\ 0a_1 + 11a_2 + a_3 \\ -a_1 + 3a_2 + 2a_3 \\ 6a_1 + 5a_2 + 7a_3 \end{bmatrix} = \begin{bmatrix} 1 & -2 & 0 \\ 0 & 11 & 1 \\ -1 & 3 & 2 \\ 6 & 5 & 7 \end{bmatrix} \begin{bmatrix} a_1 \\ a_2 \\ a_3 \end{bmatrix} = [\mathbf{x}_1 \ \mathbf{x}_2 \ \mathbf{x}_3] \begin{bmatrix} a_1 \\ a_2 \\ a_3 \end{bmatrix} = \mathbf{Xa},$$

on defining $\mathbf{X}$ as the matrix having columns $\mathbf{x}_1$, $\mathbf{x}_2$ and $\mathbf{x}_3$, and $\mathbf{a}$ as the vector of as. This is true in general: define

$$\mathbf{X} = [\mathbf{x}_1 \ \mathbf{x}_2 \ \cdots \ \mathbf{x}_n] \quad \text{and} \quad \mathbf{a}' = [a_1 \ a_2 \ \cdots \ a_n]. \tag{6.4}$$

Then the linear combination (6.1) is always a vector which can be written in the alternative forms as

$$a_1\mathbf{x}_1 + a_2\mathbf{x}_2 + \cdots + a_n\mathbf{x}_n = \sum_{i=1}^{n} a_i\mathbf{x}_i = \mathbf{Xa}. \tag{6.5}$$

Conversely, given a matrix $\mathbf{X}$ and a vector $\mathbf{a}$ for which the vector $\mathbf{Xa}$ exists, we see from (6.4) and (6.5) that

$\mathbf{Xa}$ is a column vector: a linear combination of the columns of $\mathbf{X}$.

$$(6.6)$$

Similarly,

$\mathbf{b'X}$ is a row vector: it is a linear combination of the rows of $\mathbf{X}$,

and

$\mathbf{AB}$ is a matrix:

  its rows are linear combinations of the rows of $\mathbf{B}$, and

  its columns are linear combinations of the columns of $\mathbf{A}$. (6.7)

These descriptions are a direct outcome of the definitions of a linear combination of vectors and of matrix multiplication. But they have many important uses in the development of matrix algebra, especially (6.6).

In passing, note that if we write $\mathbf{y} = \mathbf{Ax}$ then this represents the linear transformation of $\mathbf{x}$ into $\mathbf{y}$, as discussed in Section 3.1.

6.2 Linear Dependence and Independence

a. Definitions

The product $\mathbf{Xa}$ is a vector, and it is a linear combination of the column vectors in $\mathbf{X}$ as in (6.5):

$$\mathbf{Xa} = a_1\mathbf{x}_1 + a_2\mathbf{x}_2 + \cdots + a_n\mathbf{x}_n.$$

– i. Linearly dependent vectors

If there exists a vector $\mathbf{a} \neq \mathbf{0}$, such that $a_1\mathbf{x}_1 + \cdots + a_n\mathbf{x}_n = \mathbf{0}$, then provided none of $\mathbf{x}_1, \mathbf{x}_2, \ldots, \mathbf{x}_n$ is null, those vectors are said to be *linearly dependent vectors*.

An alternative statement of the definition is: if $\mathbf{Xa} = \mathbf{0}$ for some non-null $\mathbf{a}$, then the columns of $\mathbf{X}$ are linearly dependent vectors, provided none is null.

Example.

Consider the vectors

$$\mathbf{x}_1 = \begin{bmatrix} 3 \\ -6 \\ 9 \end{bmatrix}, \ \mathbf{x}_2 = \begin{bmatrix} 0 \\ 5 \\ -5 \end{bmatrix}, \ \mathbf{x}_3 = \begin{bmatrix} 2 \\ 1 \\ 1 \end{bmatrix}, \ \mathbf{x}_4 = \begin{bmatrix} -6 \\ 12 \\ -18 \end{bmatrix}$$

and

$$\mathbf{x}_5 = \begin{bmatrix} 2 \\ -3 \\ 3 \end{bmatrix}.$$

It is clear that

$$2\mathbf{x}_1 + \mathbf{x}_4 = \begin{bmatrix} \mathbf{x}_1 & \mathbf{x}_4 \end{bmatrix} \begin{bmatrix} 2 \\ 1 \end{bmatrix} = \begin{bmatrix} 6 \\ -12 \\ 18 \end{bmatrix} + \begin{bmatrix} -6 \\ 12 \\ -18 \end{bmatrix} = \begin{bmatrix} 0 \\ 0 \\ 0 \end{bmatrix} = \mathbf{0};$$

(6.8)

i.e.,

$$\mathbf{Xa} = \mathbf{0} \quad \text{for} \quad \mathbf{a} = \begin{bmatrix} a_1 \\ a_2 \end{bmatrix} = \begin{bmatrix} 2 \\ 1 \end{bmatrix},$$

which is non-null. Therefore $\mathbf{x}_1$ and $\mathbf{x}_4$ are linearly dependent. So also are $\mathbf{x}_1 \ \mathbf{x}_2$ and $\mathbf{x}_3$ because $[\mathbf{x}_1 \ \mathbf{x}_2 \ \mathbf{x}_3][2 \ 3 \ -3]' = \mathbf{0}$:

$$2\mathbf{x}_1 + 3\mathbf{x}_2 - 3\mathbf{x}_3 = \begin{bmatrix} 6 \\ -12 \\ 18 \end{bmatrix} + \begin{bmatrix} 0 \\ 15 \\ -15 \end{bmatrix} - \begin{bmatrix} 6 \\ 3 \\ 3 \end{bmatrix} = \begin{bmatrix} 0 \\ 0 \\ 0 \end{bmatrix} = \mathbf{0}.$$

(6.9)

In contrast, in

$$\mathbf{Xa} = \begin{bmatrix} \mathbf{x}_1 & \mathbf{x}_2 \end{bmatrix} \begin{bmatrix} a_1 \\ a_2 \end{bmatrix} = a_1\mathbf{x}_1 + a_2\mathbf{x}_2 = \begin{bmatrix} 3a_1 \\ -6a_1 + 5a_2 \\ 9a_1 - 5a_2 \end{bmatrix}, \quad (6.10)$$

there are no values a_1 and a_2 which make (6.10) a null vector other than $a_1 = 0 = a_2$. Therefore $\mathbf{x}_1$ and $\mathbf{x}_2$ are not linearly dependent. They are said to be linearly independent.

– ii.   Linearly independent vectors

If $\mathbf{a} = \mathbf{0}$ is the only vector for which $a_1\mathbf{x}_1 + a_2\mathbf{x}_2 + \cdots + a_n\mathbf{x}_n = \mathbf{0}$, then provided none of $\mathbf{x}_1, \mathbf{x}_2, \ldots, \mathbf{x}_n$ is null, those vectors are said to be *linearly independent vectors*.

An alternative statement of this is: if $\mathbf{Xa} = \mathbf{0}$ only for $\mathbf{a} = \mathbf{0}$, then the columns of $\mathbf{X}$ are linearly independent vectors. Equation (6.10) is an example. A difficulty with assimilating these definitions is that they are not as succinct as other definitions, for example, of $\mathbf{A}$ being idempotent. That definition is summarized by the simple statement $\mathbf{A}^2 = \mathbf{A}$. There is no such simplicity for the definitions of linear dependence and independence, because they involve not just an equation but statements about conditions for an equation being true: $\mathbf{Xa} = \mathbf{0}$ being true for some $\mathbf{a} \neq \mathbf{0}$ means the columns of $\mathbf{X}$ are linearly dependent, whereas it being true only for $\mathbf{a} = \mathbf{0}$ means they are linearly independent.

b. General properties

The linear dependence or linear independence of vectors is a property pertaining to a set of vectors of the same order. It is not a property of individual vectors.

– i. Sets of vectors

The property of vectors being linearly dependent or independent belongs to the whole set of vectors being considered in the equation $a_1\mathbf{x}_1 + a_2\mathbf{x}_2 + \cdots + a_n\mathbf{x}_n = \mathbf{0}$ and not just to subsets of them. Thus although (6.9) shows $\mathbf{x}_1$, $\mathbf{x}_2$ and $\mathbf{x}_3$ to be linearly dependent, there is no confusion of this with $\mathbf{x}_1$ and $\mathbf{x}_2$ being linearly independent as in (6.10) even though $\mathbf{x}_1$, $\mathbf{x}_2$ and $\mathbf{x}_3$ include $\mathbf{x}_1$ and $\mathbf{x}_2$. The property of linear dependence applies to the set of vectors $\mathbf{x}_1$, $\mathbf{x}_2$ and $\mathbf{x}_3$ and that of linear independence applies to the set $\mathbf{x}_1$ and $\mathbf{x}_2$.

– ii. Some coefficients zero

The definitions of linear dependence and of linear independence depend on whether $\mathbf{Xa} = \mathbf{0}$ for some $\mathbf{a} \neq \mathbf{0}$ or only for $\mathbf{a} = \mathbf{0}$. As a result, the definitions rely on whether none (or only some) of the coefficients (the a_is in $\mathbf{a}$) are zero, or on whether all of them are. Thus, with $2\mathbf{x}_1 + \mathbf{x}_4 = \mathbf{0}$ of (6.8), we say that $\mathbf{x}_1$ and $\mathbf{x}_4$ are linearly dependent; and because $2\mathbf{x}_1 + 0\mathbf{x}_2 + \mathbf{x}_4 = \mathbf{0}$, so also are $\mathbf{x}_1$, $\mathbf{x}_2$ and $\mathbf{x}_4$; in this case $a_1\mathbf{x}_1 + a_2\mathbf{x}_2 + a_3\mathbf{x}_4 = \mathbf{0}$ with some (but not all) the as being zero. Hence the conditions are satisfied for $\mathbf{x}_1$, $\mathbf{x}_2$ and $\mathbf{x}_4$ being linearly dependent. If the sole value of $\mathbf{a}$ for which $\mathbf{Xa} = \mathbf{0}$ is true is $\mathbf{a} = \mathbf{0}$ (i.e., all a_is zero), only then do we say that the columns of $\mathbf{X}$ are linearly independent.

– iii.   Existence and non-uniqueness of non-zero coefficients

Discussion of examples relies heavily on actual values of as. In practice, it is not individual values of as that are important but whether or not $\mathbf{Xa} = \mathbf{0}$ is true only for all the a_is being zero. If it is, then the vectors are linearly independent. If not, if non-zero as exist such that $\mathbf{Xa} = \mathbf{0}$ is true, then the vectors are linearly dependent and the important thing is not the individual values of those as but the fact that they do exist; i.e., it is the *existence* of non-zero as that is important. It is their existence and not their actual values that is utilized.

One reason for not being concerned with individual values of non-zero as when they exist is that they are not unique; for example, with (6.8) being true, so also is $4\mathbf{x}_1 + 2\mathbf{x}_4 = \mathbf{0}$. In general, if $\mathbf{Xa} = \mathbf{0}$ is true for some $\mathbf{a}$, then so is $\mathbf{X}\lambda\mathbf{a} = \mathbf{0}$ for scalar λ.

– iv.   Null vectors

Suppose $\mathbf{y}_1$ and $\mathbf{y}_2$ are any vectors of order 3. Then

$$a_1\mathbf{y}_1 + a_2\mathbf{y}_2 + a_3 \begin{bmatrix} 0 \\ 0 \\ 0 \end{bmatrix} = \mathbf{0}$$

is true for $a_1 = 0 = a_2$ and for any non-zero a_3. Therefore $\mathbf{y}_1, \mathbf{y}_2$ and the null vector (of order 3) are linearly dependent. This is true generally: vectors that include a null vector are always linearly dependent. This is why null vectors are excluded from the definitions of linearly dependent and linearly independent vectors. Henceforth, whenever linearly dependent or linearly independent vectors are mentioned, they never include a null vector.

Notation. The abbreviation LIN shall be used to mean "linear independence" or "linearly independent", depending on context.

6.3   Linearly Dependent Vectors

a.   At least two coefficients are non-zero

Vectors $\mathbf{x}_1, \mathbf{x}_2, \ldots, \mathbf{x}_p$ are linearly dependent when

$$a_1\mathbf{x}_1 + a_2\mathbf{x}_2 + \cdots + a_p\mathbf{x}_p = \mathbf{0} \text{ for not all the } a\text{s being zero.} \quad (6.11)$$

Suppose only one a is non-zero—suppose it is a_2, say. Then (6.11) is $a_2\mathbf{x}_2 = \mathbf{0}$, and because null vectors are excluded, $\mathbf{x}_2$ is not null and so

$a_2 = 0$. This contradicts the supposition. Therefore more than one a is non-zero, i.e., at least two of them are. Also, of course, the definition of linearly dependent vectors demands that there be at least two of them, and so at least two as are non-zero.

b. Vectors are linear combinations of others

Suppose in (6.11) that a_1 and a_2 are non-zero. (There is no loss of generality in identifying any two non-zero as as a_1 and a_2.) Then from (6.11)

$$\mathbf{x}_1 + (a_2/a_1)\mathbf{x}_2 + \cdots + (a_p/a_1)\mathbf{x}_p = \mathbf{0} \qquad (6.12)$$

so that

$$\mathbf{x}_1 = -(a_2/a_1)\mathbf{x}_2 - (a_3/a_1)\mathbf{x}_3 - \cdots - (a_p/a_1)\mathbf{x}_p;$$

i.e., $\mathbf{x}_1$ can be expressed as a linear combination of the other $\mathbf{x}$s. And this can always be done because at least two as are always non-zero. For example, in (6.9) we have $2\mathbf{x}_1 + 3\mathbf{x}_2 - 3\mathbf{x}_3 = \mathbf{0}$, so that

$$\mathbf{x}_1 = -1\tfrac{1}{2}\mathbf{x}_2 + 1\tfrac{1}{2}\mathbf{x}_3. \qquad (6.13)$$

And, of course, this can also be stated in other equivalent ways such as

$$\mathbf{x}_3 = \tfrac{2}{3}\mathbf{x}_1 + \mathbf{x}_2. \qquad (6.14)$$

Similarly, (6.8) gives $\mathbf{x}_1 = -\tfrac{1}{2}\mathbf{x}_4$ and

$$\mathbf{x}_4 = -2\mathbf{x}_1, \qquad (6.15)$$

In Section 6.2b-iii the existence rather than the individual values of non-zero as was emphasized. This emphasis carries over here to the existence (among dependent vectors) of linear combinations of some of them being equal to others: it is the existence of such combinations that is important rather than the specific scalars that multiply the vectors in those combinations.

c. Zero determinants

Suppose p linearly dependent vectors of order p are used as columns of a matrix. Then the determinant is zero. This is because linear dependence of vectors implies that one vector (column of the matrix) can always be expressed as a linear combination of the others. Therefore subtracting from that column the linear combination of the other columns which it equals gives a zero column.

Example   (*continued*)

$$|\mathbf{x}_1 \ \mathbf{x}_2 \ \mathbf{x}_3| = \begin{vmatrix} 3 & 0 & 2 \\ -6 & 5 & 1 \\ 9 & -5 & 1 \end{vmatrix}.$$

Since subtracting $(-1\frac{1}{2}\mathbf{x}_2 + 1\frac{1}{2}\mathbf{x}_3)$ from $\mathbf{x}_1$, i.e., $\mathbf{x}_1 - (-1\frac{1}{2}\mathbf{x}_2 + 1\frac{1}{2}\mathbf{x}_3)$, is by (6.13) just subtracting $\mathbf{x}_1$ from itself, carrying out this subtraction in the determinant gives it a column of zeros, and so it is zero:

$$|\mathbf{x}_1 \ \mathbf{x}_2 \ \mathbf{x}_3| = \begin{vmatrix} 0 & 0 & 2 \\ 0 & 5 & 1 \\ 0 & -5 & 1 \end{vmatrix} = 0.$$

d.   Inverse matrices

The preceding result applies to any set of p linearly dependent vectors of order p: the determinant of the matrix having those vectors as columns is always zero. (The same is true, of course, if the vectors used are rows.) Consequently, the matrix so formed has no inverse. Hence when the columns (or rows) of a square matrix are linearly dependent, that matrix has no inverse. It is singular.

e.   Testing for dependence (simple cases)

A simple test for linear dependence among p vectors of order p is to evaluate the determinant of the matrix formed from using the vectors as columns. A zero determinant means the vectors are linearly dependent, otherwise they are LIN.

Example   (*continued*)

$$|\mathbf{x}_1 \ \mathbf{x}_2 \ \mathbf{x}_3| = \begin{vmatrix} 3 & 0 & 2 \\ -6 & 5 & 1 \\ 9 & -5 & 1 \end{vmatrix} = \begin{vmatrix} 3 & 0 & 2 \\ -6 & 5 & 1 \\ 3 & 0 & 2 \end{vmatrix} = 0$$

and so $\mathbf{x}_1$, $\mathbf{x}_2$ and $\mathbf{x}_3$ are linearly dependent, whereas

$$|\mathbf{x}_1 \ \mathbf{x}_2 \ \mathbf{x}_5| = \begin{vmatrix} 3 & 0 & 2 \\ -6 & 5 & -3 \\ 9 & -5 & 3 \end{vmatrix} = \begin{vmatrix} 3 & 0 & 2 \\ -6 & 5 & -3 \\ 3 & 0 & 0 \end{vmatrix} = -30 \neq 0$$

and so x_1, x_2 and x_5 are LIN.

The definitions of linear dependence and independence of vectors represented as columns of a matrix X are based on whether the equations $Xa = 0$ are satisfied for some non-null a (dependence) or are satisfied only for $a = 0$ (independence). Given a set of vectors, their dependence or independence can therefore be ascertained by attempting to solve $Xa = 0$. If a solution can be found other than $a = 0$ (which is always a solution), then it will be a non-null solution, and the vectors will be dependent. Otherwise they are LIN.

Examples.

For $X = [x_1 \ x_2 \ x_3]$, equations $Xa = 0$ are

$$\begin{array}{rcrcrcl} 3a_1 & & & + & 2a_3 & = & 0 \\ -6a_1 & + & 5a_2 & + & a_3 & = & 0 \\ 9a_1 & - & 5a_2 & + & a_3 & = & 0 \end{array} \quad \text{with solution} \quad a = \begin{bmatrix} 2 \\ 3 \\ -3 \end{bmatrix}.$$

Since $a \neq 0$, the vectors x_1, x_2 and x_3 are linearly dependent. But for $X = [x_1 \ x_2 \ x_5]$, equations $Xa = 0$ are

$$\begin{array}{rcrcrcl} 3a_1 & & & + & 2a_3 & = & 0 \\ -6a_1 & + & 5a_2 & - & 3a_3 & = & 0 \\ 9a_1 & - & 5a_2 & + & 3a_3 & = & 0 \end{array}.$$

Adding the last two of these equations gives $3a_1 = 0$, i.e., $a_1 = 0$. Hence from the first, $a_3 = 0$ and so in the second, $a_2 = 0$. Thus $a = [0 \ 0 \ 0]' = 0$ is the only solution and so x_1, x_2 and x_5 are LIN.

This principle can also be used with rectangular X: for $X = [x_1 \ x_2 \ x_3 \ x_4]$ equations $Xa = 0$ are

$$\begin{array}{rcrcrcrcl} 3a_1 & & & + & 2a_3 & - & 6a_4 & = & 0 \\ -6a_1 & + & 5a_2 & + & a_3 & + & 12a_4 & = & 0 \\ 9a_1 & - & 5a_2 & + & a_3 & - & 18a_4 & = & 0 \end{array} \quad \text{with solution } a = \begin{bmatrix} a_4 \\ -1\frac{1}{2}a_4 \\ 1\frac{1}{2}a_4 \\ a_4 \end{bmatrix},$$

which is clearly non-null for any non-zero a_4. Hence x_1, x_2, x_3 and x_4 are linearly dependent. Similarly, for x_1 and x_3 it will be found that $[x_1 \ x_3]a = 0$ has only $a = 0$ as a solution, and so x_1 and x_3 are LIN. But $[x_1 \ x_5]a = 0$ has $a = [2 \ 1]' \neq 0$ as a solution and so x_1 and x_5 are linearly dependent.

Solving equations is easy when only a few vectors of small order are involved, as in these examples. But for more extensive situations the arithmetic can be tedious. Furthermore, since easier procedures are available for ascertaining whether or not $\mathbf{Xa} = \mathbf{0}$ has a non-null solution other than solving the equations explicitly, we proceed towards developing those procedures. Before doing so, notice the equivalence of the following statements for square $\mathbf{X}$ that has no null columns:

(i) Columns of $\mathbf{X}$ are linearly dependent.

(ii) $\mathbf{Xa} = \mathbf{0}$ can be satisfied for a non-null $\mathbf{a}$.

(iii) $\mathbf{X}$ is singular, i.e., $\mathbf{X}^{-1}$ does not exist.

(iv) $|\mathbf{X}| = 0$.

The definition of linear dependence assures the equivalence of (i) and (ii), and Section 6.3d shows the equivalence of these to (iii).

6.4   Linearly Independent (LIN) Vectors

a.   Nonzero determinants and inverse matrices

The only value of $\mathbf{a}$ for which $\mathbf{Xa} = \mathbf{0}$ when the columns of $\mathbf{X}$ are LIN is $\mathbf{a} = \mathbf{0}$, in which case no equation like (6.12) exists. Therefore no column in $\mathbf{X}$ can be expressed as a linear combination of others in $\mathbf{X}$. Hence when $\mathbf{X}$ has p LIN columns of order p, i.e., when $\mathbf{X}$ is square, its determinant is non-zero, and so its inverse does exist. Hence for square $\mathbf{X}$ the following four statements are equivalent:

(i) Columns of $\mathbf{X}$ are LIN.

(ii) $\mathbf{Xa} = \mathbf{0}$ only for $\mathbf{a} = \mathbf{0}$.

(iii) $\mathbf{X}$ is non-singular, i.e., $\mathbf{X}^{-1}$ exists.

(iv) $|\mathbf{X}| \neq 0$.

b.   Linear combinations of LIN vectors

When p vectors of order p are LIN, any other vector of order p can be expressed as a linear combination of those p vectors. Suppose $\mathbf{X}$ is the matrix of the p vectors used as columns, and let $\mathbf{v}$ be some other vector of the same order. Then this result states that there exists a vector $\mathbf{m}$ which has the property $\mathbf{Xm} = \mathbf{v}$. This is so because with

$\mathbf{X}$ being square and having LIN columns, $\mathbf{X}^{-1}$ exists and so $\mathbf{Xm} = \mathbf{v}$ implies $\mathbf{m} = \mathbf{X}^{-1}\mathbf{v}$.

Example

Any vector $\mathbf{v}$ of order 2 can be expressed as a linear combination of the two LIN vectors $\mathbf{x}_1 = \begin{bmatrix} 2 \\ 1 \end{bmatrix}$ and $\mathbf{x}_2 = \begin{bmatrix} 3 \\ 5 \end{bmatrix}$. Let $\mathbf{v} = \begin{bmatrix} a \\ b \end{bmatrix}$ and $\mathbf{X} = [\mathbf{x}_1 \ \mathbf{x}_2] = \begin{bmatrix} 2 & 3 \\ 1 & 5 \end{bmatrix}$. Then

$$\mathbf{m} = \mathbf{X}^{-1}\mathbf{v} = (1/7)\begin{bmatrix} 5 & -3 \\ -1 & 2 \end{bmatrix}\begin{bmatrix} a \\ b \end{bmatrix} = \begin{bmatrix} (5a - 3b)/7 \\ (-a + 2b)/7 \end{bmatrix};$$

and so

$$\mathbf{v} = \begin{bmatrix} a \\ b \end{bmatrix} = \mathbf{Xm} = \begin{bmatrix} 2 & 3 \\ 1 & 5 \end{bmatrix}\begin{bmatrix} (5a - 3b)/7 \\ (-a + 2b)/7 \end{bmatrix}$$

$$= \frac{5a - 3b}{7}\begin{bmatrix} 2 \\ 1 \end{bmatrix} + \frac{-a + 2b}{7}\begin{bmatrix} 3 \\ 5 \end{bmatrix}. \quad (6.16)$$

Thus $\mathbf{v}$ has been expressed as a linear combination of $\mathbf{x}_1$ and $\mathbf{x}_2$.

c. Maximum number of LIN vectors

A question that might be asked about independent vectors is "How many independent vectors are there?" The answer is contained in the following theorem. We label it as a theorem because it is a very important and useful result.

Theorem. A set of LIN vectors of order n never contains more than n such vectors.

Proof. Let $\mathbf{u}_1, \mathbf{u}_2, \ldots, \mathbf{u}_n$ be n LIN vectors of order n. Let $\mathbf{u}^*$ be any other non-null vector of order n. We show that it and $\mathbf{u}_1, \mathbf{u}_2, \ldots, \mathbf{u}_n$ are linearly dependent.

Since $\mathbf{U} = [\mathbf{u}_1, \mathbf{u}_2, \ldots, \mathbf{u}_n]$ has LIN columns, $|\mathbf{U}| \neq 0$ and $\mathbf{U}^{-1}$ exists. Let $\mathbf{q} = -\mathbf{U}^{-1}\mathbf{u}^* \neq \mathbf{0}$, because $\mathbf{u}^* \neq \mathbf{0}$; i.e., not all elements of q are zero. Then $\mathbf{Uq} + \mathbf{u}^* = \mathbf{0}$, which can be rewritten as

$$[\mathbf{U} \ \mathbf{u}^*]\begin{bmatrix} \mathbf{q} \\ 1 \end{bmatrix} = \mathbf{0} \quad \text{with} \quad \begin{bmatrix} \mathbf{q} \\ 1 \end{bmatrix} \neq \mathbf{0}.$$

Therefore $\mathbf{u}_1, \mathbf{u}_2, \ldots, \mathbf{u}_n, \mathbf{u}^*$ are linearly dependent; i.e., with $\mathbf{u}_1, \ldots, \mathbf{u}_n$ being LIN it is impossible to put another vector $\mathbf{u}$ with them and have all $n+1$ vectors be LIN.   Q.E.D

Corollary. When p vectors of order n are LIN then $p \leq n$.

It is important to note that this theorem is *not* stating that there is only a single set of n vectors of order n that are LIN. What it is saying is that if we do have a set of n LIN vectors of order n, then there is no larger set of LIN vectors of order n (i.e., there are no sets of $n+1, n+2, n+3, \ldots$ vectors of order n) that are LIN. Although there are many sets of n LIN vectors of order n, an infinite number of them in fact, for each of them it is impossible to put another vector (or vectors) with them and have the set, which then contains more than n vectors, still be LIN.

Example

$$\begin{bmatrix} 2 \\ 1 \end{bmatrix} \quad \text{and} \quad \begin{bmatrix} 3 \\ 5 \end{bmatrix}$$

are a set of two LIN vectors of order 2. Put any other vector of order 2 with these two vectors and the set will be linearly dependent. This is because, as (6.16) shows, $\begin{bmatrix} a \\ b \end{bmatrix}$ is a linear combination of $\begin{bmatrix} 2 \\ 1 \end{bmatrix}$ and $\begin{bmatrix} 3 \\ 5 \end{bmatrix}$. But the theorem is not saying that $\begin{bmatrix} 2 \\ 1 \end{bmatrix}$ and $\begin{bmatrix} 3 \\ 5 \end{bmatrix}$ is the only set of two LIN vectors of order 2. For example, $\begin{bmatrix} 5 \\ 7 \end{bmatrix}$ and $\begin{bmatrix} 9 \\ 13 \end{bmatrix}$ also form such a set; and any other vector of order 2 put with them forms a linearly dependent set of three vectors. This is so because that third vector is a linear combination of these two; e.g.,

$$\begin{bmatrix} a \\ b \end{bmatrix} = \tfrac{1}{2}(13a - 9b) \begin{bmatrix} 5 \\ 7 \end{bmatrix} + \tfrac{1}{2}(-7a + 5b) \begin{bmatrix} 9 \\ 13 \end{bmatrix}.$$

No matter what the values of a and b are, this expression holds true— i.e., every second-order vector can be expressed as a linear combination of a set of two LIN vectors of order 2. In general, every nth-order vector can be expressed as a linear combination of any set of n independent vectors of order n. The maximum number of non-null vectors in a set of independent vectors is therefore n, the order of the vectors.

6.5 Exercises

E 6.1. Correct the following sentences by adding the prefix "in" to
"___ dependent" whenever necessary.

Vectors $\mathbf{v}_1, \mathbf{v}_2, \ldots, \mathbf{v}_n$ are said to be linearly ___dependent
if $\Sigma_{i=1}^{n} k_i \mathbf{v}_i = \mathbf{0}$ is true only for $k_i = 0$ for $i = 1, 2, \ldots, n$, except
that if one of them is null they are linearly ___dependent.
When they are ___dependent at least two of the ks are non-
zero and, because they are linearly ___dependent, at least one
of the vectors can be expressed as a linear combination of the
other vectors.

It is possible to have fourteen non-null vectors of order thirteen linearly ___dependent but only thirteen of them ___depen-
dent; in which case the matrix of those thirteen vectors is non-
singular. And indeed it is impossible to have a set of fourteen
linearly ___dependent non-null vectors of order thirteen.

Consider five vectors (of order four), two of which are linear
combinations of the other three. The five vectors then form a
linearly ___dependent set. And if the five vectors are used as
the columns of a matrix the four rows of that matrix will be
linearly ___dependent.

In a set of non-null vectors of order p there cannot be more
than p vectors if they are to be linearly ___dependent.

E 6.2. For $\mathbf{x}_1 = \begin{bmatrix} 1 \\ 2 \\ 1 \end{bmatrix}$, $\mathbf{x}_2 = \begin{bmatrix} -1 \\ 3 \\ 2 \end{bmatrix}$, $\mathbf{x}_3 = \begin{bmatrix} -13 \\ -1 \\ 2 \end{bmatrix}$ and $\mathbf{x}_4 = \begin{bmatrix} 1 \\ 1 \\ 0 \end{bmatrix}$, show the following.

(a) $\mathbf{x}_1$, $\mathbf{x}_2$ and $\mathbf{x}_3$ are linearly dependent, and find a linear
relationship among them.

(b) $\mathbf{x}_1$, $\mathbf{x}_2$ and $\mathbf{x}_4$ are LIN, and find the linear combination
of them that equals $[a \ b \ c]'$.

E 6.3. Give two reasons why the columns of $\mathbf{I}_4$ are LIN, and show
that every vector of order 4 is a linear combination of those

vectors. (Note that similar results hold for all identity matrices.)

E 6.4. For a square matrix $\mathbf{A}$, suppose there is an $\mathbf{x} \neq \mathbf{0}$ such that $\mathbf{Ax} = \mathbf{0}$. Explain why $\mathbf{A}$ is singular.

E 6.5. For $\mathbf{A}$ square of order n and $\mathbf{1}$ of order n, suppose that $\mathbf{x} \neq \mathbf{0}$ exists such that $\mathbf{Ax} = \mathbf{0}$ and $\mathbf{x'1} = 0$. Prove (a) that $\mathbf{A} + \lambda \mathbf{11'}$ is singular for any scalar λ and (b) that $\mathbf{A} + \mathbf{f1'}$ is singular for any vector $\mathbf{f}$.

E 6.6. When $\mathbf{A}$ and $\mathbf{B}$ are idempotent with columns of $\mathbf{B}$ being linear combinations of columns of $\mathbf{A}$, prove that $\mathbf{AB} = \mathbf{B}$.

E 6.7. Prove that $\mathbf{vv'} - \mathbf{v'vI}$ is singular.

E 6.8. When two vectors are orthogonal, prove that they are also LIN.

E 6.9. If $\mathbf{A}$ is singular and has non-zero row sums that are the same for every row, prove (a) that $\mathbf{A} + \lambda \mathbf{11'}$ is singular and (b) that $\mathbf{A} + \mathbf{1f'}$ is also singular for any $\mathbf{f}$.

E 6.10. If $\mathbf{A}$ is singular and $\mathbf{Av} = \lambda \mathbf{v}$, prove that $\mathbf{A} + \lambda \mathbf{vv'}$ and $\mathbf{A} + \mathbf{vf'}$ for any $\mathbf{f'}$ are singular.

Chapter 7

RANK

Rank is an extremely important and useful characteristic of every matrix. It concerns the number of LIN rows (and LIN columns) in a matrix. That phrase "number of LIN rows" is used extensively; its real meaning is "the maximum number of LIN rows". For example, in a matrix of five rows having four of them that are LIN, the "number of LIN rows" is described as four. Although any three of those four will be LIN, the phrase "number of LIN rows" would always mean four, the maximum, and not three.

7.1 Number of LIN Rows and Columns in a Matrix

In the preceding chapter we explained how a determinant is zero when any of its rows (or columns) are linear combinations of other rows (or columns). In other words, a determinant is zero when its rows (or columns) do not form a set of LIN vectors. Evidently, therefore, a determinant cannot have both its rows forming a dependent set and its columns an independent set, a statement which prompts the question "What is the relationship between the number of LIN rows of a matrix and the number of LIN columns?" The answer is simple: they are the same. But because this is so important we state it as a theorem.

Theorem. The number of LIN rows in a matrix is the same as the number of LIN columns.

Before proving this, notice that independence of rows (columns) is a property of rows that is unrelated to their sequence within a matrix.

161

For example, because the rows of

$$\begin{bmatrix} 1 & 2 & 3 \\ 6 & 9 & 14 \\ 3 & 0 & 1 \end{bmatrix} \quad \text{are LIN, so also are the rows of} \quad \begin{bmatrix} 6 & 9 & 14 \\ 3 & 0 & 1 \\ 1 & 2 & 3 \end{bmatrix}.$$

Therefore, insofar as general discussion of independence properties is concerned, there is no loss of generality for a matrix that has k LIN rows in assuming that they are the first k rows. This assumption is therefore often made when discussing properties and consequences of independence of rows (columns) of a matrix.

We now prove the theorem.

Proof of Theorem.   Let $\mathbf{A}_{p \times q}$ have k LIN rows and m LIN columns. We show that $k = m$.

Assume that the first k rows of $\mathbf{A}$ are LIN, and similarly the first m columns. Then partition $\mathbf{A}$ as

$$\mathbf{A}_{p \times q} \;=\; \begin{bmatrix} \mathbf{X} & \mathbf{Y} \\ \mathbf{Z} & \mathbf{W} \end{bmatrix} = \begin{bmatrix} \mathbf{X}_{k \times m} & \mathbf{Y}_{k \times (q-m)} \\ \mathbf{Z}_{(p-k) \times m} & \mathbf{W}_{(p-k) \times (q-m)} \end{bmatrix}. \substack{\leftarrow k \text{ LIN rows}} \quad (7.1)$$

$$\substack{\uparrow \\ m \text{ LIN columns}}$$

Thus the k rows of $\mathbf{A}$ through $\mathbf{X}$ and $\mathbf{Y}$ are LIN (as are the m columns of $\mathbf{A}$ through $\mathbf{X}$ and $\mathbf{Z}$). Since $\mathbf{A}$ has only k LIN rows the other rows of $\mathbf{A}$ (those through $\mathbf{Z}$ and $\mathbf{W}$) are linear combinations of the first k rows. In particular the rows of $\mathbf{Z}$ are linear combinations of the rows of $\mathbf{X}$. Hence these rows can be expressed as $\mathbf{Z} = \mathbf{T}\mathbf{X}$ for some matrix $\mathbf{T}$. Now assume that the columns of $\mathbf{X}$ are linearly dependent, i.e., that

$$\mathbf{X}\mathbf{a} = \mathbf{0} \quad \text{for some vector} \quad \mathbf{a} \neq \mathbf{0}. \quad (7.2)$$

Then $\mathbf{Z}\mathbf{a} = \mathbf{T}\mathbf{X}\mathbf{a} = \mathbf{0}$ and so

$$\begin{bmatrix} \mathbf{X} \\ \mathbf{Z} \end{bmatrix} \mathbf{a} = \mathbf{0} \quad \text{for that same} \quad \mathbf{a} \neq \mathbf{0}.$$

But this is a statement of the linear dependence of the columns of $\begin{bmatrix} \mathbf{X} \\ \mathbf{Z} \end{bmatrix}$, i.e., of the first m columns of $\mathbf{A}$. These columns, however, have been taken in (7.1) as being LIN. This is a contradiction, and so assumption (7.2), from which it is derived, is false; i.e., the columns of $\mathbf{X}$ are not dependent. Hence they must be LIN.

Having shown the columns of $\mathbf{X}$ to be LIN, observe from (7.1) that there are m of them and they are of order k. Hence by the theorem in Section 6.4c, $m \leq k$. A similar argument based on the rows of $\mathbf{X}$ and $\mathbf{Y}$, rather than the columns of $\mathbf{X}$ and $\mathbf{Z}$, shows that $k \leq m$. Hence $m = k$. Q.E.D

Notice that this theorem says nothing about which rows (columns) of a matrix are LIN—it is concerned solely with how many of them are LIN. This means, for example, that if there are two LIN rows in a matrix of order 5×4 then once two rows are ascertained as being LIN the other three rows can be expressed as linear combinations of those two. And there may be (there usually is) more than one set of two rows that are LIN. For example, in

$$\mathbf{A} = \begin{bmatrix} 1 & 2 & 0 & 1 \\ 1 & -1 & 3 & 2 \\ 3 & 0 & 6 & 5 \\ 2 & 1 & 3 & 3 \\ 6 & 0 & 12 & 10 \end{bmatrix}$$

rows 1 and 2 are LIN, and each of rows 3, 4 and 5 is a linear combination of those first two. Rows 1 and 3 are also LIN, and the remaining rows are linear combinations of them; but not all pairs of rows are LIN. For example, rows 3 and 5 are not LIN.

7.2 Rank of a Matrix

The preceding theorem shows that every matrix has the same number of LIN rows as it does LIN columns.

Definition. The *rank* of a matrix is the number of linearly independent rows (and columns) in the matrix.

Notation. The rank of $\mathbf{A}$ will be denoted equivalently by $r_\mathbf{A}$ or $r(\mathbf{A})$. Thus if $r_\mathbf{A} \equiv r(\mathbf{A}) = k$, then $\mathbf{A}$ has k LIN rows and k LIN columns. The symbol r is often used for rank, i.e., $r_\mathbf{A} \equiv r$.

Notice again that no specific set of LIN rows (columns) is identified by knowing the rank of a matrix. Rank indicates only how many are LIN and not where they are located in the matrix. The following properties and consequences of rank are important.

(i) $r_\mathbf{A}$ is a positive integer, except that r_0 is defined as $r_0 = 0$.

(ii) $r(\mathbf{A}_{p\times q}) \leq p$ and $\leq q$: the rank of a matrix equals or is less than the smaller of its number of rows or columns.

(iii) $r(\mathbf{A}_{n\times n}) \leq n$: a square matrix has rank not exceeding its order.

(iv) When $r_\mathbf{A} = r \neq 0$ there is at least one square submatrix of $\mathbf{A}$ having order r that is non-singular. Equation (7.2) with $k = m = r$ is

$$\mathbf{A}_{p\times q} = \left[\begin{array}{cc} \mathbf{X}_{r\times r} & \mathbf{Y}_{r\times(q-r)} \\ \mathbf{Z}_{(p-r)\times r} & \mathbf{W}_{(p-r)\times(q-r)} \end{array} \right] \qquad (7.3)$$

and $\mathbf{X}_{r\times r}$, the intersection of r LIN rows and r LIN columns, is non-singular — as is evident from the proof of the preceding theorem. All square submatrices of order greater than r are singular.

(v) When $r(\mathbf{A}_{n\times n}) = n$, then by (iv) $\mathbf{A}$ is non-singular, i.e., $\mathbf{A}^{-1}$ exists. [In (7.3) $\mathbf{A} \equiv \mathbf{X}$.]

(vi) When $r(\mathbf{A}_{n\times n}) < n$ then $\mathbf{A}$ is singular and $\mathbf{A}^{-1}$ does not exist.

(vii) When $r(\mathbf{A}_{p\times q}) = p < q$, $\mathbf{A}$ is said to have *full row rank*, or to be of full row rank. Its rank equals its number of rows.

(viii) When $r(\mathbf{A}_{p\times q}) = q < p$, $\mathbf{A}$ is said to have *full column rank*, or to be of full column rank. Its rank equals its number of columns.

(ix) When $r(\mathbf{A}_{n\times n}) = n$, $\mathbf{A}$ is said to have *full rank*, or to be of full rank. Its rank equals its order, it is non-singular, its inverse exists, and it is said to be *invertible*.

Rank is one of the most useful and important characteristics of a matrix. It occurs again and again in using matrices, and plays a vital role throughout all aspects of matrix algebra. For example, from items (vi) and (ix) we see at once that ascertaining whether $|\mathbf{A}|$ is zero or not for determining the existence of $\mathbf{A}^{-1}$ can be replaced by ascertaining whether $r_\mathbf{A} < n$ or $r_\mathbf{A} = n$. And almost always it is far easier to work with rank than with determinants.

7.3   Rank and Inverse Matrices

A square matrix has an inverse if and only if its rank equals its order. This and other equivalent statements are summarized in Table 7.1. In each half of the table any one of the statements implies all the others: the first and second statements are basically definitional and the last six are equivalences. Hence whenever assurance is needed for the existence of $\mathbf{A}^{-1}$, we need only establish any one of the last seven statements in the first column of the table. The easiest of these is usually that concerning rank: when $r_\mathbf{A} = n$, then $\mathbf{A}^{-1}$ exists, and when $r_\mathbf{A} < n$, then $\mathbf{A}^{-1}$ does not exist. The problem of ascertaining the existence of an inverse is therefore equivalent to ascertaining if the rank of a square matrix is less than its order.

More generally, there are occasions when the rank of a matrix is needed exactly. Although locating $r_\mathbf{A}$ LIN rows in $\mathbf{A}$ may not always be easy, deriving $r_\mathbf{A}$ itself is conceptually not difficult. General procedures for doing this are described subsequently.

Table 7.1. Equivalent Statements for the
Existence of $\mathbf{A}^{-1}$ of Order n

Inverse Existing	Inverse Not Existing
$\mathbf{A}^{-1}$ exists	$\mathbf{A}^{-1}$ does not exist
$\mathbf{A}$ is non-singular	$\mathbf{A}$ is singular
$|\mathbf{A}| \neq 0$	$|\mathbf{A}| = 0$
$\mathbf{A}$ has full rank	$\mathbf{A}$ has less than full rank
$r_\mathbf{A} = n$	$r_\mathbf{A} < n$
$\mathbf{A}$ has n LIN rows	$\mathbf{A}$ has fewer than n LIN rows
$\mathbf{A}$ has n LIN columns	$\mathbf{A}$ has fewer than n LIN columns
$\mathbf{Ax} = \mathbf{0}$ has sole solution $\mathbf{x=0}$	$\mathbf{Ax} = \mathbf{0}$ has many solutions

7.4 Elementary Operators

Having defined rank, listed its properties, and described how knowing the rank of a square matrix immediately indicates whether that matrix has an inverse or not, we now turn to the problem of calculating the rank of a matrix. The lead-in to this is the subject of elementary operators.

The three elementary operator matrices (also known as elementary operators) $\mathbf{E}_{ij}$, $\mathbf{R}_{ii}(\lambda)$ and $\mathbf{P}_{ij}(\lambda)$ introduced in Section 4.4 play an important role in connection with the rank of a matrix. Each elementary operator is a square matrix derived from an identity matrix. When used in a product with some other matrix it leads to the same manipulation of rows (or columns) of that matrix as is used in simplifying determinants. Furthermore, the product matrix has the same rank as the original matrix. As a result, elementary operators facilitate ascertaining the rank of a matrix.

a. Row operations

Pre-multiplication by elementary operators leads to manipulation of rows. Thus $\mathbf{E}_{ij}\mathbf{A}$ interchanges the ith row and jth row of $\mathbf{A}$:

$$\mathbf{E}_{12} = \begin{bmatrix} 0 & 1 & 0 \\ 1 & 0 & 0 \\ 0 & 0 & 1 \end{bmatrix} \begin{bmatrix} 1 & 4 & 7 \\ 2 & 5 & 8 \\ 3 & 6 & 9 \end{bmatrix} = \begin{bmatrix} 2 & 5 & 8 \\ 1 & 4 & 7 \\ 3 & 6 & 9 \end{bmatrix}.$$

$\mathbf{R}_{ii}(\lambda)$ multiplies the ith row of $\mathbf{A}$ by λ:

$$[\mathbf{R}_{22}(4)]\mathbf{A} = \begin{bmatrix} 1 & 0 & 0 \\ 0 & 4 & 0 \\ 0 & 0 & 1 \end{bmatrix} \begin{bmatrix} 1 & 4 & 7 \\ 2 & 5 & 8 \\ 3 & 6 & 9 \end{bmatrix} = \begin{bmatrix} 1 & 4 & 7 \\ 8 & 20 & 32 \\ 3 & 6 & 9 \end{bmatrix},$$

and $\mathbf{P}_{ij}(\lambda)\mathbf{A}$ adds λ times the jth row of $\mathbf{A}$ to its ith row:

$$[\mathbf{P}_{12}(\lambda)]\mathbf{A} = \begin{bmatrix} 1 & \lambda & 0 \\ 0 & 1 & 0 \\ 0 & 0 & 1 \end{bmatrix} \begin{bmatrix} 1 & 4 & 7 \\ 2 & 5 & 8 \\ 3 & 6 & 9 \end{bmatrix} = \begin{bmatrix} 1+2\lambda & 4+5\lambda & 7+8\lambda \\ 2 & 5 & 8 \\ 3 & 6 & 9 \end{bmatrix}.$$

b. Transposes

The following transposes are easily verified:

$$[\mathbf{E}_{ij}]' = \mathbf{E}_{ij}, \quad [\mathbf{R}_{ii}(\lambda)]' = \mathbf{R}_{ii}(\lambda) \quad \text{and} \quad [\mathbf{P}_{ij}(\lambda)]' = \mathbf{P}_{ji}(\lambda).$$

Thus an elementary operator and its transpose are of the same form; $\mathbf{E}$- and $\mathbf{R}$-type operators are symmetric, and although a $\mathbf{P}$-type operator is not symmetric, its transpose is a $\mathbf{P}$-type operator.

c. Column operations

Let $\mathbf{T}$ be any one of the elementary operator matrices. Then, because $\mathbf{AT} = (\mathbf{T}'\mathbf{A}')'$ and $\mathbf{T}'$ is an elementary operator, $\mathbf{T}'\mathbf{A}'$ represents row operations on the rows of $\mathbf{A}'$, which are columns of $\mathbf{A}$ written as rows. Therefore $(\mathbf{T}'\mathbf{A}')' = \mathbf{AT}$ is just elementary operations carried out on columns of $\mathbf{A}$. Thus post-multiplication by elementary operators performs similar manipulations on the columns of $\mathbf{A}$ as are performed on the rows of $\mathbf{A}$ by pre-multiplication. With $\mathbf{E}$- and $\mathbf{R}$-type operators these operations are precisely the same, because $\mathbf{E}_{ij}$ and $\mathbf{R}_{ii}(\lambda)$ are symmetric. But with $\mathbf{P}$-type operators

$$\mathbf{AP}_{ji}(\lambda) = \left[\mathbf{P}'_{ji}(\lambda)\mathbf{A}'\right]' = [\mathbf{P}_{ij}(\lambda)\mathbf{A}']',$$

so that $\mathbf{AP}_{ji}(\lambda)$ does to columns of $\mathbf{A}$ what $\mathbf{P}_{ij}(\lambda)\mathbf{A}$ does to rows of $\mathbf{A}$.

Examples.

In

$$\mathbf{A}[\mathbf{P}_{21}(\lambda)] = \begin{bmatrix} 1 & 4 & 7 \\ 2 & 5 & 8 \\ 3 & 6 & 9 \end{bmatrix} \begin{bmatrix} 1 & 0 & 0 \\ \lambda & 1 & 0 \\ 0 & 0 & 1 \end{bmatrix} = \begin{bmatrix} 1+4\lambda & 4 & 7 \\ 2+5\lambda & 5 & 8 \\ 3+6\lambda & 6 & 9 \end{bmatrix},$$

$\lambda \times$ (column 2 of $\mathbf{A}$) is added to column 1, which is the same operation as $[\mathbf{P}_{12}(\lambda)]\mathbf{A}$ did to rows. In contrast, in

$$\mathbf{A}[\mathbf{P}_{12}(\lambda)] = \begin{bmatrix} 1 & 4 & 7 \\ 2 & 5 & 8 \\ 3 & 6 & 9 \end{bmatrix} \begin{bmatrix} 1 & \lambda & 0 \\ 0 & 1 & 0 \\ 0 & 0 & 1 \end{bmatrix} = \begin{bmatrix} 1 & \lambda+4 & 7 \\ 2 & 2\lambda+5 & 8 \\ 3 & 3\lambda+6 & 9 \end{bmatrix},$$

$\lambda \times$ (column 1 of $\mathbf{A}$) is added to column 2.

d. Inverses

As indicated in Section 4.4, all three of the elementary operator matrices have non-zero determinants. Hence all three have inverses. They are as follows:

$$\mathbf{E}_{ij}^{-1} = \mathbf{E}_{ij}, \quad [\mathbf{R}_{ii}(\lambda)]^{-1} = \mathbf{R}_{ii}(1/\lambda) \quad \text{and} \quad [\mathbf{P}_{ij}(\lambda)]^{-1} = \mathbf{P}_{ij}(-\lambda).$$

Thus an $\mathbf{E}$-type operator is the same as its inverse, and the effect of inverting $\mathbf{R}$- and $\mathbf{P}$-type operators is simply that of changing the constants, replacing λ by $1/\lambda$ in the $\mathbf{R}$-type and λ by $-\lambda$ in the $\mathbf{P}$-type.

In summary, the underlying form of any elementary operator is unchanged when it is either inverted or transposed.

7.5 Rank and the Elementary Operators

a. Rank

The rank of a matrix is unaffected when it is multiplied by an elementary operator; i.e., if the matrix $\mathbf{A}$ is multiplied by an E-, P- or R-type operator the rank of the product is the same as the rank of $\mathbf{A}$. We discuss the validity of this statement for each operator in turn. If $\mathbf{A}$ is pre-multiplied by an E-type operator the product $\mathbf{EA}$ is $\mathbf{A}$ with two rows interchanged. Hence $\mathbf{EA}$ is the same as $\mathbf{A}$ except with the rows in a different sequence. The number of rows that are linearly independent will be unaltered. Therefore $r(\mathbf{EA}) = r(\mathbf{A})$.

In the product of $\mathbf{A}$ and an R-type operator every element of some row (or column) of $\mathbf{A}$ is multiplied by a constant, and the product of a P-type operator with $\mathbf{A}$ is the same as $\mathbf{A}$ except it has a multiple of one row (column) added to another. In both cases the independence of rows (columns) is unaffected and the same number will be linearly independent in the product as in $\mathbf{A}$. Thus multiplication of any matrix by an elementary operator does not alter rank.

b. Products of elementary operators

We now use $\mathbf{E}$ to represent *any* elementary operator matrix (be it of the E-, P- or R-type) and $\mathbf{P}$ and $\mathbf{Q}$ to represent products of elementary operators; e.g., $\mathbf{P} = \mathbf{E}_3\mathbf{E}_2\mathbf{E}_1$ is a product of three of them. Then, because multiplication by an elementary operator does not affect rank we know that for any $\mathbf{A}$

$$r(\mathbf{A}) = r(\mathbf{E}_1\mathbf{A}) = r[\mathbf{E}_2(\mathbf{E}_1\mathbf{A})] = r[\mathbf{E}_3(\mathbf{E}_2\mathbf{E}_1\mathbf{A})] = r(\mathbf{PA}).$$

This result is used for finding the rank of any matrix $\mathbf{A}$. It is done by using elementary operators to change $\mathbf{A}$ until its rank is obvious.

c. Equivalence

When $\mathbf{A}$ is multiplied by elementary operator matrices, the product is said to be *equivalent* to $\mathbf{A}$. Whenever $\mathbf{B} = \mathbf{PAQ}$, for $\mathbf{P}$ and $\mathbf{Q}$ each being products of elementary operators, then $\mathbf{B}$ is equivalent to $\mathbf{A}$ and we write $\mathbf{B} \cong \mathbf{A}$. Since all elementary operator matrices have inverses, $\mathbf{P}^{-1}$ and $\mathbf{Q}^{-1}$ exist and so $\mathbf{A} = \mathbf{P}^{-1}\mathbf{B}\mathbf{Q}^{-1}$; furthermore, because $\mathbf{P}$ and $\mathbf{Q}$ are products of elementary operators whose inverses are themselves also elementary operators, $\mathbf{P}^{-1}$ and $\mathbf{Q}^{-1}$ are products of elementary operators too, and so $\mathbf{A} = \mathbf{P}^{-1}\mathbf{B}\mathbf{Q}^{-1}$ implies that $\mathbf{A}$ is equivalent to $\mathbf{B}$, i.e., $\mathbf{A} \cong \mathbf{B}$. Hence equivalence is a reflexive operation. Furthermore, $r_{\mathbf{A}} = r_{\mathbf{B}}$.

7.6 Calculating the Rank of a Matrix

a. Some special LIN vectors

A set of vectors such as

$$\mathbf{x}_1 = \begin{bmatrix} 4 \\ 0 \\ 0 \\ 0 \\ 0 \end{bmatrix}, \qquad \mathbf{x}_2 = \begin{bmatrix} 2 \\ 3 \\ 0 \\ 0 \\ 0 \end{bmatrix} \qquad \text{and} \qquad \mathbf{x}_3 = \begin{bmatrix} -1 \\ 2 \\ 7 \\ 0 \\ 0 \end{bmatrix}$$

is linearly independent because

$$a_1\mathbf{x}_1 + a_2\mathbf{x}_2 + a_3\mathbf{x}_3 = \begin{bmatrix} 4a_1 + 2a_2 - a_3 \\ 3a_2 + 2a_3 \\ 7a_3 \\ 0 \\ 0 \end{bmatrix} \qquad (7.4)$$

can equal $\mathbf{0}$ only if all the as are zero. Furthermore, any vector of order 5 having its only non-zero elements in its first three elements is a linear combination of $\mathbf{x}_1$, $\mathbf{x}_2$ and $\mathbf{x}_3$. This is so because equating (7.4), to such a vector, e.g.,

$$\begin{bmatrix} 4a_1 + 2a_2 - a_3 \\ 3a_2 + 2a_3 \\ 7a_3 \\ 0 \\ 0 \end{bmatrix} = \begin{bmatrix} 9 \\ 4 \\ 6 \\ 0 \\ 0 \end{bmatrix},$$

always yields solutions for the as that are not all zero. Hence, for example, in the matrix

$$\begin{bmatrix} 4 & 2 & -1 & 9 & 3 \\ 0 & 3 & 2 & 4 & 7 \\ 0 & 0 & 7 & 6 & 1 \\ 0 & 0 & 0 & 0 & 0 \\ 0 & 0 & 0 & 0 & 0 \end{bmatrix} \qquad (7.5)$$

all columns after the first three are linear combinations of those three, and so the matrix has rank 3; and this rank is, by the nature of the occurrence of the zeros, the number of non-null rows. Matrices of the form (7.5), with non-decreasing number of zeros in their rows (until, maybe, all elements are zero), are said to be matrices of *row echelon form* .

b.   Calculating rank

Matrices like (7.5) are the foundation for calculating the rank of any matrix $\mathbf{A}$. By row operations on $\mathbf{A}$ similar to those used in evaluating determinants, we reduce elements of $\mathbf{A}$ to zero until $\mathbf{A}$ has a form like (7.5)—all elements below the "diagonal" $a_{11}, a_{22}, a_{33} \ldots$ being zero. Because these row operations are identical to those represented by elementary operator matrices, the resulting matrix $\mathbf{B}$ is $\mathbf{B} = \mathbf{PA}$ for $\mathbf{P}$ being a product of elementary operators. Therefore $r_{\mathbf{A}}$, which we seek, equals $r_{\mathbf{B}}$, which by the form of $\mathbf{B}$ is immediately obvious. An example illustrates the procedure.

Example

$$\mathbf{A} = \begin{bmatrix} 1 & 2 & 4 & 3 \\ 3 & -1 & 2 & -2 \\ 5 & -4 & 0 & -7 \end{bmatrix} .$$

Carrying out the operations

$$\begin{array}{l} \text{row 2} \quad - \quad 3(\text{row 1}) \\ \text{row 3} \quad - \quad 5(\text{row 1}) \end{array} \quad \text{gives} \quad \mathbf{A} \cong \begin{bmatrix} 1 & 2 & 4 & 3 \\ 0 & -7 & -10 & -11 \\ 0 & -14 & -20 & -22 \end{bmatrix}$$

(7.6)

and then on this matrix the operation

$$\text{row 3} - 2(\text{row 2}) \quad \text{gives} \quad \mathbf{A} \cong \begin{bmatrix} 1 & 2 & 4 & 3 \\ 0 & -7 & -10 & -11 \\ 0 & 0 & 0 & 0 \end{bmatrix} = \mathbf{B}.$$

(7.7)

$\mathbf{B}$ is of the desired form, its rank is 2 and that is the rank of $\mathbf{A}$.

Finding the rank of a matrix by the procedure exemplified here is very simple—merely manipulate the rows of the matrix by adding multiples of them to each other to reduce the subdiagonal elements to zero. A formal description for the general case is available in many places, e.g., Searle (1982, p. 189). With some matrices the manipulation of rows will not be quite as arithmetically straightforward as in the preceding example. But the important thing is that it is always possible, for any real $\mathbf{A}$, to get $\mathbf{EA}$ in a form akin to (7.7), where $\mathbf{E}$ is a product of elementary operators and where the number of LIN rows in $\mathbf{EA}$ is obvious; this number is the rank of $\mathbf{EA}$, and the rank of $\mathbf{A}$.

7.7 Permutation Matrices

Proof of the theorem in Section 7.1 begins by assuming that all the LIN rows come first in a matrix, and the LIN columns likewise. But this is not always so. For example, in

$$\mathbf{M} = \begin{bmatrix} 1 & 1 & 3 \\ 1 & 1 & 3 \\ 4 & 4 & 12 \\ 2 & 2 & 5 \end{bmatrix}$$

there are two LIN rows and two LIN columns—but not the first two. Despite this, what are called permutation matrices are available for situations like that of the theorem in Section 7.1, to provide a mechanism for resequencing rows and columns in a matrix so that a matrix having k LIN rows can be resequenced into one having its first k rows and its first k columns LIN. Properties of the permutation matrices then allow many properties concerning linear independence of the resequenced matrices to also apply to the original matrix.

In Section 7.4 we defined an elementary operator $\mathbf{E}_{ij}$ which in pre-multiplying a matrix $\mathbf{A}$ interchanges the ith and jth rows of $\mathbf{A}$. In this context $\mathbf{E}_{ij}$ is called an *elementary permutation matrix*. It is $\mathbf{I}$ with its ith and jth rows interchanged. Now consider a product, $\mathbf{P}$, of several $\mathbf{E}_{ij}$-matrices. $\mathbf{P}$ will simply be an identity matrix, $\mathbf{I}$, after several successive interchanges of its rows: so $\mathbf{P}$ will be $\mathbf{I}$ with its rows rearranged, or permuted. Thus $\mathbf{P}$ is called a *permutation matrix*. And for any permutation matrix $\mathbf{P}$ (of appropriate order) $\mathbf{PA}$ is $\mathbf{A}$ with its rows permuted. Likewise $\mathbf{AP}$ is $\mathbf{A}$ with its columns permuted.

An important feature of $\mathbf{P}$ is that it is orthogonal (but not symmetric). $\mathbf{E}_{ij}$ is symmetric (by virtue of its definition) which means that $\mathbf{E}_{ij}\mathbf{E}'_{ij}$ is $\mathbf{E}_{ij}\mathbf{E}_{ij}$ and this is $\mathbf{I}$ because $\mathbf{E}_{ij}\mathbf{E}_{ij}$ is $\mathbf{E}_{ij}$ operating on $\mathbf{E}_{ij}$, which is $\mathbf{E}_{ij}$ with *its* ith and jth rows interchanged. All this leads to $\mathbf{PP}' = \mathbf{I}$ (and hence $\mathbf{P}'\mathbf{P} = \mathbf{I}$ because $\mathbf{P}$ is square). Therefore, for example, if

$$\mathbf{P} = \mathbf{E}_{ij}\mathbf{E}_{rs}\mathbf{E}_{qp} \quad \text{then} \quad \mathbf{P}^{-1} = \mathbf{P}' = \mathbf{E}_{qp}\mathbf{E}_{rs}\mathbf{E}_{ij},$$

and thus not only is $\mathbf{P}$ a permutation matrix but so also is $\mathbf{P}^{-1} = \mathbf{P}'$.

7.8   Matrix Factorization

An immediate consequence of the notion of rank is that a $p \times q$ matrix of rank $r \neq 0$ can be partitioned into a group of r independent rows and a group of $p - r$ rows that are linear combinations of the first group. This leads to useful factorizations.

a.   Matrices with linearly dependent columns

In Sections 6.2 and 6.3 we used the numerical example of

$$\mathbf{x}_1 = \begin{bmatrix} 3 \\ -6 \\ 9 \end{bmatrix}, \quad \mathbf{x}_2 = \begin{bmatrix} 0 \\ 5 \\ -5 \end{bmatrix}, \quad \mathbf{x}_3 = \begin{bmatrix} 2 \\ 1 \\ 1 \end{bmatrix} \quad \text{and} \quad \mathbf{x}_4 = \begin{bmatrix} -6 \\ 12 \\ -18 \end{bmatrix},$$

where $\mathbf{x}_1$ and $\mathbf{x}_2$ are LIN but $\mathbf{x}_1$, $\mathbf{x}_2$ and $\mathbf{x}_3$ are not, and neither are $\mathbf{x}_1$ and $\mathbf{x}_4$. This is so because

$$2\mathbf{x}_1 + 3\mathbf{x}_2 - 3\mathbf{x}_3 = 0 \quad \text{and} \quad 2\mathbf{x}_1 + \mathbf{x}_4 = 0.$$

Therefore

$$\mathbf{x}_3 = \tfrac{2}{3}\mathbf{x}_1 + \mathbf{x}_2 = [\mathbf{x}_1 \ \ \mathbf{x}_2]\begin{bmatrix} \tfrac{2}{3} \\ 1 \end{bmatrix} \quad \text{and} \quad \mathbf{x}_4 = -2\mathbf{x}_1 = [\mathbf{x}_1 \ \ \mathbf{x}_2]\begin{bmatrix} -2 \\ 0 \end{bmatrix}.$$

As a result the matrix $\mathbf{X}$ of all four vectors can be written as

$$\begin{aligned}
\mathbf{X} &= [\mathbf{x}_1 \ \ \mathbf{x}_2 \ \ \mathbf{x}_3 \ \ \mathbf{x}_4] \\
&= \left[\mathbf{x}_1 \ \ \mathbf{x}_2 \ \ (\mathbf{x}_1 \ \ \mathbf{x}_2)\begin{pmatrix} \tfrac{2}{3} \\ 1 \end{pmatrix} \ \ (\mathbf{x}_1 \ \ \mathbf{x}_2)\begin{pmatrix} -2 \\ 0 \end{pmatrix}\right] \\
&= \left[(\mathbf{x}_1 \ \ \mathbf{x}_2) \ \ (\mathbf{x}_1 \ \ \mathbf{x}_2)\begin{pmatrix} \tfrac{2}{3} & -2 \\ 1 & 0 \end{pmatrix}\right] \\
&= [\mathbf{X}_1 \ \ \mathbf{X}_1\mathbf{R}], \quad \text{for} \quad \mathbf{X}_1 = [\mathbf{x}_1 \ \ \mathbf{x}_2] \quad \text{and} \quad \mathbf{R} = \begin{bmatrix} \tfrac{2}{3} & -2 \\ 1 & 0 \end{bmatrix}.
\end{aligned}$$

The general concept of this result, $\mathbf{X} = [\mathbf{X}_1 \quad \mathbf{X}_1\mathbf{R}]$, is important: when $\mathbf{X}$ has rank r and its first r columns, to be denoted $\mathbf{X}_1$, are LIN (and using permutation matrices can always achieve that), then $\mathbf{X}$ can be written as $\mathbf{X} = [\mathbf{X}_1 \quad \mathbf{X}_1\mathbf{R}]$ for some $\mathbf{R}$. That we might not know $\mathbf{R}$ is seldom of concern; but it always exists. Likewise if $\mathbf{Y}$ has rank s and its first s rows, denoted $\mathbf{Y}_1$, are LIN one can always write

$$\mathbf{Y} = \begin{bmatrix} \mathbf{Y}_1 \\ \mathbf{L}\mathbf{Y}_1 \end{bmatrix} \quad \text{for some } \mathbf{L}.$$

b.   Full-rank factorization

The preceding partitioning can be extended to express $\mathbf{A}$ of rank r as

$$\mathbf{A}_{p\times q} = \mathbf{K}_{p\times r}\mathbf{L}_{r\times q} \tag{7.8}$$

where $\mathbf{K}$ and $\mathbf{L}$ have, respectively, full column and full row rank r. This is the *full rank factorization*. It can be developed as follows.

Ignoring the possible need for using permutation matrices we can, from the proof of the theorem in Section 7.1, assume that $\mathbf{A}$ can be partitioned as

$$\mathbf{A} = \begin{bmatrix} \mathbf{A}_{11} & \mathbf{A}_{12} \\ \mathbf{A}_{21} & \mathbf{A}_{22} \end{bmatrix} \tag{7.9}$$

with $\mathbf{A}_{11}$ square, of order r and non-singular, and with $[\mathbf{A}_{11} \quad \mathbf{A}_{12}]$ being r LIN rows. Hence, for some matrix $\mathbf{S}$

$$[\mathbf{A}_{21} \quad \mathbf{A}_{22}] = \mathbf{S}[\mathbf{A}_{11} \quad \mathbf{A}_{12}], \quad \text{so that} \quad \mathbf{S} = \mathbf{A}_{21}\mathbf{A}_{11}^{-1}. \tag{7.10}$$

Similarly, for some $\mathbf{R}$,

$$\begin{bmatrix} \mathbf{A}_{12} \\ \mathbf{A}_{22} \end{bmatrix} = \begin{bmatrix} \mathbf{A}_{11} \\ \mathbf{A}_{21} \end{bmatrix}\mathbf{R}, \quad \text{so that} \quad \mathbf{R} = \mathbf{A}_{11}^{-1}\mathbf{A}_{12}. \tag{7.11}$$

Also from (7.10) and (7.11)

$$\mathbf{A}_{22} = \mathbf{A}_{21}\mathbf{A}_{11}^{-1}\mathbf{A}_{12} = \mathbf{S}\mathbf{A}_{12} = \mathbf{A}_{21}\mathbf{R} \tag{7.12}$$

so that

$$\mathbf{A} = \begin{bmatrix} \mathbf{A}_{11} & \mathbf{A}_{12} \\ \mathbf{A}_{21} & \mathbf{A}_{21}\mathbf{A}_{11}^{-1}\mathbf{A}_{12} \end{bmatrix} = \begin{bmatrix} \mathbf{A}_{11} \\ \mathbf{A}_{21} \end{bmatrix}[\mathbf{I} \quad \mathbf{A}_{11}^{-1}\mathbf{A}_{12}] \tag{7.13}$$

$$= \begin{bmatrix} \mathbf{I} \\ \mathbf{A}_{21}\mathbf{A}_{11}^{-1} \end{bmatrix}[\mathbf{A}_{11} \quad \mathbf{A}_{12}]. \tag{7.14}$$

In (7.13) the left-hand matrix has full column rank r, from the nature of the partitioning in (7.9); and the right-hand matrix has full row rank r, as does any matrix of the form $[\mathbf{I}_r \quad \mathbf{M}]$. Complementary conclusions apply to (7.14). Thus each of (7.13') and (7.14') is of the form (7.8).

Example

Equation (7.11) is illustrated by

$$\mathbf{A} = \begin{bmatrix} 1 & 2 & 1 & 5 \\ 2 & 5 & 1 & 14 \\ 4 & 9 & 3 & 24 \end{bmatrix}$$

$$= \begin{bmatrix} \mathbf{I}_2 \\ (4 \ 9) \begin{pmatrix} 1 & 2 \\ 2 & 5 \end{pmatrix}^{-1} \end{bmatrix} \begin{bmatrix} 1 & 2 \\ 2 & 5 \end{bmatrix} \begin{bmatrix} \mathbf{I}_2 & \begin{pmatrix} 1 & 2 \\ 2 & 5 \end{pmatrix}^{-1} \begin{pmatrix} 1 & 5 \\ 1 & 14 \end{pmatrix} \end{bmatrix}$$

which reduces to (7.10) in the form

$$\mathbf{A} = \begin{bmatrix} 1 & 0 \\ 0 & 1 \\ 2 & 1 \end{bmatrix} \begin{bmatrix} 1 & 2 \\ 2 & 5 \end{bmatrix} \begin{bmatrix} 1 & 0 & 3 & -3 \\ 0 & 1 & -1 & 4 \end{bmatrix}$$

and this in turn gives the two equivalent forms of (7.12):

$$\mathbf{A} = \begin{bmatrix} 1 & 0 \\ 0 & 1 \\ 2 & 1 \end{bmatrix} \begin{bmatrix} 1 & 2 & 1 & 5 \\ 2 & 5 & 1 & 14 \end{bmatrix} = \begin{bmatrix} 1 & 2 \\ 2 & 5 \\ 4 & 9 \end{bmatrix} \begin{bmatrix} 1 & 0 & 3 & -3 \\ 0 & 1 & -1 & 4 \end{bmatrix}.$$

c.   Using permutation matrices

Development of (7.8) rests upon the first r rows (and r columns) of $\mathbf{A}$ in (7.9) being LIN. But suppose this is not the case. It is in just such a situation that permutation matrices play their part. Let $\mathbf{M} = \mathbf{PAQ}$ where $\mathbf{P}$ and $\mathbf{Q}$ are permutation matrices and where $\mathbf{M}$ has the form (7.9). Then $\mathbf{M}$ can be expressed as $\mathbf{M} = \mathbf{KL}$ as in (7.8). Therefore, on using the orthogonality of $\mathbf{P}$ and $\mathbf{Q}$ (e.g., $\mathbf{P}^{-1} = \mathbf{P}'$),

$$\mathbf{A} = \mathbf{P}^{-1}\mathbf{M}\mathbf{Q}^{-1} = \mathbf{P}'\mathbf{KL}\mathbf{Q}' = (\mathbf{P}'\mathbf{K})(\mathbf{L}\mathbf{Q}')$$

where $\mathbf{P}'\mathbf{K}$ and $\mathbf{L}\mathbf{Q}'$ here play the same roles as $\mathbf{K}$ and $\mathbf{L}$ do in (7.8). Hence (7.8) can be derived quite generally, for any matrix.

7.9 Results on Rank

We list here a number of results on rank, labeled for subsequent reference as theorems. Section 8.7 has proofs, some of which need canonical forms, the subject of Chapter 8.

Theorem 1: $r(\mathbf{AB}) \leq$ lesser of $r(\mathbf{A})$ and $r(\mathbf{B})$.

Theorem 2: For $\mathbf{A}$ non-singular, $r(\mathbf{AB}) = r(\mathbf{B})$.

Theorem 3: $r([\mathbf{A} \quad \mathbf{B}]) \leq r(\mathbf{A}) + r(\mathbf{B})$.

Theorem 4: $r(\mathbf{A} + \mathbf{B}) \leq r([\mathbf{A} \quad \mathbf{B}])$.

Theorem 5: $r(\mathbf{AB}) \geq r(\mathbf{A}) + r(\mathbf{B}) - n$, for $\mathbf{A}_{n \times n}$.

Theorem 6: For $\mathbf{M}_{n \times n}$ idempotent, $r(\mathbf{I} - \mathbf{M}) = n - r(\mathbf{M})$.

7.10 Vector Spaces

This section introduces the reader to ideas which are used extensively in many books on linear and matrix algebra. They involve extensions of geometry that we choose not to use, but which are nevertheless quite widespread in current matrix literature. We therefore provide a brief account of the underlying concepts as a convenient introductory reference for the reader. We confine ourselves, as usual, to real numbers and to vectors whose elements are real numbers.

a. Euclidean space

A vector $[x_0 \quad y_0]'$ of order 2 can be thought of as representing a point on a plane using familiar Cartesian x, y coordinates, as in Figure 7.1(a). Similarly, a vector $[x_0 \quad y_o \quad z_0]'$ of order 3 can represent a point in three-dimensional space: Figure 7.1(b). In similar manner a vector of order n can be said to represent a point in what is called *n-dimensional space*, or *n-space*. This appeal to geometry invites conceptualization of space as we ordinarily know it to more than three dimensions, a conceptualization that is not easy for some people. However, inasmuch as the set of all vectors of order 3 can be thought of as defining (all points in) 3-space, we use the word *n-space* to mean a representation of the set of all vectors of order n. Since it is a natural extension of familiar lines (1-space), planes (2-space) and the three-dimensional space in which we live, it is called *Euclidean n-space*. It will be denoted by R^n. Thus R^3 is the space of the world we live in.

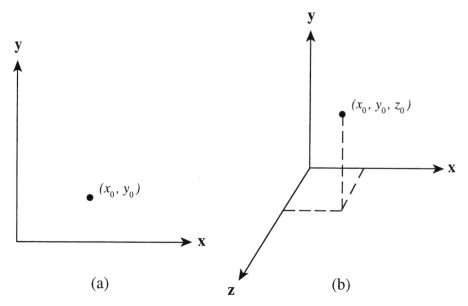

Figure 7.1: Representations of Euclidean 2-space and 3-space

b.   Vector spaces

R^n is a special case of a wider concept called a *vector space*, which we now define. Suppose x_i and x_j are two vectors in a set of vectors to be called S:

$$x_i \text{ and } x_j \text{ are in } S. \tag{7.15}$$

If, for every such pair x_i and x_j we also find that

$$x_i + x_j \text{ is in } S \tag{7.16}$$

and

$$ax_i \text{ is in } S, \text{ for all real scalars } a, \tag{7.17}$$

then S is said to be a *vector space*. When the vectors have order n, S is a vector space of order n.

Examples

Euclidean n-space is a vector space of order n: for every pair of vectors in R^n their sum is in R^n, and so is the product of a scalar and any vector that is in R^n.

Viewed as a set of vectors, $\mathbf{0}$ is a set with but a single vector in it and so (7.16) and (7.17) need be checked for the only possible choices,

$\mathbf{x}_i = \mathbf{x}_j = \mathbf{0}$; obviously they are satisfied. Hence $\mathbf{0}$ constitutes a vector space. More than that, $\mathbf{0}$ belongs to every vector space in R^n.

c. Spanning sets and bases

The defining conditions (7.15), (7.16) and (7.17) of a vector space ensure that all linear combinations of vectors that are in a vector space are also in the same vector space. Consider $\mathbf{x}_1$, $\mathbf{x}_2$ and $\mathbf{x}_3$ in some vector space S. Then by (7.17), for real values a_1, a_2 and a_3, we know that $a_1\mathbf{x}_1$, $a_2\mathbf{x}_2$ and $a_3\mathbf{x}_3$ are in S, as is $a_1\mathbf{x}_1 + a_2\mathbf{x}_2$ from (7.16); and similarly $(a_1\mathbf{x}_1 + a_2\mathbf{x}_2) + a_3\mathbf{x}_3 = a_1\mathbf{x}_1 + a_2\mathbf{x}_2 + a_3\mathbf{x}_3$ is also in S.

Suppose that every vector in a vector space S can be expressed as a linear combination of the t vectors $\mathbf{x}_1$, $\mathbf{x}_2, \ldots, \mathbf{x}_t$. Then this set of vectors is said to *span* or *generate* the vector space S. It is called a *spanning set* for S. When the vectors of a spanning set are also LIN then the set is said to be a *basis* for S. It is *a* basis and not *the* basis because for any S there are many bases. All bases of a space S have the same number of vectors and that number is called the *dimension* of S, or dim(S).

Example

$$\mathbf{x}_1 = \begin{bmatrix} 1 \\ 2 \\ 0 \end{bmatrix}, \qquad \mathbf{x}_2 = \begin{bmatrix} 1 \\ -1 \\ 0 \end{bmatrix} \qquad \text{and} \qquad \mathbf{x}_3 = \begin{bmatrix} 3 \\ 0 \\ 0 \end{bmatrix}$$

are in R^3. Any two of them form a basis for the vector space whose typical vector is $[\alpha \quad \beta \quad 0]'$ for α and β real. The dimension of the space is 2. (The space in this case is, of course, a subspace of R^3.)

A basis for any vector space S consists of LIN vectors, dim(S) of them in number; and all vectors in S are linear combinations of those vectors in the basis.

d. Many spaces of order n

It is important to realize that although there is only one Euclidean space of order n, there are many vector spaces of order n.

Example

Consider the set S_1 of vectors $\mathbf{x} = [\alpha \quad \beta \quad \gamma \quad 0 \quad 0]'$ for all real values α, β and γ. They satisfy (7.15), (7.16) and (7.17) for S_1. Hence S_1 is a vector space of order 5.

Clearly every vector in S_1 is also in R^5; but there are also vectors in R^5 that are not in S_1, e.g., $[2 \quad 1 \quad 7 \quad 3 \quad -4]'$. Thus S_1 and R^5 are different vector spaces, although both are of order 5.

e. Sub-spaces

Because all vectors of S_1 are also in R^5, and R^5 has vectors that are not in S_1, the set of vectors in S_1 is clearly a subset of those in R^5. Since S_1 is also a vector space it is therefore called a *vector sub-space* of R^5. All vector spaces of order n (involving vectors of real numbers) are sub-spaces of R^n, and some of them are also sub-spaces of others.

Example

Let S_2, S_3 and S_4 be the sets of vectors having the forms **x**, **y** and **z**, respectively:

$$\mathbf{x} = \begin{bmatrix} \alpha \\ 0 \\ 0 \end{bmatrix}, \quad \mathbf{y} = \begin{bmatrix} 0 \\ 0 \\ \beta \end{bmatrix} \quad \text{and} \quad \mathbf{z} = \begin{bmatrix} \gamma \\ 0 \\ \delta \end{bmatrix}, \quad \text{for real } \alpha, \beta, \gamma \text{ and } \delta.$$

Then S_2, S_3 and S_4 each define a vector space, and they are all sub-spaces of R^3. Furthermore, S_2 and S_3 are each a sub-space of S_4.

f. Range and null space of a matrix

A matrix **A** of rank $r_\mathbf{A}$ has $r_\mathbf{A}$ LIN columns. All vectors that are linear combinations of those independent columns form a vector space. It is known as the *column space* of **A**, *range* of **A** or *manifold* of **A**, and is often denoted by $\mathcal{R}(\mathbf{A})$. It has dimension $r_\mathbf{A}$:

$$r_\mathbf{A} = \dim[\mathcal{R}(\mathbf{A})].$$

Any set of $r_\mathbf{A}$ LIN columns of **A** is a basis for $\mathcal{R}(\mathbf{A})$.

The equations

$$\begin{bmatrix} 2 & 1 & 3 & -1 \\ 3 & 5 & 1 & 2 \end{bmatrix} \mathbf{x} = \mathbf{0}$$

have many solutions for $\mathbf{x}' = [x_1 \quad x_2 \quad x_3 \quad x_4]$; for example, $[-1 \quad 0 \quad 1 \quad 1]$ and $[-3 \quad 1 \quad 2 \quad 1]$ are solutions. More generally, there are numerous situations when $\mathbf{Ax} = \mathbf{0}$ has many solutions for **x**. When this is so, the space defined by those solution vectors **x** is called the *null space* of **A**, denoted by $\mathcal{N}(\mathbf{A})$. Its dimension is called the *nullity* of **A**:

$$\text{nullity}(\mathbf{A}) \equiv \dim[\mathcal{N}(\mathbf{A})].$$

Suppose $\mathbf{X'y} = \mathbf{0}$, i.e., $\mathbf{y'X} = \mathbf{0'}$. Then $\mathbf{y'x} = 0$ for every column **x** of **X**; i.e., $\mathbf{y'}$ and **x** are orthogonal. Let **Y** be the matrix of all vectors **y** satisfying $\mathbf{X'y} = \mathbf{0}$. Then for every **x** in $\mathcal{R}(\mathbf{X})$ and every **y** in $\mathcal{R}(\mathbf{Y})$ the vectors **x** and **y** are orthogonal and we say that $\mathcal{R}(\mathbf{X})$ is orthogonal to $\mathcal{R}(\mathbf{Y})$. This is sometimes written as $\mathcal{R}(\mathbf{X}) \perp \mathcal{R}(\mathbf{Y})$. Furthermore, because columns of **X** (i.e., rows of **X'**) and columns of **Y** are orthogonal, they are also LIN (see E 6.8). Therefore, if these columns have n elements, the theorem of Section 6.4c indicates that $r_\mathbf{Y} = n - r_\mathbf{X}$; i.e., $\dim(\mathbf{Y}) = n - r_\mathbf{X}$. Hence

$$\text{nullity}(\mathbf{X}) = n - r_\mathbf{X}.$$

7.11 Exercises

E 7.1. For $\mathbf{A} = \begin{bmatrix} 1 & 2 & 3 \\ 4 & 5 & 7 \\ 9 & 8 & 6 \end{bmatrix}$

and $\mathbf{E} = \begin{bmatrix} 0 & 1 & 0 \\ 1 & 0 & 0 \\ 0 & 0 & 1 \end{bmatrix}$, $\mathbf{R} = \begin{bmatrix} 1 & 0 & 0 \\ 0 & 1 & 0 \\ 0 & 0 & 2 \end{bmatrix}$ and $\mathbf{P} = \begin{bmatrix} 1 & 3 & 0 \\ 0 & 1 & 0 \\ 0 & 0 & 1 \end{bmatrix}$,

show that

(a) $\mathbf{EA} = \begin{bmatrix} 4 & 5 & 7 \\ 1 & 2 & 3 \\ 9 & 8 & 6 \end{bmatrix}$, $\mathbf{AE} = \begin{bmatrix} 2 & 1 & 3 \\ 5 & 4 & 7 \\ 8 & 9 & 6 \end{bmatrix}$, $\mathbf{EAE} = \begin{bmatrix} 5 & 4 & 7 \\ 2 & 1 & 3 \\ 8 & 9 & 6 \end{bmatrix}$;

(b) $\mathbf{PA} = \begin{bmatrix} 13 & 17 & 24 \\ 4 & 5 & 7 \\ 9 & 8 & 6 \end{bmatrix}$, $\mathbf{AP} = \begin{bmatrix} 1 & 5 & 3 \\ 4 & 17 & 7 \\ 9 & 35 & 6 \end{bmatrix}$, $\mathbf{PAP} = \begin{bmatrix} 13 & 56 & 24 \\ 4 & 17 & 7 \\ 9 & 35 & 6 \end{bmatrix}$;

(c) $\mathbf{REA} = \begin{bmatrix} 4 & 5 & 7 \\ 1 & 2 & 3 \\ 18 & 16 & 12 \end{bmatrix}$, $\mathbf{REAP} = \begin{bmatrix} 4 & 17 & 7 \\ 1 & 5 & 3 \\ 18 & 70 & 12 \end{bmatrix}$;

(d) $-|\mathbf{E}| = 1 = |\mathbf{P}| = \frac{1}{2}|\mathbf{R}|$;

(e) $\mathbf{E} = \mathbf{E}^{-1} = \mathbf{E'}$; and $\mathbf{R} = \mathbf{R'}$

(f) $\mathbf{R}^{-1} = \begin{bmatrix} 1 & 0 & 0 \\ 0 & 1 & 0 \\ 0 & 0 & \frac{1}{2} \end{bmatrix}$ and $\mathbf{P}^{-1} = \begin{bmatrix} 1 & -3 & 0 \\ 0 & 1 & 0 \\ 0 & 0 & 1 \end{bmatrix}$.

E 7.2. Find the rank of

$$\mathbf{A} = \begin{bmatrix} 1 & 6 \\ 2 & 9 \\ 4 & 3 \end{bmatrix}, \qquad \mathbf{B} = \begin{bmatrix} 3 & 7 & 6 \\ 2 & 1 & 7 \\ 4 & 6 & 3 \\ -1 & -1 & 0 \\ 6 & 8 & 3 \end{bmatrix},$$

$$\mathbf{C} = \begin{bmatrix} 1 & 1 & 0 & 2 \\ -1 & 3 & 6 & 0 \\ 1 & 5 & 6 & 4 \\ 6 & 4 & -3 & 11 \end{bmatrix}, \qquad \mathbf{D} = \begin{bmatrix} 1 & -1 & 0 \\ 3 & 2 & -4 \\ 5 & 0 & -4 \\ 1 & -6 & 4 \end{bmatrix}.$$

E 7.3. Express each matrix $\mathbf{A}$, $\mathbf{C}$ and $\mathbf{D}$ of E 7.2 as a product $\mathbf{XY}$, where the number of rows in $\mathbf{Y}$ is the rank of the matrix.

E 7.4. Define

$$\mathbf{A} = \begin{bmatrix} 1 & 2 & 4 & 0 \\ -2 & -3 & -1 & 1 \\ 0 & 1 & 7 & 1 \\ -2 & -2 & 6 & 2 \\ -3 & -6 & -12 & 0 \end{bmatrix}.$$

(a) Find the rank of $\mathbf{A}$.

(b) Define $\mathbf{B}$ as the matrix of two LIN rows of $\mathbf{A}$. Find $\mathbf{K}$ such that $\mathbf{KB}$ is the matrix of the other three rows of $\mathbf{A}$.

(c) Develop the results

$$\mathbf{A} = \begin{bmatrix} 1 & 0 \\ 0 & 1 \\ 2 & 1 \\ 2 & 2 \\ -3 & 0 \end{bmatrix} \begin{bmatrix} 1 & 2 & 4 & 0 \\ -2 & -3 & -1 & 1 \end{bmatrix} = \begin{bmatrix} -1 & \frac{1}{2} \\ 1 & 0 \\ -1 & 1 \\ 0 & 1 \\ 3 & -\frac{3}{2} \end{bmatrix} \begin{bmatrix} -2 & -3 & -1 & 1 \\ -2 & -2 & 6 & 2 \end{bmatrix}.$$

E 7.5. With the aid of permutation matrices calculate (7.8) for

$$\mathbf{A} = \begin{bmatrix} 2 & 2 & 3 & 1 \\ 2 & 2 & 3 & 1 \\ 4 & 4 & 6 & 2 \\ 1 & 1 & 1 & 1 \\ -1 & -1 & -1 & 3 \end{bmatrix}.$$

E 7.6. Express $\begin{bmatrix} 2 & -2 & 4 \\ 3 & -3 & 6 \end{bmatrix}$ in full rank factoring form.

E 7.7. Explain why a matrix of full row rank can always be written as a product of matrices one of which has the partitioned form $[\mathbf{I} \quad \mathbf{H}]$ for some matrix $\mathbf{H}$.

E 7.8. Prove, for $\mathbf{X}$ of full column rank, that $(\mathbf{X}'\mathbf{X})^{-1}$ exists.

E 7.9. For $\mathbf{A} = \mathbf{X}\mathbf{Y}'$ with $\mathbf{X}$ and $\mathbf{Y}$ of full column rank $r_{\mathbf{A}}$, prove that $\mathbf{A}$ and $\mathbf{A}^2$ have the same rank if and only if $\mathbf{Y}'\mathbf{X}$ is non-singular.

E 7.10. For $\mathbf{K}$ of full column rank, prove that $\mathbf{T}$ and $\mathbf{K}\mathbf{T}$ have the same rank.

E 7.11. Prove that the difference between two non-singular matrices has the same rank as the difference between their inverses.

E 7.12. When $\mathbf{A}\mathbf{B}$ is non-singular, prove that $\mathbf{A}$ has full row rank, and $\mathbf{B}$ has full column rank.

E 7.13. (Business cycle) Samuelson (1939) considers the interaction of the multiplier and accelerator in relation to business cycle theory. He presents the following model, where the symbols are scalars:

$$\begin{aligned} Y_t &= g_t + C_t + I_t, \\ C_t &= \alpha Y_{t-1}, \end{aligned}$$

$$I_t = \beta(C_t - C_{t-1}) = \alpha\beta Y_{t-1} - \alpha\beta Y_{t-2}$$

and

$$g_t = 1,$$

where, in period t, Y_t is national income, g_t is government expenditures, C_t is consumption expenditures and I_t is induced private investment.

(a) Find $\mathbf{A}$ such that

$$\mathbf{A}\mathbf{x}_t = \mathbf{y}_t$$

where

$$\mathbf{x}_t = \begin{bmatrix} Y_t \\ g_t \\ C_t \\ I_t \end{bmatrix} \quad \text{and} \quad \mathbf{y}_t = \begin{bmatrix} 0 \\ \alpha Y_{t-1} \\ \alpha\beta Y_{t-1} - \alpha\beta y_{t-2} \\ 1 \end{bmatrix}.$$

(b) What is the rank of $\mathbf{A}$?

(c) Solve the system for $\mathbf{x}_t$ if it is possible. If you can't solve for $\mathbf{x}_t$ explain why.

E 7.14. (Management) A hotel chain must establish uniform prices for its suite, double, queen and king accommodations. Expected sales, where columns represent accommodations and rows represent the four hotels, are shown in the following matrix:

$$\mathbf{A} = \begin{bmatrix} 335 & 2250 & 3345 & 4100 \\ 345 & 213 & 512 & 905 \\ 225 & 845 & 1250 & 540 \\ 450 & 1690 & 2500 & 1080 \end{bmatrix}.$$

The hotel chain revenue is

$$\mathbf{R} = \begin{bmatrix} 776,575 \\ 167,110 \\ 223,100 \\ 128,740 \end{bmatrix}.$$

(a) Assuming that sales are unaffected by price, use a computer to derive the prices the chain should charge for each type of room. If you are unable to do so, explain why.

(b) Suppose the fourth hotel had sales of 152, 125, 450 and 850 for its suite, double, queen and king accommodations, respectively. Using a computer, derive the prices the chain should charge for each type of room.

Chapter 8

CANONICAL FORMS

8.1 Introduction

This chapter deals with being able to take a matrix $\mathbf{A}$ and from it develop matrices $\mathbf{P}$ and $\mathbf{Q}$ such that

$$\mathbf{PAQ} = \begin{bmatrix} \mathbf{I}_r & \mathbf{0} \\ \mathbf{0} & \mathbf{0} \end{bmatrix} \equiv \mathbf{C} \tag{8.1}$$

where r is the rank of $\mathbf{A}$, and $\mathbf{C}$ of the form shown in (8.1) is known as a canonical form. There are several variations on the form of (8.1) (e.g., $\mathbf{Q}$ can be $\mathbf{P}'$ for symmetric $\mathbf{A}$), two of which are described here and two in Chapter 11. For calculating (8.1) one sometimes gets embroiled in fussy arithmetic, depending on the nature of the elements of $\mathbf{A}$, but that is of little concern in practice because in using (8.1) numerical values of $\mathbf{P}$ and $\mathbf{Q}$ are seldom needed. The important features of $\mathbf{P}$ and $\mathbf{Q}$ are that they *always* exist and that they are always non-singular, i.e., $\mathbf{P}^{-1}$ and $\mathbf{Q}^{-1}$ always exist. It is the certainty of the existence and non-singularity of $\mathbf{P}$ and $\mathbf{Q}$ which has been, and continues to be, so useful in the development of a variety of matrix properties that are valuable in applications.

8.2 Equivalent Canonical Form

a. Equivalent matrices

In Section 7.4 we described the three elementary operators and showed how using them as pre-multipliers of $\mathbf{A}$ affected the rows of $\mathbf{A}$, and post-multiplication affected the columns of $\mathbf{A}$. Denoting by $\mathbf{B}$ any product of $\mathbf{A}$ with elementary operators, we say that $\mathbf{A}$ is *equivalent* to $\mathbf{B}$ and write $\mathbf{A} \cong \mathbf{B}$. Thus if $\mathbf{B} = \mathbf{E}_1\mathbf{A}\mathbf{E}_2$ where $\mathbf{E}_1$ and $\mathbf{E}_2$ are products of elementary operators we have $\mathbf{A} \cong \mathbf{B}$. But $\mathbf{E}_1$ and $\mathbf{E}_2$, being products of elementary operators have inverses (Section 7.4d) which are products of elementary operators and so $\mathbf{A} = \mathbf{E}_1^{-1}\mathbf{B}\mathbf{E}_2^{-1}$, thus showing that $\mathbf{B} \cong \mathbf{A}$, i.e., $\mathbf{B}$ is equivalent to $\mathbf{A}$. In this way we see that equivalence is reflexive; i.e., $\mathbf{A} \cong \mathbf{B}$ and $\mathbf{B} \cong \mathbf{A}$ always go together. We therefore write $\mathbf{A} \cong \mathbf{B}$ to mean that $\mathbf{A}$ and $\mathbf{B}$ are equivalent, namely that simultaneously $\mathbf{A}$ is equivalent to $\mathbf{B}$ and $\mathbf{B}$ is equivalent to $\mathbf{A}$.

b.   Reduction to equivalent canonical form

– i.   Example

Section 7.6 shows how to use elementary row operations on a matrix to get it in a form where rank is easily established. Similarly, column operations can then yield the equivalent canonical form $\mathbf{C}$ of (8.1). We illustrate for the same matrix used in Section 7.6, namely

$$\mathbf{A} = \begin{bmatrix} 1 & 2 & 4 & 3 \\ 3 & -1 & 2 & -2 \\ 5 & -4 & 0 & -7 \end{bmatrix}. \qquad (8.2)$$

To develop $\mathbf{C}$ we first carry out on $\mathbf{A}$ the row operations (i) and (ii) shown in Table 8.1. That yields the rank of $\mathbf{A}$ as $r(\mathbf{A}) = 2$. Then we carry out on the row-amended $\mathbf{A}$, namely the matrix shown in (ii) of Table 8.1, the column operations (iii) and (iv) of Table 8.1, using "col" as an abbreviation for "column". After that $\mathbf{A}$ has been amended to be of the form

$$\mathbf{A} \cong \Delta = \begin{bmatrix} \mathbf{D}_r & \mathbf{0} \\ \mathbf{0} & \mathbf{0} \end{bmatrix} \qquad (8.3)$$

where $\mathbf{D}_r$ is a diagonal matrix of r non-zero elements. Δ is called the *diagonal form* of $\mathbf{A}$. It can easily be reduced to the canonical form $\mathbf{C}$, as shown in (v) of Table 8.1. Thus

$$\mathbf{A} \cong \mathbf{C} = \begin{bmatrix} 1 & 0 & 0 & 0 \\ 0 & 1 & 0 & 0 \\ 0 & 0 & 0 & 0 \end{bmatrix} = \begin{bmatrix} \mathbf{I}_2 & \mathbf{0} \\ \mathbf{0} & \mathbf{0} \end{bmatrix}, \qquad (8.4)$$

the expected result, since $\mathbf{A}$ has rank 2.

Table 8.1. Reducing the **A** of (8.2) to Equivalent Canonical Form

	Row Operations	**Effect on A**
(i)	row 2 − 3(row 1) row 3 − 5(row 1)	$\begin{bmatrix} 1 & 2 & 4 & 3 \\ 0 & -7 & -10 & -11 \\ 0 & -14 & -20 & -22 \end{bmatrix}$
(ii)	row 3 − 2(row 2)	$\begin{bmatrix} 1 & 2 & 4 & 3 \\ 0 & -7 & -10 & -11 \\ 0 & 0 & 0 & 0 \end{bmatrix}$

Column Operations

(iii)	col 2 − 2(col 1) col 3 − 4(col 1) col 4 − 3(col 1)	$\begin{bmatrix} 1 & 0 & 0 & 0 \\ 0 & -7 & -10 & -11 \\ 0 & 0 & 0 & 0 \end{bmatrix}$
(iv)	col 3 − (-10/-7)(col 2) col 4 − (-11/-7)(col 2)	$\begin{bmatrix} 1 & 0 & 0 & 0 \\ 0 & -7 & 0 & 0 \\ 0 & 0 & 0 & 0 \end{bmatrix} = \boldsymbol{\Delta}$
(v)	(col 2)×(1/-7)	$\begin{bmatrix} 1 & 0 & 0 & 0 \\ 0 & 1 & 0 & 0 \\ 0 & 0 & 0 & 0 \end{bmatrix} = \begin{bmatrix} \mathbf{I}_2 & \mathbf{0} \\ \mathbf{0} & \mathbf{0} \end{bmatrix} = \mathbf{C}$

In general, for $\mathbf{A}_{m \times n}$ of rank r, operations of this nature lead to

$$\mathbf{A}_{m \times n} \cong \begin{bmatrix} \mathbf{I}_r & \mathbf{0}_{r \times n-r} \\ \mathbf{0}_{m-r \times r} & \mathbf{0}_{m-r \times n-r} \end{bmatrix},$$

the zeros being null matrices of the orders shown. If $r = m < n$ the form is

$$[\mathbf{I}_m \quad \mathbf{0}]; \text{ it is } \begin{bmatrix} \mathbf{I}_n \\ \mathbf{0} \end{bmatrix} \text{ if } r = n < m \text{ and } \mathbf{I}_n \text{ if } r = m = n.$$

Implied in the above result is the existence of matrices $\mathbf{P}$ and $\mathbf{Q}$ such that $\mathbf{PAQ} = \mathbf{C}$ where $\mathbf{P}$ and $\mathbf{Q}$ are products of elementary operators.

If $\mathbf{A}$ is $m \times n$, $\mathbf{P}$ is square of order m and $\mathbf{Q}$ is square of order n, and

$$\mathbf{P}_{m \times m} \mathbf{A}_{m \times n} \mathbf{Q}_{n \times n} = \mathbf{C} = \begin{bmatrix} \mathbf{I}_r & \mathbf{0} \\ \mathbf{0} & \mathbf{0} \end{bmatrix}.$$

The matrix $\mathbf{C}$ is usually known as the *equivalent canonical form* of $\mathbf{A}$, or the *canonical form under equivalence*. The procedure for obtaining it is often called *reduction to canonical form under equivalence*.

When only the $\mathbf{C}$ matrix is required it can be found, as above, by carrying out elementary operations on $\mathbf{A}$ without explicitly deriving $\mathbf{P}$ and $\mathbf{Q}$. In situations when these are needed, however, they can be obtained from noting that $\mathbf{PAQ} = (\mathbf{PI})\mathbf{A}(\mathbf{IQ})$. Hence performing on identity matrices the same operations as performed on $\mathbf{A}$ to reduce it to canonical form produces the matrices $\mathbf{P}$ and $\mathbf{Q}$, grouping row operations together to get $\mathbf{P}$ and column operations to get $\mathbf{Q}$. Thus in the example we obtain $\mathbf{P}$ by applying operations (i) and (ii) to $\mathbf{I}_3$. Hence

$$\mathbf{I}_3 = \begin{bmatrix} 1 & 0 & 0 \\ 0 & 1 & 0 \\ 0 & 0 & 1 \end{bmatrix} \cong \begin{bmatrix} 1 & 0 & 0 \\ -3 & 1 & 0 \\ -5 & 0 & 1 \end{bmatrix} \cong \begin{bmatrix} 1 & 0 & 0 \\ -3 & 1 & 0 \\ 1 & -2 & 1 \end{bmatrix} = \mathbf{P}; \quad (8.5)$$

using $\mathbf{P}$ and $\mathbf{A}$ of (8.2) it will be found that $\mathbf{PA}$ is the matrix shown after operations (ii). Similarly, carrying out operations (iii), (iv) and (v) on $\mathbf{I}_4$ yields, after a little arithmetic,

$$\mathbf{Q} = \begin{bmatrix} 1 & 2/7 & -8/7 & 1/7 \\ 0 & -1/7 & -10/7 & -11/7 \\ 0 & 0 & 1 & 0 \\ 0 & 0 & 0 & 1 \end{bmatrix} \quad (8.6)$$

with which it will be found that

$$\mathbf{PAQ} = \begin{bmatrix} \mathbf{I}_2 & \mathbf{0} \\ \mathbf{0} & \mathbf{0} \end{bmatrix} = \mathbf{C} \quad (8.7)$$

of (8.4).

− ii.   The general case

Result (8.7) is an example of the quite general result given in the following theorem.

Theorem. Any non-null matrix **A** of rank r is equivalent to

$$\mathbf{PAQ} = \begin{bmatrix} \mathbf{I}_r & \mathbf{0} \\ \mathbf{0} & \mathbf{0} \end{bmatrix} = \mathbf{C}, \tag{8.8}$$

where $\mathbf{I}_r$ is the identity matrix of order r, and the null submatrices are of appropriate order to make **C** the same order as **A**. For **A** of order $m \times n$, **P** and **Q** are non-singular matrices of order m and n, respectively, being products of elementary operators.

The proof of (8.8), which is somewhat lengthy and couched largely in general discursive terms, is available in Searle (1982, Section 7.4c).

In the example of the preceding subsection, the arithmetic was easy. With some matrices, that is not always the case. Nevertheless, even with tricky arithmetic there are straightforward ways, using the elementary operators, to achieve (8.8). The exact manner in which they get used depends on the specific elements of **A**. That is why the proof of the theorem is so discursive. But neither that nor the specificity of **A** and the arithmetic values of **P** and **Q** really matter because, as has already been stated, it is the existence and non-singularity of **P** and **Q** that are so important, not their values for a specific **A**.

c. Non-uniqueness of P and Q

An important property of **P** and **Q** should be noted: for any **A**, the matrices **P** and **Q** are not unique. For example, column operation v could be eliminated and replaced by the row operation (row 2)$\times(1/\text{-}7)$. This would give, from (8.5),

$$\mathbf{P} = \begin{bmatrix} 1 & 0 & 0 \\ 3/7 & -1/7 & 0 \\ 1 & -2 & 1 \end{bmatrix}. \tag{8.9}$$

Also, with operation (v) on columns no longer being used, **Q** will be (8.6) but without the denominator 7 in its second column. With **Q** so amended and with **P** of (8.9) we will still have $\mathbf{PAQ} = \mathbf{C}$.

d. Existence and non-singularity of P and Q

The existence and non-singularity of **P** and **Q** are assured by the preceding theorem, the non-singularity arising from **P** and **Q** each being products of elementary operators. The existence of relationships among linearly dependent vectors rather than the actual values of those relationships was emphasized in Section 6.4c. The same is true here. It is the existence of $\mathbf{P}$ and $\mathbf{Q}$ that is much more important than their specific values. It is knowing that they always exist, without needing to know their specific values, that permits us to establish useful results. An example follows.

e. Full rank factorization

The full-rank factoring of a matrix developed in Section 7.8b is now stated formally as a theorem, and proved more easily by using the equivalent canonical form.

Theorem. $\mathbf{A}_{p \times q}$ of rank r can always be factored as

$$\mathbf{A} = \mathbf{K}_{p \times r} \mathbf{L}_{r \times q} \tag{8.10}$$

where $\mathbf{K}$ and $\mathbf{L}$ have full column and full row rank, respectively.

Proof. There exist non-singular matrices $\mathbf{P}$ and $\mathbf{Q}$ such that

$$\mathbf{PAQ} = \begin{bmatrix} \mathbf{I}_r & \mathbf{0} \\ \mathbf{0} & \mathbf{0} \end{bmatrix} \quad \text{and, equivalently,} \quad \mathbf{A} = \mathbf{P}^{-1} \begin{bmatrix} \mathbf{I}_r & \mathbf{0} \\ \mathbf{0} & \mathbf{0} \end{bmatrix} \mathbf{Q}^{-1}. \tag{8.11}$$

Partition $\mathbf{P}^{-1}$ and $\mathbf{Q}^{-1}$ as

$$\mathbf{P}^{-1} = \begin{bmatrix} \mathbf{K}_{p \times r} & \mathbf{W}_{p \times (p-r)} \end{bmatrix} \quad \text{and} \quad \mathbf{Q}^{-1} = \begin{bmatrix} \mathbf{L}_{r \times q} \\ \mathbf{Z}_{(q-r) \times q} \end{bmatrix}. \tag{8.12}$$

Then

$$\mathbf{A} = \begin{bmatrix} \mathbf{K} & \mathbf{W} \end{bmatrix} \begin{bmatrix} \mathbf{I}_r & \mathbf{0} \\ \mathbf{0} & \mathbf{0} \end{bmatrix} \begin{bmatrix} \mathbf{L} \\ \mathbf{Z} \end{bmatrix} = \begin{bmatrix} \mathbf{K} & \mathbf{0} \end{bmatrix} \begin{bmatrix} \mathbf{L} \\ \mathbf{Z} \end{bmatrix} = \mathbf{KL} = \mathbf{K}_{p \times r} \mathbf{L}_{r \times q}. \tag{8.13}$$

Since $\mathbf{P}$ is non-singular, so is $\mathbf{P}^{-1}$, and so the columns of $\mathbf{P}^{-1}$ are LIN. In particular, the r columns of $\mathbf{K}$ are LIN, and so $r_{\mathbf{K}} = r$; i.e., $\mathbf{K}$ has full column rank. (Of course, with $\mathbf{A}$ being $p \times q$ of rank r, it is clear that $r \leq p$, the number of rows in $\mathbf{K}$.) Similar reasoning for rows of $\mathbf{L}$ shows that $\mathbf{L}$ has full row rank.   Q.E.D.

Example

The example in Section 7.8b has

$$\mathbf{A} = \begin{bmatrix} 1 & 2 & 1 & 5 \\ 2 & 5 & 1 & 14 \\ 4 & 9 & 3 & 24 \end{bmatrix}, \quad \text{with} \quad \mathbf{PAQ} = \begin{bmatrix} \mathbf{I}_2 & \mathbf{0} \\ \mathbf{0} & \mathbf{0} \end{bmatrix}$$

for which

$$\mathbf{P} = \begin{bmatrix} 1 & 0 & 0 \\ -2 & 1 & 0 \\ -2 & -1 & 1 \end{bmatrix} \quad \text{with} \quad \mathbf{P}^{-1} = \begin{bmatrix} 1 & 0 & 0 \\ 2 & 1 & 0 \\ 4 & 1 & 1 \end{bmatrix}$$

and

$$\mathbf{Q} = \begin{bmatrix} 1 & -2 & -3 & 3 \\ 0 & 1 & 1 & -4 \\ 0 & 0 & 1 & 0 \\ 0 & 0 & 0 & 1 \end{bmatrix} \quad \text{with} \quad \mathbf{Q}^{-1} = \begin{bmatrix} 1 & 2 & 1 & 5 \\ 0 & 1 & -1 & 4 \\ 0 & 0 & 1 & 0 \\ 0 & 0 & 0 & 1 \end{bmatrix}.$$

Therefore by (8.13), using the first two columns of $\mathbf{P}^{-1}$ and the first two rows of $\mathbf{Q}^{-1}$,

$$\mathbf{A} = \mathbf{KL} = \begin{bmatrix} 1 & 0 \\ 2 & 1 \\ 4 & 1 \end{bmatrix} \begin{bmatrix} 1 & 2 & 1 & 5 \\ 0 & 1 & -1 & 4 \end{bmatrix} = \begin{bmatrix} 1 & 2 & 1 & 5 \\ 2 & 5 & 1 & 14 \\ 4 & 9 & 3 & 24 \end{bmatrix} = \mathbf{A}.$$

It is the product of a full column rank matrix pre-multiplying one of full row rank.

An important special case of this $\mathbf{A} = \mathbf{KL}$ factoring is when $\mathbf{A}$ is non-singular so that $\mathbf{C} = \mathbf{I}$ and hence

$$\mathbf{PAQ} = \mathbf{I} \quad \text{and so} \quad \mathbf{A} = \mathbf{P}^{-1}\mathbf{Q}^{-1};$$

this shows that every non-singular matrix is a product of elementary operators.

A very important consequence of this is that multiplication by a non-singular matrix does not alter rank; i.e., if $\mathbf{A}$ is non-singular, $r(\mathbf{AB}) = r(\mathbf{B})$. Why? See E 8.8.

8.3 Congruent Reduction

a. Symmetric matrices

Derivation of the canonical form is simplified for symmetric matrices, which frequently occur in situations where the canonical form is of particular use, notably the simplification of quadratic forms.

In working towards the reduction of $\mathbf{A}$ to canonical form we get $\mathbf{A}$ reduced to what is called the diagonal form of (8.3),

$$\boldsymbol{\Delta} = \begin{bmatrix} \mathbf{D}_r & \mathbf{0} \\ \mathbf{0} & \mathbf{0} \end{bmatrix}, \tag{8.14}$$

where $\mathbf{D}_r$ is a diagonal matrix of r non-zero diagonal elements. And, in view of the theorem in Section 8.2b-ii, this occurs for every $\mathbf{A}$ where $\mathbf{D}_r$ is a diagonal matrix of r non-zero elements. Consider the operations for deriving $\mathbf{\Delta}$ for a symmetric $\mathbf{A}$. Since each column of $\mathbf{A}$ is the same as the corresponding row, the same operations as are made on the rows of $\mathbf{A}$ to reduce the sub-diagonal elements to zero will, if performed on the columns of $\mathbf{A}$, reduce the elements above the diagonal to zeros also. Therefore, when $\mathbf{A}$ is symmetric, performing the same operations on columns as on rows reduces $\mathbf{A}$ to the diagonal form $\mathbf{\Delta}$. This means that in $\mathbf{PAQ}$ the matrix $\mathbf{Q}$ equals the transpose of $\mathbf{P}$, i.e., $\mathbf{Q} = \mathbf{P}'$. Hence we have $\mathbf{PAP}' = \mathbf{\Delta}$.

Suppose all diagonal elements of $\mathbf{D}_r$ are positive and from $\mathbf{D}_r$ we form the diagonal matrix $\mathbf{R}_r$, whose elements are the reciprocals of the square roots of the diagonal elements of $\mathbf{D}_r$, i.e., $\mathbf{R}_r^2 = \mathbf{D}_r^{-1}$. If $\mathbf{A}$ is of order n,

$$\mathbf{F} = \begin{bmatrix} \mathbf{R}_r & \mathbf{0} \\ \mathbf{0} & \mathbf{I}_{n-r} \end{bmatrix}, \tag{8.15}$$

then

$$(\mathbf{FP})\mathbf{A}(\mathbf{FP})' = \mathbf{FPAP}'\mathbf{F} = \mathbf{C} = \begin{bmatrix} \mathbf{I}_r & \mathbf{0} \\ \mathbf{0} & \mathbf{0} \end{bmatrix} \quad \text{since} \quad \mathbf{F}' = \mathbf{F}. \tag{8.16}$$

This means that, provided all elements of $\mathbf{D}_r$ are positive, a symmetric matrix can be reduced to canonical form under equivalence by pre-multiplying by a matrix $\mathbf{FP}$ and post-multiplying by its transpose $\mathbf{P}'\mathbf{F}$. This is called the *congruent reduction* of the symmetric matrix $\mathbf{A}$, and $\mathbf{C}$ is known as the *canonical form under congruence*.

Suppose now that not all diagonal elements of $\mathbf{D}_r$ are positive, but that q of them are negative. Then pre- and post-multiply $\mathbf{PAP}'$ by an E-type elementary operator so that the first $r - q$ diagonal elements of $\mathbf{D}_r$ are positive and the last q are negative. Denote the new product $\mathbf{P}^*\mathbf{AP}^{*'}$. Now define $\mathbf{F}$ as before except that $\mathbf{R}_r$ has as elements the reciprocals of the square roots of the elements of $\mathbf{D}_r$ disregarding sign. We then have

$$(\mathbf{FP}^*)\mathbf{A}(\mathbf{FP}^*)' = \mathbf{FP}^*\mathbf{AP}^{*'}\mathbf{F} = \begin{bmatrix} \mathbf{I}_{r-q} & \mathbf{0} & \mathbf{0} \\ \mathbf{0} & -\mathbf{I}_q & \mathbf{0} \\ \mathbf{0} & \mathbf{0} & \mathbf{0} \end{bmatrix}. \tag{8.17}$$

This is the general canonical form under congruence, for the symmetric matrix $\mathbf{A}$. When $q = 0$ it reduces to $\mathbf{C}$. The difference between the order of $\mathbf{I}_{r-q}$ and that of $\mathbf{I}_q$ is known as the *signature* of $\mathbf{A}$; i.e., signature $= r - 2q$. Retaining the negative signs in the diagonal means that the reduction to this form is entirely in terms of real numbers. If we are prepared to use imaginary numbers, involving $i = \sqrt{-1}$, then reduction to $\mathbf{C}$ can be made, using an $\mathbf{F}$ that involves imaginary numbers.

b. Example

For the symmetric matrix

$$\mathbf{A} = \begin{bmatrix} 4 & 12 \\ 12 & 27 \end{bmatrix}$$

the following operations reduce $\mathbf{A}$ to diagonal form.

Operations	**Effect on A**
(i) row 2 $-$ 3(row 1)	$\mathbf{A} \cong \begin{bmatrix} 4 & 12 \\ 0 & -9 \end{bmatrix}$
(ii) column 2 $-$ 3(column 1)	$\mathbf{A} \cong \begin{bmatrix} 4 & 0 \\ 0 & -9 \end{bmatrix}$

$\mathbf{P}$ is obtained by carrying out operation (i) on $\mathbf{I}_2$:

$$\mathbf{I}_2 = \begin{bmatrix} 1 & 0 \\ 0 & 1 \end{bmatrix} \cong \begin{bmatrix} 1 & 0 \\ -3 & 1 \end{bmatrix} = \mathbf{P}$$

and

$$\mathbf{PAP}' = \begin{bmatrix} 1 & 0 \\ -3 & 1 \end{bmatrix} \begin{bmatrix} 4 & 12 \\ 12 & 27 \end{bmatrix} \begin{bmatrix} 1 & -3 \\ 0 & 1 \end{bmatrix} = \begin{bmatrix} 4 & 0 \\ 0 & -9 \end{bmatrix}.$$

We then have

$$\mathbf{R} = \begin{bmatrix} 1/\sqrt{4} & 0 \\ 0 & 1/\sqrt{9} \end{bmatrix} = \begin{bmatrix} \frac{1}{2} & 0 \\ 0 & \frac{1}{3} \end{bmatrix}$$

and in this case $\mathbf{F} = \mathbf{R}$ so that

$$\mathbf{FP} = \mathbf{RP} = \begin{bmatrix} \frac{1}{2} & 0 \\ 0 & \frac{1}{3} \end{bmatrix} \begin{bmatrix} 1 & 0 \\ -3 & 1 \end{bmatrix} = \begin{bmatrix} \frac{1}{2} & 0 \\ -1 & \frac{1}{3} \end{bmatrix}.$$

Then

$$
\mathbf{FPA(FP)'} = \begin{bmatrix} \frac{1}{2} & 0 \\ -1 & \frac{1}{3} \end{bmatrix} \begin{bmatrix} 4 & 12 \\ 12 & 27 \end{bmatrix} \begin{bmatrix} \frac{1}{2} & -1 \\ 0 & \frac{1}{3} \end{bmatrix} = \begin{bmatrix} 1 & 0 \\ 0 & -1 \end{bmatrix},
$$

the desired form.

Result (8.17) is a method for dealing with negative elements of $\mathbf{D}_r$ separately from its positive elements. But then (8.17) is not of the form of (8.16). But if $\mathbf{D}_r$ has some negative elements, and the consequent occurrence of complex numbers (involving $i = \sqrt{-1}$) in $\mathbf{R}_r$ of $\mathbf{R}_r^2 = \mathbf{D}_r^{-1}$, is of no concern, then (8.16) will be achieved. In contrast, (8.17) avoids complex numbers.

8.4   Quadratic Forms

Quadratic forms $\mathbf{x'Ax}$ are discussed at some length in Section 3.6, wherein at equation (3.18) it is shown that $\mathbf{A}$ can always be taken as symmetric. A feature of $\mathbf{x'Ax}$, which is particularly important in statistics, is whether $\mathbf{x'Ax}$ (and hence $\mathbf{A}$) is n.n.d. (non-negative definite— see Section 3.6f) or not; and this is important in terms of properties of $\mathbf{A}$. The clue to this decision is whether or not $\mathbf{x'Ax}$ can be expressed as a sum of squares. Section 3.6f shows this for some easy examples but, with the aid of the congruent reduction $\mathbf{PAP'}$, this decision can easily be made for any quadratic form. We begin with an example.

Suppose our quadratic form is

$$
q = 4x_1^2 + 45x_2^2 + 136x_3^2 + 24x_1x_2 + 40x_1x_3 + 156x_2x_3 .
$$

With the aid of

$$
\mathbf{x} = \begin{bmatrix} x_1 \\ x_2 \\ x_3 \end{bmatrix} \quad \text{and} \quad \mathbf{A} = \begin{bmatrix} 4 & 12 & 20 \\ 12 & 45 & 78 \\ 20 & 78 & 136 \end{bmatrix} \tag{8.18}
$$

it can be written as

$$
q = \mathbf{x'Ax}. \tag{8.19}
$$

We now convert q to a sum of squares (of linear combinations of the

xs) by using the congruent canonical form of $\mathbf{A}$. With $\mathbf{P}$ being

$$\mathbf{P} = \begin{bmatrix} \frac{1}{2} & 0 & 0 \\ -1 & \frac{1}{3} & 0 \\ 1 & -2 & 1 \end{bmatrix} \quad \text{we get} \quad \mathbf{PAP}' = \begin{bmatrix} 1 & 0 & 0 \\ 0 & 1 & 0 \\ 0 & 0 & 0 \end{bmatrix}. \quad (8.20)$$

Now comes the crucial step. We convert (transform, in the language of Section 3.1) $\mathbf{x}$ to $\mathbf{y}$ taking

$$\mathbf{y} = (\mathbf{P}')^{-1}\mathbf{x} \quad (8.21)$$

equivalent to

$$\mathbf{x} = \mathbf{P}'\mathbf{y}. \quad (8.22)$$

Substituting (8.22) into (8.19) then gives

$$q = \mathbf{x}'\mathbf{A}\mathbf{x} = \mathbf{y}'\mathbf{PAP}'\mathbf{y} = \begin{bmatrix} y_1 & y_2 & y_3 \end{bmatrix} \begin{bmatrix} 1 & 0 & 0 \\ 0 & 1 & 0 \\ 0 & 0 & 0 \end{bmatrix} \begin{bmatrix} y_1 \\ y_2 \\ y_3 \end{bmatrix} = y_1^2 + y_2^2.$$

Hence with

$$\mathbf{y} = \begin{bmatrix} y_1 \\ y_2 \\ y_3 \end{bmatrix} = \mathbf{P}'^{-1}\mathbf{x} = \begin{bmatrix} 2 & 6 & 10 \\ 0 & 3 & 6 \\ 0 & 0 & 1 \end{bmatrix} \begin{bmatrix} x_1 \\ x_2 \\ x_3 \end{bmatrix} = \begin{bmatrix} 2x_1 + 6x_2 + 10x_3 \\ 3x_2 + 6x_3 \\ x_3 \end{bmatrix},$$

the quadratic form becomes

$$q = \mathbf{x}'\mathbf{A}\mathbf{x} = y_1^2 + y_2^2 = (2x_1 + 6x_2 + 10x_3)^2 + (3x_2 + 6x_3)^2. \quad (8.23)$$

Note that the rank of $\mathbf{A}$, the number of non-zero elements in $\mathbf{PAP}'$, is the number of squared terms in the final form of $\mathbf{x}'\mathbf{A}\mathbf{x}$.

8.5 Full Rank Factoring a Symmetric Matrix

Because in (8.16) the matrices $\mathbf{F}$ and $\mathbf{P}$ are non-singular,

$$\mathbf{A} = \mathbf{A}' = (\mathbf{FP})^{-1} \begin{bmatrix} \mathbf{I}_r & \mathbf{0} \\ \mathbf{0} & \mathbf{0} \end{bmatrix} [(\mathbf{FP})']^{-1}. \quad (8.24)$$

Partitioning $(\mathbf{FP})^{-1}$ as $[\mathbf{K}_{p \times r} \quad \mathbf{W}_{p \times (p-r)}]$, similar to the partitioning of $\mathbf{P}^{-1}$ in (8.12), therefore leads, as in (8.13) only with $\mathbf{L}$ now being $\mathbf{K}'$, to

$$\mathbf{A} = \mathbf{A}' = \mathbf{KK}' \quad (8.25)$$

for $\mathbf{K}_{p \times r}$ with $r = r(\mathbf{A})$.

But note: in (8.24) matrix $\mathbf{F}$ involves $\mathbf{R}$ of $\mathbf{R}^2 = \mathbf{D}^{-1}$. Therefore if $\mathbf{D}$ of (8.14) has negative elements, $\mathbf{R}$ will contain imaginary numbers, multiples of $i = \sqrt{-1}$. And thus $\mathbf{F}$ and $\mathbf{K}$ will also. Despite this, $\mathbf{A} = \mathbf{KK}'$ is real. But when $\mathbf{D}$ has only positive elements, $\mathbf{K}$ is real.

8.6   Non-negative Definite Matrices

Non-negative definite (n.n.d.) matrices, defined in Section 3.6, are symmetric matrices $\mathbf{A}$ such that $\mathbf{x}'\mathbf{Ax} \geq 0$ for all $\mathbf{x}$. If $\mathbf{x}'\mathbf{Ax} = 0$ only for $\mathbf{x} = \mathbf{0}$, then $\mathbf{A}$ is positive definite (p.d.), but if $\mathbf{x}'\mathbf{Ax} = 0$ for $\mathbf{x} = \mathbf{0}$ and for some $\mathbf{x} \neq \mathbf{0}$, then $\mathbf{A}$ is positive semi-definite (p.s.d.). We limit discussion to real matrices and vectors, for which $\mathbf{x}'\mathbf{x} > 0$ for $\mathbf{x} \neq \mathbf{0}$. Under these conditions n.n.d. matrices have a variety of properties, helpful in applications to statistics.

a.   Diagonal elements and principal minors

An important property of a p.(s.)d. matrix is that all its leading principal minors are positive (positive or zero). Proofs of this result, which for p.d. matrices is both a necessary and sufficient condition, are in many texts, but since most are lengthy none is given here. An elegant inductive proof in Seelye (1958) is commended to the interested reader.

Non-negative definiteness of $\mathbf{A}$ is a property that is not affected by changing the sequence of rows and columns in $\mathbf{A}$, because that merely amounts to relabeling the elements of $\mathbf{x}$ in $\mathbf{x}'\mathbf{Ax}$. Therefore the non-negativity of principal minors of a n.n.d. matrix $\mathbf{A}$ applies to all its diagonal elements (which, after suitable resequencing of rows and columns, are principal minors of order 1) and to $|\mathbf{A}|$ (which is the principal minor of largest order). Hence

$$\text{if } \mathbf{A} \text{ is p.d. }, \quad a_{ii} > 0 \quad \text{and} \quad |\mathbf{A}| > 0;$$
$$\text{if } \mathbf{A} \text{ is p.s.d.}, \quad a_{ii} \geq 0 \quad \text{and} \quad |\mathbf{A}| = 0 \qquad (8.26)$$

b.   Congruent canonical form

The diagonal form of an n.n.d. matrix never has negative elements, and so reduction to congruent canonical form involves only real numbers. This is so because $\mathbf{x}'\mathbf{Ax} = \mathbf{y}'\mathbf{PAP}'\mathbf{y}$ for $\mathbf{y} = \mathbf{Px}$, so that for any x which makes $\mathbf{x}'\mathbf{A}\mathbf{x}$ be zero, so also will $\mathbf{y}'\mathbf{P}\mathbf{A}\mathbf{P}'\mathbf{y}$ be zero; i.e., for $\mathbf{A}$ being n.n.d. so also is $\mathbf{P}\mathbf{A}\mathbf{P}'$. Therefore by (8.26) $\mathbf{D}_r$ has non-negative elements in

$$\mathbf{P}\mathbf{A}\mathbf{P}' = \begin{bmatrix} \mathbf{D}_r & \mathbf{0} \\ \mathbf{0} & \mathbf{0} \end{bmatrix}. \tag{8.27}$$

All this means that when $\mathbf{A}$ is n.n.d., the congruent canonical form will lead to $\mathbf{x}'\mathbf{A}\mathbf{x}$ being a sum of squares, as exemplified by (8.23).

c. Full-rank factorization

Because n.n.d. matrices are symmetric, they have the full-rank factorization $\mathbf{A} = \mathbf{K}\mathbf{K}'$ of (8.25). Since $\mathbf{D}_r$ of (8.27) has non-negative elements for n.n.d. matrices, $\mathbf{K}$ is always real for them too. Hence when

$$\mathbf{A} \text{ is n.n.d.}, \quad \mathbf{A} = \mathbf{K}\mathbf{K}' \text{ for } \mathbf{K} \text{ real and of full column rank.} \tag{8.28}$$

Furthermore, when $\mathbf{A}$ is p.d., $\mathbf{K}$ is non-singular.

These results have important applications in statistics where sums of squares are written as quadratic forms in n.n.d. and p.d. matrices. It is to be noted, however, that $\mathbf{D}_r$ having non-negative elements is a consequence of $\mathbf{A}$ being p.s.d., but the converse is not true.

d. Useful products

In both $\mathbf{A}_{p\times q} = \mathbf{K}\mathbf{L}$ for non-symmetric $\mathbf{A}$, and $\mathbf{A}_{p\times p} = \mathbf{K}\mathbf{K}'$ when $\mathbf{A}$ is symmetric, $\mathbf{K}$ is of full column rank $r = r(\mathbf{A})$. We show that $(\mathbf{K}'\mathbf{K})^{-1}$ exists.

With $\mathbf{K}$ being $p \times r$ of rank r (and ignoring the possible need for using permutation matrices) $\mathbf{K}$ can be partitioned as

$$\mathbf{K} = \begin{bmatrix} \mathbf{M}_{r\times r} \\ \mathbf{N}_{(p-r)\times r} \end{bmatrix} \quad \text{for non-singular } \mathbf{M}. \tag{8.29}$$

Thus

$$\mathbf{K} = \begin{bmatrix} \mathbf{I} \\ \mathbf{N}\mathbf{M}^{-1} \end{bmatrix} \mathbf{M} = \begin{bmatrix} \mathbf{I} \\ \mathbf{S} \end{bmatrix} \mathbf{M} \quad \text{for } \mathbf{S} = \mathbf{N}\mathbf{M}^{-1}.$$

Therefore

$$\mathbf{K}'\mathbf{K} = \mathbf{M}'(\mathbf{I} + \mathbf{S}'\mathbf{S})\mathbf{M}.$$

Now with $(\mathbf{I} + \mathbf{S}'\mathbf{S})$ being positive definite (see E 3.17c), and hence by (8.26) having a non-zero determinant, we have $(\mathbf{I} + \mathbf{S}'\mathbf{S})^{-1}$ existing.

Therefore with $\mathbf{M}^{-1}$ existing from (8.29)

$$(\mathbf{K}'\mathbf{K})^{-1} = \mathbf{M}^{-1}(\mathbf{I} + \mathbf{S}'\mathbf{S})^{-1}\mathbf{M}'^{-1}$$

exists. This result is, of course, true for any full column rank matrix
$\mathbf{K}$; and it is also true for $\mathbf{L}$ of full row rank, using $\mathbf{L}$ in place of $\mathbf{K}'$; i.e.,
$\mathbf{L}\mathbf{L}'$ is non-singular.

8.7 Proofs of Results on Rank

Section 7.9 lists without proof a number of theorems on rank.
Proofs are provided here.

Theorem 1. $r_{\mathbf{AB}} \leq$ the lesser of $r_{\mathbf{A}}$ and $r_{\mathbf{B}}$.

Proof. For $r_{\mathbf{A}} = r$ we know that non-singular $\mathbf{P}$ and $\mathbf{Q}$ exist such
that

$$\mathbf{PAQ} = \mathbf{C} = \begin{bmatrix} \mathbf{I}_r & \mathbf{0} \\ \mathbf{0} & \mathbf{0} \end{bmatrix} \quad \text{and equivalently} \quad \mathbf{PA} = \begin{bmatrix} \mathbf{I}_r & \mathbf{0} \\ \mathbf{0} & \mathbf{0} \end{bmatrix} \mathbf{Q}^{-1}.$$

Post-multiply both sides of the second equation by $\mathbf{B}$, and for $\mathbf{B}$ having
n columns partition $\mathbf{Q}^{-1}\mathbf{B}$ as $\mathbf{Q}^{-1}\mathbf{B} = \begin{bmatrix} \mathbf{T}_{r \times n} \\ \mathbf{S} \end{bmatrix}$ so that

$$\mathbf{PAB} = \begin{bmatrix} \mathbf{I}_r & \mathbf{0} \\ \mathbf{0} & \mathbf{0} \end{bmatrix} \mathbf{Q}^{-1}\mathbf{B} = \begin{bmatrix} \mathbf{I}_r & \mathbf{0} \\ \mathbf{0} & \mathbf{0} \end{bmatrix} \begin{bmatrix} \mathbf{T}_{r \times n} \\ \mathbf{S} \end{bmatrix} = \begin{bmatrix} \mathbf{T}_{r \times n} \\ \mathbf{0} \end{bmatrix}.$$

Then, because $\mathbf{P}$ is a product of elementary operators,

$$r_{\mathbf{AB}} = r_{\mathbf{PAB}} = r \begin{bmatrix} \mathbf{T}_{r \times n} \\ \mathbf{0} \end{bmatrix} \leq r \quad \text{because } \mathbf{T} \text{ has } r \text{ rows;}$$

i.e., $r_{\mathbf{AB}} \leq r_{\mathbf{A}}$, since $r = r_{\mathbf{A}}$. Similar reasoning, based on the canonical
form of $\mathbf{B}$, gives $r_{\mathbf{AB}} = r_{\mathbf{B}}$.   Q.E.D.

Even though this theorem does not involve an equality statement, its
utility is that it can often be applied twice to yield inequalities of the
form rank $\leq m$ and $m \leq$ rank and hence rank $= m$. This is illustrated
in some of the following proofs.

Theorem 2. Multiplication by a non-singular matrix does not alter rank.

Proof. Let $\mathbf{Q} = \mathbf{TA}$ for $\mathbf{T}$ non-singular. Then, by Theorem 1, $r_{\mathbf{Q}} \leq r_{\mathbf{A}}$. But $\mathbf{A} = \mathbf{T}^{-1}\mathbf{Q}$ so that Theorem 1 also gives $r_{\mathbf{A}} \leq r_{\mathbf{Q}}$. Hence $r_{\mathbf{Q}} = r_{\mathbf{A}}$. Q.E.D.

This theorem is used repeatedly where problems of rank are concerned.

Theorem 3. $r([\mathbf{A}\ \ \mathbf{B}]) \leq r(\mathbf{A}) + r(\mathbf{B})$.

Proof.

$$
\begin{aligned}
r([\mathbf{A}\ \ \mathbf{B}]) &= \text{number of LIN columns in } [\mathbf{A}\ \ \mathbf{B}] \\
&\leq \ (\text{number of LIN columns in } \mathbf{A}) \\
&\quad + \ (\text{number of LIN columns in } \mathbf{B}) \\
&\leq \ r(\mathbf{A}) + r(\mathbf{B}). \qquad \text{Q.E.D.}
\end{aligned}
$$

This theorem does not involve Theorem 1, but it is part of the next one, which does.

Theorem 4. $r([\mathbf{A} + \mathbf{B}]) \leq r(\mathbf{A}\ \ \mathbf{B}) \leq r(\mathbf{A}) + r(\mathbf{B})$.

Proof. Applying Theorem 1 to the identity $\mathbf{A} + \mathbf{B} \equiv [\mathbf{A}\ \ \mathbf{B}]\begin{bmatrix} \mathbf{I} \\ \mathbf{I} \end{bmatrix}$ gives $r(\mathbf{A} + \mathbf{B}) \leq r([\mathbf{A}\ \ \mathbf{B}])$. Q.E.D.

Theorem 4 is used to prove Theorem 5, which is used in Theorem 6.

Theorem 5. For $\mathbf{A}$ square, of order n, $r(\mathbf{AB}) \geq r(\mathbf{A}) + r(\mathbf{B}) - n$.

Proof. For non-singular $\mathbf{P}$ and $\mathbf{Q}$ such that

$$
\mathbf{PAQ} = \begin{bmatrix} \mathbf{I}_{r_{\mathbf{A}}} & 0 \\ 0 & 0 \end{bmatrix}, \quad \text{define} \quad \mathbf{X} = \mathbf{P}^{-1}\begin{bmatrix} 0 & 0 \\ 0 & \mathbf{I}_{n-r_{\mathbf{A}}} \end{bmatrix}\mathbf{Q}^{-1}.
$$

Then $\mathbf{A} + \mathbf{X} = \mathbf{P}^{-1}\mathbf{Q}^{-1}$ and so, starting with Theorem 2,

$$
\begin{aligned}
r(\mathbf{B}) &= r(\mathbf{P}^{-1}\mathbf{Q}^{-1}\mathbf{B}) = r(\mathbf{AB} + \mathbf{XB}) \\
&\leq \ r(\mathbf{AB}) + r(\mathbf{XB}) \ \text{ by Theorem 4} \\
&\leq \ r(\mathbf{AB}) + r(\mathbf{X}) \ \text{ because } r(\mathbf{XB}) \leq r(\mathbf{X}) \\
&\leq \ r(\mathbf{AB}) + n - r_{\mathbf{A}} \ \text{ by the nature of } \mathbf{X}. \qquad \text{Q.E.D.}
\end{aligned}
$$

This theorem is known as Sylvester's law of nullity.

Theorem 6. For $\mathbf{M}$ idempotent, of order n, $r(\mathbf{I} - \mathbf{M}) = n - r(\mathbf{M})$.

Proof. Because $\mathbf{I} = \mathbf{I} - \mathbf{M} + \mathbf{M}$, Theorem 4 gives $r(\mathbf{I}) \leq r(\mathbf{I} - \mathbf{M}) + r(\mathbf{M})$, so that $n - r(\mathbf{M}) \leq r(\mathbf{I} - \mathbf{M})$. But because $\mathbf{0} = (\mathbf{I} - \mathbf{M})\mathbf{M}$, Theorem 5 gives $r(\mathbf{0}) \leq r(\mathbf{I} - \mathbf{M}) + r(\mathbf{M}) - n$, which is $n - r(\mathbf{M}) \leq r(\mathbf{I} - \mathbf{M})$. Therefore $r(\mathbf{I} - \mathbf{M}) = n - r(\mathbf{M})$.   Q.E.D.

A useful result for left and right inverses (Section 5.11) is the following.

Theorem 7. A matrix has left inverses only if it has full column rank.

Proof. Let $\mathbf{L}$ be a left inverse of $\mathbf{A}_{r \times c}$. Then $\mathbf{LA} = \mathbf{I}$ or, equivalently, $\mathbf{I}_{c \times c} = \mathbf{L}_{c \times r}\mathbf{A}_{r \times c}$, so that Theorem 1 gives $r_{\mathbf{I}} = c \leq r_{\mathbf{A}}$. But, by definition of rank, $r_{\mathbf{A}} = r(\mathbf{A}_{r \times c}) \leq c$ and therefore $r_{\mathbf{A}} = c$, i.e., $\mathbf{A}$ has full column rank.   Q.E.D.

Converse. A matrix has right inverses only if it has full row rank.

Corollary. A matrix has both a left inverse and a right inverse only if it is non-singular; and each is the regular inverse.

These results on left and right inverses are consonant with Section 5.11 on the existence of such inverses. One form of the inverses is given in Section 10.3.

8.8 Exercises

E 8.1. Reduce the following matrices to diagonal form and thence to equivalent canonical form.

$$\mathbf{A}_1 = \begin{bmatrix} 7 & 13 \\ 2 & 9 \end{bmatrix}, \qquad \mathbf{A}_2 = \begin{bmatrix} 3 & 6 & 48 \\ 1 & 9 & 2 \\ 4 & 1 & 3 \end{bmatrix},$$

$$\mathbf{A}_3 = \begin{bmatrix} 1 & 0 & -1 \\ 3 & -4 & 2 \\ 5 & -4 & 0 \\ 1 & 4 & -6 \end{bmatrix}, \qquad \mathbf{A}_4 = \begin{bmatrix} 3 & 6 & 2 & 4 \\ 9 & 1 & 3 & 2 \\ 6 & -5 & 1 & -2 \end{bmatrix}.$$

E 8.2. Reduce the following symmetric matrices to diagonal form:

$$\mathbf{B}_1 = \begin{bmatrix} 4 & 4 & 10 \\ 4 & 20 & 18 \\ 10 & 18 & 29 \end{bmatrix}, \qquad \mathbf{B}_2 = \begin{bmatrix} 4 & 6 & 12 \\ 6 & 8 & 1 \\ 12 & 1 & 5 \end{bmatrix},$$

$$\mathbf{B}_3 = \begin{bmatrix} 4 & -2 & 0 \\ -2 & 3 & -2 \\ 0 & -2 & 2 \end{bmatrix}, \qquad \mathbf{B}_4 = \begin{bmatrix} 1 & 8 & 6 & 7 \\ 8 & 65 & 99 & 40 \\ 6 & 99 & 81 & 78 \\ 7 & 40 & 78 & 21 \end{bmatrix}.$$

E 8.3. For $\mathbf{A} = \mathbf{KL}$ of (8.10) explain why $\mathbf{KS(S^{-1}L)}$ yields, for different values of $\mathbf{S}$, different factorings similar to $\mathbf{KL}$.

E 8.4. Show that if the operations used on rows to reduce the elements below the diagonal of

$$\mathbf{A} = \begin{bmatrix} 1 & 2 & 3 \\ 2 & 1 & 2 \\ 3 & 2 & 4 \end{bmatrix}$$

to zero are then performed on columns of the resultant equivalent matrix, a diagonal matrix results. What are $\mathbf{P}$ and $\mathbf{Q}$ such that $\mathbf{PAQ} = \mathbf{\Delta}$ in this case?

E 8.5. Verify (8.16).

E 8.6. (a) Express

$$q = 3x_1^2 + 16x_2^2 + 139x_3^2 + 12x_1x_2 + 30x_1x_3 + 92x_2x_3$$

as a quadratic form $\mathbf{x'Ax}$ with $\mathbf{A} = \mathbf{A'}$. Find the congruent reduction of $\mathbf{A}$ and from it express q as a sum of squares of linear combinations of x_1, x_2 and x_3.

(b) Repeat (a) for

$$q = 4x_1^2 + 21x_2^2 + 116x_3^2 + 16x_1x_2 + 24x_1x_3 + 88x_2x_3$$

E 8.7. Prove, for real $\mathbf{X}$, that $r(\mathbf{X'X}) = r(\mathbf{X})$. (Hint: refer to Sections 7.8b and 8.6d.)

E 8.8. For non-singular $\mathbf{A}$, why is $r(\mathbf{AB}) = r(\mathbf{B})$?

E 8.9. If $\mathbf{A}$ is p.d., prove that (a) $\mathbf{A}^{-1}$ exists (b) $\mathbf{A}^{-1}$ is p.d.

Chapter 9

GENERALIZED INVERSES

A variation of inverse matrices for non-singular matrices is generalized inverses for rectangular and for singular matrices. Although the basic idea of generalized inverses seems to have originated with Moore (1920), it was Penrose (1955) and Rao (1962) who prompted wide interest in them, particularly in regard to their value in solving linear equations $\mathbf{Ax} = \mathbf{y}$ when $\mathbf{A}^{-1}$ does not exist. That widespread interest led to numerous papers and a spate of books in the 1970s. Indeed, 1971 saw publication of three books with almost identical titles, namely Boullion and Odell (1971), Pringle and Rayner (1971) and Rao and Mitra (1971). Another is Ben-Israel and Greville (1974). All four have titles starting with "Generalized Inverse"! Before that era the literature contains little on generalized inverses. But a reader wanting more than this chapter would do well to refer to those books.

9.1 Moore-Penrose Inverse

The starting point is what is known as the Moore-Penrose inverse, defined as follows.

Given any matrix $\mathbf{A}$, there is a unique matrix $\mathbf{M}$ such that

$$
\begin{array}{lll}
\text{(i)} & \mathbf{AMA} = \mathbf{A} & \text{(ii)} \quad \mathbf{MAM} = \mathbf{M} \\
\text{(iii)} & \mathbf{AM} \text{ is symmetric} & \text{(iv)} \quad \mathbf{MA} \text{ is symmetric.}
\end{array}
\tag{9.1}
$$

Penrose (1955), in developing (9.1) on foundations laid by Moore (1920), established not only the existence of $\mathbf{M}$ but also its uniqueness for a given $\mathbf{A}$. One way of writing $\mathbf{M}$ is based on the factoring of $\mathbf{A}_{p \times q}$ as $\mathbf{A} = \mathbf{KL}$ as in Section 7.8b, where $\mathbf{K}$ and $\mathbf{L}$ have full column and row rank, respectively, equal to $r_{\mathbf{A}}$. Then $\mathbf{M}$ of (9.1) is

$$\mathbf{M} = \mathbf{L}'(\mathbf{K}'\mathbf{AL}')^{-1}\mathbf{K}'. \tag{9.2}$$

Readers should confirm for themselves that (9.2) satisfies (9.1), using the non-singularity of $\mathbf{PP}'$ where $\mathbf{P}$ is any matrix of full row rank.

$\mathbf{M}$ is generally known as the *Moore-Penrose inverse* of $\mathbf{A}$; and the four conditions in (9.1) are usually called the *Penrose conditions*. For $\mathbf{A}$ of order $p \times q$, the order of $\mathbf{M}$ is $q \times p$. When $\mathbf{A}$ is non-singular, $\mathbf{K}$ and $\mathbf{L}$ are likewise, and the Moore-Penrose inverse is the regular inverse, $\mathbf{M} = \mathbf{A}^{-1}$.

9.2   Generalized Inverses

The matrix $\mathbf{M}$ defined by the four Penrose conditions in (9.1) is unique for a given $\mathbf{A}$. But there are many matrices $\mathbf{G}$ which satisfy just the first Penrose condition:

$$\mathbf{AGA} = \mathbf{A}. \tag{9.3}$$

Nevertheless, they are of such importance in solving linear equations that we direct most attention to those matrices $\mathbf{G}$ rather than to the Moore-Penrose inverse $\mathbf{M}$.

Any matrix $\mathbf{G}$ which satisfies (9.3) is called a *generalized inverse* of $\mathbf{A}$; it applies whether $\mathbf{A}$ is square or rectangular. In the latter case, $\mathbf{A}_{r \times c}$ leads to $\mathbf{G}_{c \times r}$. The name generalized inverse is now accepted almost universally, but that was not so in the early writings on generalized inverses.

Notice that $\mathbf{G}$ satisfying $\mathbf{AGA} = \mathbf{A}$ is described as "a" (not "the") generalized inverse of $\mathbf{A}$, because for any given $\mathbf{A}$ there is an infinite number of such matrices; with but one exception. That exception is when $\mathbf{A}$ is non-singular there is only one $\mathbf{G}$, namely $\mathbf{A}^{-1}$; thus

$$\text{for } \mathbf{A} \text{ non-singular: } \mathbf{G} = \mathbf{A}^{-1} = \mathbf{M}. \tag{9.4}$$

A common notation for a generalized inverse of $\mathbf{A}$ is $\mathbf{A}^-$, read as "$\mathbf{A}$ to the minus" or as "$\mathbf{A}$ minus" (*not* "minus one"). And $\mathbf{A}^+$ is commonly used for the Moore-Penrose inverse. Moreover, suppose $\mathbf{A}^*$ is a generalized inverse satisfying the first two Penrose conditions:

$$\mathbf{AA}^*\mathbf{A} = \mathbf{A} \qquad \text{and} \qquad \mathbf{A}^*\mathbf{AA}^* = \mathbf{A}^*. \tag{9.5}$$

Then $\mathbf{A}^*$ is said to be a *reflexive generalized inverse*. Given any generalized inverse $\mathbf{A}^-$ it is easy to create from it a reflexive generalized inverse in the form

$$\mathbf{A}^* = \mathbf{A}^- \mathbf{A} \mathbf{A}^-. \tag{9.6}$$

Occasionally one comes across the notation g_{12}-inverse for $\mathbf{A}^*$. The subscript on g indicates that $\mathbf{A}^*$ is a generalized inverse satisfying the first and second of the four Penrose conditions. Likewise g_{1234} is sometimes seen for $\mathbf{A}^+$, satisfying all four of the Penrose conditions.

9.3 Deriving a Generalized Inverse

Although there are quite a number of methods for deriving a generalized inverse of a known matrix $\mathbf{A}$, we give here but two of those methods—two which are easy to use.

a. Using the diagonal form

It is always possible, for any real matrix $\mathbf{A}$, to obtain $\mathbf{P}$ and $\mathbf{Q}$ as products of elementary operators such that the product $\mathbf{PAQ}$ is diagonal; i.e.,

$$\mathbf{PAQ} = \mathbf{\Delta} = \begin{bmatrix} \mathbf{D}_r & \mathbf{0} \\ \mathbf{0} & \mathbf{0} \end{bmatrix} \tag{9.7}$$

where $\mathbf{D}_r$ is diagonal of order $r = r_{\mathbf{A}}$ with all diagonal elements nonzero. Then

$$\mathbf{\Delta}^- = \begin{bmatrix} \mathbf{D}_r^{-1} & \mathbf{0} \\ \mathbf{0} & \mathbf{0} \end{bmatrix}$$

is a generalized inverse of $\mathbf{\Delta}$ with $\mathbf{\Delta}\mathbf{\Delta}^-\mathbf{\Delta} = \mathbf{\Delta}$; and

$$\mathbf{G} = \mathbf{Q} \begin{bmatrix} \mathbf{D}_r^{-1} & \mathbf{X} \\ \mathbf{Y} & \mathbf{Z} \end{bmatrix} \mathbf{P} \tag{9.8}$$

for any matrices $\mathbf{X}$, $\mathbf{Y}$ and $\mathbf{Z}$ of appropriate orders is a generalized inverse of $\mathbf{A}$ satisfying $\mathbf{AGA} = \mathbf{A}$. The reader should verify this (using $\mathbf{A} = \mathbf{P}^{-1}\mathbf{\Delta}\mathbf{Q}^{-1}$).

Example

$$\mathbf{PAQ} = \begin{bmatrix} 1 & 0 & 0 \\ 0 & 1 & 0 \\ 1 & -2 & 1 \end{bmatrix} \begin{bmatrix} 1 & 0 & -1 & 1 \\ 0 & 2 & 2 & 2 \\ -1 & 4 & 5 & 3 \end{bmatrix} \begin{bmatrix} 1 & 0 & 1 & -1 \\ 0 & 1 & -1 & -1 \\ 0 & 0 & 1 & 0 \\ 0 & 0 & 0 & 1 \end{bmatrix}$$

$$
= \begin{bmatrix} 1 & 0 & 0 & 0 \\ 0 & 2 & 0 & 0 \\ 0 & 0 & 0 & 0 \end{bmatrix} = \boldsymbol{\Delta}
$$

illustrates (9.7), so that from (9.8)

$$
\mathbf{G} = \begin{bmatrix} 1 & 0 & 1 & -1 \\ 0 & 1 & -1 & -1 \\ 0 & 0 & 1 & 0 \\ 0 & 0 & 0 & 1 \end{bmatrix} \begin{bmatrix} 1 & 0 & x_1 \\ 0 & \frac{1}{2} & x_2 \\ y_{11} & y_{12} & z_1 \\ y_{21} & y_{22} & z_2 \end{bmatrix} \begin{bmatrix} 1 & 0 & 0 \\ 0 & 1 & 0 \\ 1 & -2 & 1 \end{bmatrix} \tag{9.9}
$$

for any values of the xs, ys and zs is a generalized inverse of $\mathbf{A}$, meaning that $\mathbf{AGA} = \mathbf{A}$. The reader can easily verify this.

Note that the *only* way of establishing that $\mathbf{G}$ of (9.9) is a generalized inverse of $\mathbf{A}$ is to multiply out the product $\mathbf{AGA}$ and see that it yields $\mathbf{A}$. And this applies not only for $\mathbf{G}$ of (9.9) but for any matrix which is purported to be a generalized inverse of $\mathbf{A}$.

Note the occurrence of arbitrary $\mathbf{X}$, $\mathbf{Y}$ and $\mathbf{Z}$ in (9.8) illustrated with xs, ys and zs as elements in (9.9). These can be any values at all and the defining relationship $\mathbf{AGA} = \mathbf{A}$ still holds. The presence of these arbitrary values emphasizes the existence of many generalized inverses $\mathbf{G}$ for any particular $\mathbf{A}$. The nature of this arbitrariness is characterized further in Section 9.4. Only when $\mathbf{A}$ is square of full rank is there but one $\mathbf{G}$, and that is the regular inverse, $\mathbf{A}^{-1}$. The easiest form of $\mathbf{G}$ in (9.8) is, of course, with $\mathbf{X}$, $\mathbf{Y}$, $\mathbf{Z}$ null, giving

$$
\mathbf{G} = \mathbf{Q} \begin{bmatrix} \mathbf{D}^{-1} & \mathbf{0} \\ \mathbf{0} & \mathbf{0} \end{bmatrix} \mathbf{P}. \tag{9.10}
$$

b.   Inverting a sub-matrix

Suppose that in $\mathbf{A}_{p \times q}$ the leading principal sub-matrix $\mathbf{A}_{11}$ is non-singular of rank $r_\mathbf{A}$. Then a generalized inverse obtainable from partitioning $\mathbf{A}$ as

$$
\mathbf{A}_{p \times q} = \begin{bmatrix} \mathbf{A}_{11} & \mathbf{A}_{12} \\ \mathbf{A}_{21} & \mathbf{A}_{22} \end{bmatrix} \quad \text{is} \quad \mathbf{G}_{q \times p} = \begin{bmatrix} \mathbf{A}_{11}^{-1} & \mathbf{0} \\ \mathbf{0} & \mathbf{0} \end{bmatrix}, \tag{9.11}
$$

where the null matrices in $\mathbf{G}$ have appropriate order to make $\mathbf{G}$ be of order $q \times p$. Then verification of $\mathbf{AGA} = \mathbf{A}$ involves using $\mathbf{A}_{22} = \mathbf{A}_{21} \mathbf{A}_{11}^{-1} \mathbf{A}_{12}$, which comes from Section 7.8b.

Example

A generalized inverse of

$$\mathbf{A} = \begin{bmatrix} 1 & 2 & 4 & 3 \\ 3 & -1 & 2 & -2 \\ 5 & -4 & 0 & -7 \end{bmatrix} \quad \text{is} \quad \mathbf{G} = \tfrac{1}{7} \begin{bmatrix} 1 & 2 & 0 \\ 3 & -1 & 0 \\ 0 & 0 & 0 \\ 0 & 0 & 0 \end{bmatrix}.$$

c. A more general inversion

There is no need for the non-singular sub-matrix of order r to be in the principal leading position. It can be anywhere. Let $\mathbf{R}$ and $\mathbf{S}$ be permutation matrices (Section 7.7) such that $\mathbf{RAS}$ brings a non-singular sub-matrix of order r to the leading position; i.e.,

$$\mathbf{RAS} = \mathbf{B} = \begin{bmatrix} \mathbf{B}_{11} & \mathbf{B}_{12} \\ \mathbf{B}_{21} & \mathbf{B}_{22} \end{bmatrix} \tag{9.12}$$

with $\mathbf{B}_{11}$ being non-singular of order r. Then a generalized inverse of $\mathbf{B}$ is

$$\mathbf{F} = \begin{bmatrix} \mathbf{B}_{11}^{-1} & \mathbf{0} \\ \mathbf{0} & \mathbf{0} \end{bmatrix}, \quad \text{and} \quad \mathbf{G} = \mathbf{SFR} \tag{9.13}$$

is a generalized inverse of $\mathbf{A}$. Further details of this derivation are in Searle (1982, Sections 8.4 and 8.6).

9.4 Arbitrariness in a Generalized Inverse

The non-uniqueness of matrices $\mathbf{G}$ satisfying $\mathbf{AGA} = \mathbf{A}$ is evident in (9.8), which holds for any $\mathbf{X}$, $\mathbf{Y}$ and $\mathbf{Z}$ of appropriate order. More useful characterizations of arbitrariness, based on Urquhart (1969), are as follows. First, a generalization of (9.11) is that

$$\mathbf{G} = \begin{bmatrix} \mathbf{A}_{11}^{-1} - \mathbf{U}\mathbf{A}_{21}\mathbf{A}_{11}^{-1} - \mathbf{A}_{11}^{-1}\mathbf{A}_{12}\mathbf{V} - \mathbf{A}_{11}^{-1}\mathbf{A}_{12}\mathbf{W}\mathbf{A}_{21}\mathbf{A}_{11}^{-1} & \mathbf{U} \\ \mathbf{V} & \mathbf{W} \end{bmatrix} \tag{9.14}$$

is a generalized inverse of $\mathbf{A}$ for arbitrary $\mathbf{U}$, $\mathbf{V}$ and $\mathbf{W}$. An important consequence of (9.14) is the evidence it provides for the existence of non-singular generalized inverses of singular matrices. For example, suppose $\mathbf{U} = -\mathbf{A}_{11}^{-1}\mathbf{A}_{12}\mathbf{W}$, with $\mathbf{V} = \mathbf{0}$ and $\mathbf{W}$ non-singular. Then (9.14) gives

$$\mathbf{G} = \begin{bmatrix} \mathbf{A}_{11}^{-1} & \mathbf{A}_{11}^{-1}\mathbf{A}_{12}\mathbf{W} \\ \mathbf{0} & \mathbf{W} \end{bmatrix}, \tag{9.15}$$

which is non-singular. More generally, when $\mathbf{A}$ is rectangular and $\mathbf{W}$ has full row rank, then $\mathbf{G}$ of (9.15) has full row rank also, i.e., $r_{\mathbf{G}} > r_{\mathbf{A}}$.

A further characterization of the arbitrariness is that when $\mathbf{G}$ is a generalized inverse of $\mathbf{A}$ so also is

$$\mathbf{G}^* = \mathbf{GAG} + (\mathbf{I} - \mathbf{GA})\mathbf{T} + \mathbf{S}(\mathbf{I} - \mathbf{AG}) \tag{9.16}$$

for any matrices $\mathbf{T}$ and $\mathbf{S}$ of appropriate order. A consequence of (9.16) is the very useful result that given one particular generalized inverse $\mathbf{G}$, all others can be generated from it by using (9.16) for all possible arbitrary matrices $\mathbf{T}$ and $\mathbf{S}$. To see that this is so, suppose $\dot{\mathbf{G}}$ is some generalized inverse different from $\mathbf{G}$. Then putting $\mathbf{T} = \dot{\mathbf{G}}$ and $\mathbf{S} = \mathbf{GA}\dot{\mathbf{G}}$ into (9.16) yields $\mathbf{G}^*$ as $\dot{\mathbf{G}}$. The existence of a $\mathbf{T}$ and an $\mathbf{S}$ that achieve this shows that no matter what generalized inverse we have, it can always be written as (9.16) with an appropriate $\mathbf{T}$ and $\mathbf{S}$; i.e., (9.16) generates all generalized inverses of $\mathbf{A}$.

9.5   Symmetric Matrices

a.   Two general results

An all-important result for symmetric matrices is that if $\mathbf{G}$ is a generalized inverse of a symmetric matrix then so is $\mathbf{G}'$. This is so because for $\mathbf{A} = \mathbf{A}'$, transposing $\mathbf{AGA} = \mathbf{A}$ gives

$$\mathbf{AG}'\mathbf{A} = \mathbf{A}. \tag{9.17}$$

And, of course, this applies whether or not $\mathbf{G}$ is symmetric, and it does not have to be. This is evident from using the diagonal form

$$\mathbf{PAP}' = \begin{bmatrix} \mathbf{D}_r & \mathbf{0} \\ \mathbf{0} & \mathbf{0} \end{bmatrix} \tag{9.18}$$

to obtain a generalized inverse as

$$\mathbf{G} = \mathbf{P}' \begin{bmatrix} \mathbf{D}_r^{-1} & \mathbf{X} \\ \mathbf{Y} & \mathbf{Z} \end{bmatrix} \mathbf{P}. \tag{9.19}$$

We see at once that a generalized inverse of a symmetric matrix does not have to be symmetric.

b. Non-negative definite matrices

Section 8.5 shows that when $\mathbf{A}$ is symmetric, $\mathbf{A} = \mathbf{KK}'$ for $\mathbf{K}$ of full column rank. Then Section 8.6d shows that $\mathbf{K}'\mathbf{K}$ is non-singular, and so by using $\mathbf{K}$ in (9.2) in place of $\mathbf{L}$ the Moore-Penrose inverse of $\mathbf{A} = \mathbf{A}'$ is

$$\mathbf{A}^+ = \mathbf{K}(\mathbf{K}'\mathbf{AK})^{-1}\mathbf{K}' = \mathbf{K}(\mathbf{K}'\mathbf{K})^{-2}\mathbf{K}' \qquad (9.20)$$

where $(\mathbf{K}'\mathbf{K})^{-2}$ is, of course, $(\mathbf{K}'\mathbf{K})^{-1}$ squared.

Confining attention to n.n.d. matrices ensures that $\mathbf{K}$ is real and that $(\mathbf{K}'\mathbf{K})^{-1}$ exists. And, as in Section 8.3, the canonical form exists for all symmetric matrices, so leading to $\mathbf{A} = \mathbf{A}' = \mathbf{KK}'$. But for matrices which are not n.n.d., $\mathbf{K}$ may have $i = \sqrt{-1}$ in some elements. In that case $\mathbf{K}$ of $\mathbf{A} = \mathbf{KK}'$ will still have full column rank but the $\sqrt{-1}$ factor in $\mathbf{K}$ may lead to $\mathbf{K}'\mathbf{K}$ being singular. For example,

$$\mathbf{K} = \begin{bmatrix} i & 0 \\ 0 & i \\ 0 & 1 \\ 1 & 0 \end{bmatrix}$$

has full column rank but $\mathbf{K}'\mathbf{K} = \mathbf{0}$ has no inverse and so (9.20) would not exist.

c. The matrix $\mathbf{X}'\mathbf{X}$

The matrix $\mathbf{X}'\mathbf{X}$ has an important role in statistics where it arises in least squares equations $\mathbf{X}'\mathbf{Xb} = \mathbf{X}'\mathbf{y}$. Properties of generalized inverses of $\mathbf{X}'\mathbf{X}$ are therefore of interest.

Theorem. When $\mathbf{G}$ is any generalized inverse of $\mathbf{X}'\mathbf{X}$:

(i) $\mathbf{G}'$ is also a generalized inverse of $\mathbf{X}'\mathbf{X}$.

(ii) $\mathbf{XGX}'\mathbf{X} = \mathbf{X}$; i.e., $\mathbf{GX}'$ is a generalized inverse of $\mathbf{X}$.

(iii) $\mathbf{XGX}'$ is invariant to $\mathbf{G}$, i.e., is the same for every $\mathbf{G}$.

(iv) $\mathbf{XGX}'$ is symmetric, whether $\mathbf{G}$ is or not.

Proof. (i) is a consequence of (9.17). Applying (3.5) of Section 3.2c yields (ii). To prove (iii) suppose $\mathbf{F}$ is a generalized inverse of $\mathbf{X}'\mathbf{X}$ possibly different from $\mathbf{G}$. Then (ii) gives $\mathbf{XGX}'\mathbf{X} = \mathbf{XFX}'\mathbf{X}$ and (3.5) of Section 3.2c applied to this yields (iii). Finally, suppose $\mathbf{T}$ is a symmetric generalized inverse of $\mathbf{X}'\mathbf{X}$. ($\mathbf{T} = \mathbf{GX}'\mathbf{XG}'$ is always one possibility.) Then $\mathbf{XTX}'$ is symmetric. But $\mathbf{XGX}' = \mathbf{XTX}'$ by (iii) and so $\mathbf{XGX}'$ is symmetric too.   Q.E.D.

Corollary. $\mathbf{XG}'\mathbf{X}'\mathbf{X} = \mathbf{X}$ and $\mathbf{X}'\mathbf{XGX}' = \mathbf{X}'\mathbf{XG}'\mathbf{X}' = \mathbf{X}'$; and $\mathbf{XGX}' = \mathbf{XG}'\mathbf{X}'$. In the latter, note that both expressions are symmetric, even without $\mathbf{G}$ equaling $\mathbf{G}'$.

d.   Moore-Penrose inverses

Using (ii) of the preceding theorem enables one to show that for any real matrix $\mathbf{A}$,

$$\mathbf{A}^+ = \mathbf{A}'(\mathbf{AA}')^-\mathbf{A}(\mathbf{A}'\mathbf{A})^-\mathbf{A}'. \tag{9.21}$$

Then using (9.19) we can show that

$$\mathbf{XX}^+ = \mathbf{X}(\mathbf{X}'\mathbf{X})^-\mathbf{X}'. \tag{9.22}$$

Note that although $(\mathbf{X}'\mathbf{X})^-\mathbf{X}'$ is a generalized inverse of $\mathbf{X}$, it is not necessarily $\mathbf{X}^+$ as might be (wrongly) concluded from (9.22).

9.6   Exercises

E 9.1. Confirm that $\mathbf{M}$ of (9.2) satisfies (9.1).

E 9.2. Show that (9.6) satisfies (9.5).

E 9.3. Show that using $\mathbf{Z} = \mathbf{YDX}$ in (9.8) gives $\mathbf{G}$ as a reflexive generalized inverse.

E 9.4. Find a generalized inverse of each of the following matrices.

(a) $\begin{bmatrix} 2 & 1 & 4 \\ 6 & 9 & 3 \\ 4 & 4 & 5 \end{bmatrix}$
(b) $\begin{bmatrix} 1 & 0 & -1 & 2 \\ 3 & 1 & 2 & 1 \\ 4 & 3 & -2 & 1 \\ 13 & 11 & 3 & -5 \end{bmatrix}$

(c) $\begin{bmatrix} 1 & 2 & 1 & 2 \\ 1 & 3 & 2 & 1 \\ 0 & 1 & 1 & 1 \\ -1 & 2 & 3 & 1 \end{bmatrix}$
(d) $\begin{bmatrix} 4 & 3 \\ 1 & 2 \end{bmatrix}$

E 9.5. Determine which conditions in (9.1) are satisfied by each of your generalized inverses of the matrices in E 9.4.

E 9.6. Show that (9.14) is a generalized inverse of $\mathbf{A}$ as partitioned in (9.11). [Hint: use $\mathbf{A}_{22}$ of (7.12).]

E 9.7. Show that (9.19) is a generalized inverse of $\mathbf{A}$ occurring in (9.18).

E 9.8. Show that $\mathbf{M} = \mathbf{A}^+$ of (9.21) satisfies (9.1); and for symmetric $\mathbf{A}$ the $\mathbf{A}^+$ in (9.20) satisfies (9.1).

E 9.9. If $\mathbf{G}$ is a generalized inverse of $\mathbf{X}'\mathbf{X}$, show that $\mathbf{G}^*$, based on (9.6), is a symmetric, reflexive generalized inverse of $\mathbf{X}'\mathbf{X}$.

E 9.10. For $\mathbf{P} = \mathbf{I} - \mathbf{X}(\mathbf{X}'\mathbf{X})^-\mathbf{X}'$ prove that

(a) $\mathbf{P}$ is symmetric and idempotent.

(b) $\mathbf{PX} = \mathbf{0}$ and $\mathbf{X}'\mathbf{P} = \mathbf{0}$.

(c) For $\mathbf{X}$ of order $n \times k$, the rank of $\mathbf{P}$ is $n - r_{\mathbf{X}}$.

(d) Columns of $\mathbf{P}$ are orthogonal to, and linearly independent of, columns of $\mathbf{X}$.

(e) Row sums of $\mathbf{P}$ are zero when row sums of $\mathbf{X}$ are all the same and zero.

(f) Row sums of $\mathbf{P}$ are zero when $\mathbf{X}$ has $\mathbf{1}$ as a column.

Chapter 10

SOLVING LINEAR EQUATIONS

Linear equations in several unknowns have been represented in earlier chapters as $\mathbf{Ax} = \mathbf{y}$, where $\mathbf{x}$ is the vector of unknowns, $\mathbf{y}$ is a vector of known values and $\mathbf{A}$ is the matrix of coefficients. So long as $\mathbf{A}$ has an inverse the solution is $\mathbf{x} = \mathbf{A}^{-1}\mathbf{y}$, a situation demanding that $\mathbf{A}$ be square and that its rank equal its order (see Chapter 5). The more general case is now considered, that of having p equations in q unknowns,

$$\mathbf{A}_{p \times q}\mathbf{x}_{q \times 1} = \mathbf{y}_{p \times 1},$$

with the rank of $\mathbf{A}$ being $r_\mathbf{A} = r$. The instances when the solution is $\mathbf{x} = \mathbf{A}^{-1}\mathbf{y}$ are then particular cases of this, namely when $p = q = r$.

Throughout this chapter we deal with linear equations only, of the kind $3x_1 + 7x_2 + 2x_3 = 13$, i.e., linear in scalar unknowns, x_1, x_2 and x_3. At no time are equations involving powers of the unknowns considered. Thus by "equations" we always mean "linear equations".

10.1 Equations Having Many Solutions

Illustration (Sales)

In Table 2.1 the number of sales of three kinds of product to domestic professionals are shown as $\begin{bmatrix} 22 \\ 30 \\ 40 \end{bmatrix}$. And in equation (2.19) two sets of prices are matched up with these sales. Suppose there was a further increase in prices to \$10, \$40 and \$90 for the three kinds of products. Then the three possible sales revenues would be an extension of (2.19) to

$$\begin{bmatrix} 5 & 10 & 20 \\ 7 & 12 & 23 \\ 10 & 40 & 90 \end{bmatrix} \begin{bmatrix} 22 \\ 30 \\ 40 \end{bmatrix} = \begin{bmatrix} 1210 \\ 1434 \\ 5020 \end{bmatrix} \qquad (10.1)$$

where the first two rows are exactly (2.19).

Although not immediately obvious, it does not take much arithmetic to figure out that in the 3×3 matrix of (10.1), the third row is $16 \times$ (first row) $- 10 \times$ (second row); i.e.,

$$[10 \quad 40 \quad 90] = 16[5 \quad 10 \quad 20] - 10[7 \quad 12 \quad 23].$$

Therefore the equations

$$\begin{bmatrix} 5 & 10 & 20 \\ 7 & 12 & 23 \\ 10 & 40 & 90 \end{bmatrix} \begin{bmatrix} x_1 \\ x_2 \\ x_3 \end{bmatrix} = \begin{bmatrix} 1210 \\ 1434 \\ 5020 \end{bmatrix} \qquad (10.2)$$

have many solutions; and those solutions will be different values for the product sales which, at the same prices as shown in the 3×3 matrix, yield the same total sales as now. It turns out that four possible solutions are as shown in Table 10.1, where the first is as in (10.1) and the second is impossible because it contains a negative value, and negative product sales do not make sense (unless they represent returned product). By direct substitution it will be found that each set of values $\tilde{x}_1$, $\tilde{x}_2$, $\tilde{x}_3$ and $\tilde{x}_4$ is a solution.

Table 10.1. Four Solutions to Equations (10.1)

Unknowns	Solutions*			
$\mathbf{x}$	$\tilde{\mathbf{x}}_1$	$\tilde{\mathbf{x}}_2$	$\tilde{\mathbf{x}}_3$	$\tilde{\mathbf{x}}_4$
x_1	22	-6	4	13
x_2	30	100	75	52.5
x_3	40	12	22	31

*The notation $\tilde{\mathbf{x}}$ is used for a solution, to distinguish it from $\mathbf{x}$, the vector of unknowns.

Readers accustomed to thinking of a set of linear equations as having only a single solution may find this idea of many solutions somewhat novel. Nevertheless, it is a situation of wide occurrence, especially in the analysis of linear statistical models.

10.2 Consistent Equations

a. Definition

Consider the two equations

$$\begin{array}{rcrcl} x_1 & + & x_2 & = & 5 \\ 2x_1 & + & 2x_2 & = & 11. \end{array}$$

Obviously if one is true the other cannot be, for the second is incompatible, or inconsistent, with the first. Considered as a set of linear equations they are said to be *inconsistent*. Similarly, consider

$$\begin{bmatrix} 5 & 10 & 20 \\ 7 & 12 & 23 \\ 10 & 40 & 90 \end{bmatrix} \begin{bmatrix} x_1 \\ x_2 \\ x_3 \end{bmatrix} = \begin{bmatrix} 1210 \\ 1434 \\ 7888 \end{bmatrix}. \tag{10.3}$$

As noted earlier the third row of the 3×3 matrix is $16 \times$ (first row) $- 10 \times$ (second row). But this is not true of the right-hand side of equation (10.3). The third element there, 7888, is equal to $16 \times (1210) - 8(1434)$ and not $16 \times (1210) - 10(1434)$. This incompatibility with the relationship among the rows of the matrix leads to the equations in (10.3) being described as *inconsistent*. Equations in which linear relationships existing among the rows of $\mathbf{A}$ also hold among the corresponding elements of $\mathbf{y}$ are said to be *consistent*. A formal definition is as follows.

Definition. The linear equations $\mathbf{Ax} = \mathbf{y}$ are defined as consistent if linear relationships existing among the rows of $\mathbf{A}$ also exist among the corresponding elements of $\mathbf{y}$.

The definition does not require that linear relationships must exist among the rows of $\mathbf{A}$; but if they do, the same relationships must also exist among the corresponding elements of $\mathbf{y}$ for the equations to be consistent. When $\mathbf{A}$ has full row rank there are no linear relationships among the rows of $\mathbf{A}$ and therefore none that the elements of $\mathbf{y}$ must satisfy. Therefore equations $\mathbf{Ax} = \mathbf{y}$ with $\mathbf{A}$ of full row rank (including $\mathbf{A}$ non-singular) are always consistent.

b. Existence of solutions

The importance of the concept of consistency is that consistent equations can be solved and inconsistent equations cannot. This fact represents the starting point for a series of theorems on the solution of linear equations.

Theorem 1. A set of linear equations can be solved if, and only if, they are consistent.

Proof. Suppose $\mathbf{A}_{p \times q}$ has rank r. Rewrite $\mathbf{A}$ so that r LIN rows become the first r rows to be denoted by $\mathbf{B}$, and the remaining rows will then be $\mathbf{KB}$ for some $\mathbf{K}$. Also rewrite $\mathbf{y}$ so that its r elements corresponding to the rows of $\mathbf{B}$ are brought to be the first r elements of $\mathbf{y}$ and denote them by $\mathbf{y}_r$. Then, if the equations are consistent, the remaining elements will be $\mathbf{Ky}_r$. Thus the equations $\mathbf{Ax} = \mathbf{y}$ come to be

$$\mathbf{Ax} = \left[\begin{array}{c} \mathbf{B} \\ \mathbf{KB} \end{array} \right] \mathbf{x} = \left[\begin{array}{c} \mathbf{y}_r \\ \mathbf{Ky}_r \end{array} \right] = \mathbf{y}. \tag{10.4}$$

Clearly if $\tilde{\mathbf{x}}$ satisfies $\mathbf{B}\tilde{\mathbf{x}} = \mathbf{y}_r$ it also satisfies $\mathbf{KB}\tilde{\mathbf{x}} = \mathbf{Ky}_r$ and thus satisfies (10.4). Therefore it is only necessary to consider

$$\mathbf{Bx} = \mathbf{y}_r.$$

Now partition $\mathbf{B}$ as $[\mathbf{B}_1 \quad \mathbf{B}_2]$ with $\mathbf{B}_1$ non-singular, and $\mathbf{x}$ as $\mathbf{x}' = [\mathbf{x}_1' \quad \mathbf{x}_2']$ so that

$$\mathbf{Bx} = \mathbf{B}_1\mathbf{x}_1 + \mathbf{B}_2\mathbf{x}_2 = \mathbf{y}_r.$$

Then

$$\mathbf{x}_1 = \mathbf{B}_1^{-1}(\mathbf{y}_r - \mathbf{B}_2\mathbf{x}_2). \tag{10.5}$$

Therefore any vector $\mathbf{x}_2$, together with the corresponding $\mathbf{x}_1$ given by using that $\mathbf{x}_2$ in (10.5), constitute a solution $\mathbf{x}$ that satisfies $\mathbf{Ax} = \mathbf{y}$; i.e., if the equations are consistent they have a solution.

Conversely, suppose $\mathbf{x}_0$ is a solution. Then $\mathbf{x}_0$ satisfies (10.4) and the equations are consistent.   Q.E.D.

c.   Testing for consistency

Since it is only consistent equations which can be solved, it is essential to be able to ascertain if equations $\mathbf{Ax} = \mathbf{y}$ are consistent. They are if the matrix $[\mathbf{A} \quad \mathbf{y}]$ (known as the augmented matrix) has the same rank as $\mathbf{A}$. A proof of this is to be found in Searle (1982, Section 9.2c).

Alternatively, one can test for consistency using the diagonal matrix $\mathbf{PAQ}$. Specifically, if $\mathbf{A}$ is a matrix of p rows and rank r, and if $\mathbf{PAQ}$ is a diagonal form of $\mathbf{A}$, then the equations $\mathbf{Ax} = \mathbf{y}$ are consistent if and only if the last $p - r$ elements of $\mathbf{Py}$ are zero. To determine whether or not equations are consistent we calculate the rank of $\mathbf{A}$, namely r, by determining the number of non-zero elements in $\mathbf{PAQ}$. Then, knowing p and r, we investigate the last $p - r$ elements of $\mathbf{Py}$. If they are zero the equations are consistent. If any of those elements are non-zero the equations are not consistent.

10.3   Equations with Only One Solution

Consistent equations $\mathbf{Ax} = \mathbf{y}$ have just one solution when $\mathbf{A}$ has full column rank (for $\mathbf{A}$ being rectangular; or for $\mathbf{A}$ square and non-singular, for which the equations are always consistent, with solution $\mathbf{A}^{-1}\mathbf{y}$). Otherwise, for $\mathbf{A}$ of less than full column rank, there are many solutions. We begin with $\mathbf{A}$ of full column rank.

An easily obtained solution of $\mathbf{Ax} = \mathbf{y}$ when $\mathbf{A}$ has full column rank and the equations are partitioned as

$$\begin{bmatrix} \mathbf{A}_{11} \\ \mathbf{KA}_{11} \end{bmatrix} \mathbf{x} = \begin{bmatrix} \mathbf{y}_1 \\ \mathbf{Ky}_1 \end{bmatrix} \tag{10.6}$$

with $\mathbf{A}_{11}$ non-singular is

$$\tilde{\mathbf{x}} = \mathbf{A}_{11}^{-1}\mathbf{y}. \tag{10.7}$$

An alternative form of the *same* solution is

$$\tilde{\mathbf{x}} = \mathbf{Ly} \tag{10.8}$$

for $\mathbf{L}$ being any left inverse (see Section 5.11) of $\mathbf{A}$. And using $(\mathbf{A}'\mathbf{A})^{-1}\mathbf{A}'$ as $\mathbf{L}$ also gives

$$\tilde{\mathbf{x}} = (\mathbf{A}'\mathbf{A})^{-1}\mathbf{A}'\mathbf{y}. \tag{10.9}$$

Then, because $\tilde{\mathbf{x}}$ satisfies $\mathbf{Ax} = \mathbf{y}$ we have $\mathbf{A}\tilde{\mathbf{x}} = \mathbf{y}$ and so

$$\mathbf{A}(\mathbf{A}'\mathbf{A})^{-1}\mathbf{A}'\mathbf{y} = \mathbf{y}. \tag{10.10}$$

This result holds for any vector $\mathbf{y}$ which is a linear combination of columns of $\mathbf{A}$.

Notice that $\mathbf{A}$ being square and non-singular is a particular case of $\mathbf{A}$ having full column rank, in which case all three of (10.7), (10.8) and (10.9) reduce to $\mathbf{x} = \mathbf{A}^{-1}\mathbf{y}$, as one would expect. With this in mind we can say that whenever $\mathbf{A}$ has full column rank, consistent equations $\mathbf{Ax} = \mathbf{y}$ have just one solution; otherwise they have infinitely many solutions. If the equations are inconsistent they have no solution.

10.4   Solutions Using a Generalized Inverse

When there are many solutions to $\mathbf{Ax} = \mathbf{y}$ we make use of generalized inverse matrices. This occurs when $\mathbf{A}$ has less than full column rank, i.e., $\mathbf{A}$ square and singular, or $\mathbf{A}$ rectangular and rank less than the number of columns. The case of $\mathbf{A}$ having full row rank does not need to be treated separately.

a.   Obtaining one solution

There is a simple relationship between a generalized inverse of $\mathbf{A}$ and solution of consistent equations $\mathbf{Ax} = \mathbf{y}$, as in the following theorem.

Theorem 2. If $\mathbf{AGA} = \mathbf{A}$, then consistent equations $\mathbf{Ax} = \mathbf{y}$ with $\mathbf{y} \neq \mathbf{0}$ have $\tilde{\mathbf{x}} = \mathbf{Gy}$ as a solution.

This result is easily verified, as follows. If $\mathbf{AGA} = \mathbf{A}$ then $\mathbf{AGAx} = \mathbf{Ax}$; and $\mathbf{Ax} = \mathbf{y}$ means $\mathbf{AGy} = \mathbf{y}$, which shows that $\mathbf{Gy}$ is a solution.

Example

For the earlier equations (10.1),

$$\mathbf{Ax} = \mathbf{y} \equiv \begin{bmatrix} 5 & 10 & 20 \\ 7 & 12 & 23 \\ 10 & 40 & 90 \end{bmatrix} \mathbf{x} = \begin{bmatrix} 1210 \\ 1434 \\ 5020 \end{bmatrix} \qquad (10.11)$$

a generalized inverse of $\mathbf{A}$ is, from (9.11) of Section 9.3c,

$$\mathbf{G} = \begin{bmatrix} \left(\begin{matrix} 5 & 10 \\ 7 & 12 \end{matrix} \right)^{-1} & \mathbf{0} \\ \mathbf{0} & \mathbf{0} \end{bmatrix} = \begin{bmatrix} -1.2 & 1.0 & 0 \\ 0.7 & -0.5 & 0 \\ 0 & 0 & 0 \end{bmatrix}. \qquad (10.12)$$

Hence

$$\tilde{\mathbf{x}} = \mathbf{Gy} = \begin{bmatrix} -18 \\ 130 \\ 0 \end{bmatrix} \qquad (10.13)$$

is a solution, as is easily verified by direct substitution. Note that for our software company example this might not be a viable solution as there is a negative volume of sales.

Theorem 2 is the part of a theorem from Rao (1962) that is of great use: if (we calculate $\mathbf{G}$ such that) $\mathbf{AGA} = \mathbf{A}$, then $\mathbf{Gy}$ is a solution of consistent equations $\mathbf{Ax} = \mathbf{y}$. The full theorem includes the necessity statement that when $\mathbf{Gy}$ is a solution, then we have $\mathbf{AGA} = \mathbf{A}$; this is clearly less useful than Theorem 2, because we seldom have a solution $\mathbf{Gy}$ and want to know that $\mathbf{AGA} = \mathbf{A}$. Therefore we concentrate on Theorem 2. Readers interested in proving the necessity statement can find a proof of it in Searle (1982, p. 236).

Two results of occasional use are

$$\mathbf{AGy} = \mathbf{y} \qquad \text{and} \qquad \mathbf{GA\tilde{x}} = \mathbf{Gy}. \tag{10.14}$$

The first comes from pre-multiplying $\mathbf{Ax} = \mathbf{y}$ by $\mathbf{AG}$, and the second from pre-multiplying $\mathbf{A\tilde{x}} = \mathbf{y}$ by $\mathbf{G}$.

b. Obtaining many solutions

In $\mathbf{\tilde{x}} = \mathbf{Gy}$ of Theorem 2 we have, for any given $\mathbf{G}$, a single solution. From it

$$\mathbf{\tilde{x}} = \mathbf{Gy} + (\mathbf{GA} - \mathbf{I})\mathbf{z} \tag{10.15}$$

is a solution for any (and every) arbitrary vector $\mathbf{z}$ of appropriate order. Pre-multiplying (10.15) by $\mathbf{A}$ and using (10.14) shows that $\mathbf{A\tilde{x}} = \mathbf{y}$.

Illustration (Sales)

To calculate from (10.15) the values in Table 10.1 we use from (10.11), (10.12) and (10.13)

$$\mathbf{A} = \begin{bmatrix} 5 & 10 & 20 \\ 7 & 12 & 23 \\ 10 & 40 & 90 \end{bmatrix}, \mathbf{G} = \begin{bmatrix} \begin{pmatrix} 5 & 10 \\ 7 & 12 \end{pmatrix}^{-1} & \mathbf{0} \\ \mathbf{0} & \mathbf{0} \end{bmatrix} = \begin{bmatrix} -1.2 & 1.0 & 0 \\ 0.7 & -0.5 & 0 \\ 0 & 0 & 0 \end{bmatrix}$$

so that (10.15) gives

$$\begin{aligned}
\mathbf{\tilde{x}} &= \begin{bmatrix} -1.2 & 1.0 & 0 \\ 0.7 & -0.5 & 0 \\ 0 & 0 & 0 \end{bmatrix} \begin{bmatrix} 1210 \\ 1434 \\ 5020 \end{bmatrix} + \left[\begin{pmatrix} -1.2 & 1.0 & 0 \\ 0.7 & -0.5 & 0 \\ 0 & 0 & 0 \end{pmatrix} \begin{pmatrix} 5 & 10 & 20 \\ 7 & 12 & 23 \\ 10 & 40 & 90 \end{pmatrix} - \mathbf{I}_3 \right] \begin{bmatrix} z_1 \\ z_2 \\ z_3 \end{bmatrix} \\
&= \begin{bmatrix} -18 \\ 130 \\ 0 \end{bmatrix} + \begin{bmatrix} 0 & 0 & -1 \\ 0 & 0 & 2.5 \\ 0 & 0 & -1 \end{bmatrix} \begin{bmatrix} z_1 \\ z_2 \\ z_3 \end{bmatrix} \\
&= \begin{bmatrix} -18 - z_3 \\ 130 + 2.5z_3 \\ -z_3 \end{bmatrix}. \tag{10.16}
\end{aligned}$$

Using $z_3 = -40, -12, -22$ and -31 then gives the four solutions shown in Table 10.1.

Up to this point we have maintained that $\mathbf{Gy}$ is a solution of $\mathbf{Ax} = \mathbf{y}$ for $\mathbf{y} \neq \mathbf{0}$. Well, of course, for $\mathbf{y} = \mathbf{0}$ it is clear that $\mathbf{Gy} = \mathbf{0}$ is also a solution of $\mathbf{Ax} = \mathbf{0}$; but then in (10.15) we see that $(\mathbf{GA} - \mathbf{I})\mathbf{z}$ for arbitrary $\mathbf{z}$ is also a solution which provides as many non-null solutions as we want. Thus in the preceding example, if $\mathbf{y}$ were to be taken as $\mathbf{0}$ the solution (10.16) would be

$$\begin{bmatrix} -z_3 \\ 2.5z_3 \\ -z_3 \end{bmatrix}$$

for any values of z_3.

Armed with the ability to calculate an infinite number of solutions of consistent equations $\mathbf{Ax} = \mathbf{y}$, we need to be aware of the following properties those solutions have.

(i) $\tilde{\mathbf{x}} = \mathbf{Gy} + (\mathbf{GA} - \mathbf{I})\mathbf{z}$, for any $\mathbf{G}$, will generate all possible solutions by using all possible values of $\mathbf{z}$.

(ii) $\tilde{\mathbf{x}} = \mathbf{Gy}$ will generate all possible solutions by using all possible values for $\mathbf{G}$ satisfying $\mathbf{AGA} = \mathbf{A}$.

(iii) For any number of solutions $\tilde{\mathbf{x}}_t$, the sum $\Sigma_t \lambda_t \tilde{\mathbf{x}}_t$ is a solution when $\mathbf{y} = \mathbf{0}$; and when $\mathbf{y} \neq \mathbf{0}$ it is a solution if and only if $\Sigma_t \lambda_t = 1$.

(iv) When $\mathbf{x}$ has order q, the maximum number of linearly independent solutions is $q - r_\mathbf{A}$ when $\mathbf{y} = \mathbf{0}$, and $q - r_\mathbf{A} + 1$ when $\mathbf{y} \neq \mathbf{0}$.

(v) For every solution $\tilde{\mathbf{x}}$, the value of $\mathbf{k}'\tilde{\mathbf{x}}$ is the same if $\mathbf{k}' = \mathbf{k}'\mathbf{GA}$. This holds for each such $\mathbf{k}'$. It means for two solutions, $\tilde{\mathbf{x}}_1$ and $\tilde{\mathbf{x}}_2$, say, that if $\mathbf{k}' = \mathbf{k}'_1\mathbf{GA}$, then $\mathbf{k}_1\tilde{\mathbf{x}}_1 = \mathbf{k}'_1\tilde{\mathbf{x}}_2$. And for some other such $\mathbf{k}'$, say $\mathbf{k}'_2$, then $\mathbf{k}'_2\tilde{\mathbf{x}}_1 = \mathbf{k}'_2\tilde{\mathbf{x}}_2$. It does not mean that $\mathbf{k}'_1\tilde{\mathbf{x}}_1$ equals $\mathbf{k}'_2\tilde{\mathbf{x}}_1$.

These five properties are given as theorems 4-9 (with proofs and numerical examples) in Searle (1982, Sections 9.4-9.7).

10.5 Complete Example

Here we give a complete example. Consider $\mathbf{Ax} = \mathbf{y}$ as

$$\begin{bmatrix} 5 & 2 & -1 & 2 \\ 2 & 2 & 3 & 1 \\ 1 & 1 & 4 & -1 \\ 2 & -1 & -3 & -1 \\ 3 & 0 & 1 & -2 \end{bmatrix} \mathbf{x} = \begin{bmatrix} 7 \\ 9 \\ 5 \\ -6 \\ -1 \end{bmatrix}.$$

Row operations represented by

$$\mathbf{P} = \begin{bmatrix} 0 & 0 & 1 & 0 & 0 \\ 1 & 0 & -5 & 0 & 0 \\ 0 & 1 & -2 & 0 & 0 \\ -1 & 2 & -1 & 1 & 0 \\ -1 & 2 & -2 & 0 & 1 \end{bmatrix} \quad \text{give } \mathbf{PA} = \begin{bmatrix} 1 & 1 & 4 & -1 \\ 0 & -3 & -21 & 7 \\ 0 & 0 & -5 & 3 \\ 0 & 0 & 0 & 0 \\ 0 & 0 & 0 & 0 \end{bmatrix}$$

so that $r_{\mathbf{A}} = 3$, and the last two elements of $\mathbf{Py}$ are $-7+18-5-6 = 0$ and $-7 + 18 - 10 - 1 = 0$. Hence by Section 10.2c the equations are consistent, and can be solved. A generalized inverse of $\mathbf{A}$, which can be obtained from Section 9.3, is

$$\mathbf{G} = \tfrac{1}{15} \begin{bmatrix} 5 & -9 & 8 & 0 & 0 \\ -5 & 21 & -17 & 0 & 0 \\ 0 & -3 & 6 & 0 & 0 \\ 0 & 0 & 0 & 0 & 0 \end{bmatrix} \quad \text{with } \mathbf{GA} = \begin{bmatrix} 1 & 0 & 0 & -7/15 \\ 0 & 1 & 0 & 28/15 \\ 0 & 0 & 1 & -9/15 \\ 0 & 0 & 0 & 0 \end{bmatrix}. \quad (10.17)$$

Solutions given by (10.15) are therefore

$$\tilde{\mathbf{x}} = \tfrac{1}{15} \begin{bmatrix} 5 & -9 & 8 & 0 & 0 \\ -5 & 21 & -17 & 0 & 0 \\ 0 & -3 & 6 & 0 & 0 \\ 0 & 0 & 0 & 0 & 0 \end{bmatrix} \begin{bmatrix} 7 \\ 9 \\ 5 \\ -6 \\ -1 \end{bmatrix} + \begin{bmatrix} 0 & 0 & 0 & -7/15 \\ 0 & 0 & 0 & 28/15 \\ 0 & 0 & 0 & -9/15 \\ 0 & 0 & 0 & -1 \end{bmatrix} z$$

$$= \tfrac{1}{15} \begin{bmatrix} -6 - 7z \\ 69 + 28z \\ 3 - 9z \\ - 15z \end{bmatrix} \qquad (10.18)$$

where z is an arbitrary scalar.

Because $q = 4$ and $r_A = 3$, there are $q - r_A + 1 = 4 - 3 + 1 = 2$ linearly independent solutions: two possibilities are obtained by putting $z = 0$ and $z = -3$ in (10.18), giving

$$\tilde{x}_1 = \begin{bmatrix} -0.4 \\ 4.6 \\ 0.2 \\ 0 \end{bmatrix} \quad \text{and} \quad \tilde{x}_2 = \begin{bmatrix} 1 \\ -1 \\ 2 \\ 3 \end{bmatrix},$$

respectively. Any other solution will be a linear combination of these two with coefficients summing to unity. For example,

$$\tilde{x}_3 = 2\tilde{x}_1 - \tilde{x}_2 = \begin{bmatrix} -1.8 \\ 10.2 \\ -1.6 \\ -3.0 \end{bmatrix}$$

is a solution.

With an arbitrary vector $\mathbf{w}'$, values $\mathbf{k}'\tilde{x}$ invariant to $\tilde{x}$ are available using property (v) at the end of Section 10.4, for $\mathbf{k}'$ of the form

$$\mathbf{k}' = \mathbf{w}'\mathbf{G}\mathbf{A} = [w_1 \quad w_2 \quad w_3 \quad (-7w_1 + 28w_2 - 9w_3)/15]; \quad (10.19)$$

and from (10.18) the value of $\mathbf{k}'\tilde{x}$ is then

$$\mathbf{k}'\tilde{x} = \mathbf{k}'\mathbf{G}\mathbf{y} = (-6w_1 + 69w_2 + 3w_3)/15. \quad (10.20)$$

There are $r = 3$ LIN vectors of the form (10.19), examples being $\mathbf{k}'_1 = [3 \quad 0 \quad 1 \quad -2]$, $\mathbf{k}'_2 = [4 \quad 1 \quad 5 \quad -3]$ and $\mathbf{k}'_3 = [1 \quad 4 \quad 0 \quad 7]$, the corresponding invariant values from (10.20) being $\mathbf{k}'_1\tilde{x} = -1$, $\mathbf{k}'_2\tilde{x} = 4$ and $\mathbf{k}'_3\tilde{x} = 18$. Any other $\mathbf{k}'$ of the form (10.19) will be a linear combination of $\mathbf{k}'_1$, $\mathbf{k}'_2$ and $\mathbf{k}'_3$ and the corresponding value of $\mathbf{k}'\tilde{x}$ will be the same linear combination of $\mathbf{k}'_1\tilde{x}$, $\mathbf{k}'_2\tilde{x}$ and $\mathbf{k}'_3\tilde{x}$. For example, for

$$\mathbf{k}'_4 = [11 \quad -1 \quad 30 \quad -25] = -5\mathbf{k}'_1 + 7\mathbf{k}'_2 - 2\mathbf{k}'_3,$$

$$\mathbf{k}'_4\tilde{x} = -5(\mathbf{k}'_1\tilde{x}) + 7(\mathbf{k}'_2\tilde{x}) - 2(\mathbf{k}'_3\tilde{x}) = -5(-1) + 7(4) - 2(18) = -3.$$

The procedures described here may appear somewhat lengthy for the small examples chosen, but advantages become apparent when dealing with a large number of equations, say 20 or more. At all stages, the matrix operations are readily amenable to computer processing.

10.6 Exercises

E 10.1. Find a set of LIN solutions for each of the following sets of
equations and in each case show that any other solution is a
linear combination of the solutions you have obtained.

(a) $\begin{bmatrix} 6 & 2 & 0 \\ -1 & 0 & 3 \\ 3 & 2 & 9 \end{bmatrix} \mathbf{x} = \begin{bmatrix} 8 \\ 2 \\ 14 \end{bmatrix}$ (b) $\begin{bmatrix} 1 & 0 & 2 \\ 4 & 1 & 7 \\ 3 & 2 & 3 \end{bmatrix} \mathbf{x} = \begin{bmatrix} 12 \\ 46 \\ 27 \end{bmatrix}$

(c) $\begin{bmatrix} 1 & 1 & 0 & 1 \\ -1 & -1 & 1 & 1 \\ 1 & 0 & 0 & 1 \\ -1 & 0 & 1 & 1 \end{bmatrix} \mathbf{x} = \begin{bmatrix} 8 \\ -1 \\ 6 \\ 1 \end{bmatrix}$

(d) $\begin{bmatrix} 4 & -9 & -1 & 2 \\ 3 & 1 & 0 & 1 \\ 10 & -7 & -1 & 4 \\ 25 & -2 & -1 & 9 \end{bmatrix} \mathbf{x} = \begin{bmatrix} 7 \\ 5 \\ 17 \\ 42 \end{bmatrix}$

(e) $\begin{bmatrix} 6 & 1 & 4 & 2 & 1 \\ 3 & 0 & 1 & 4 & 2 \\ -3 & -2 & -5 & 8 & 4 \end{bmatrix} \mathbf{x} = \begin{bmatrix} 11 \\ 6 \\ -4 \end{bmatrix}$

(f) $\begin{bmatrix} 1 & 2 & 3 & 4 \\ 5 & 6 & 7 & 8 \end{bmatrix} \mathbf{x} = \begin{bmatrix} 10 \\ 26 \end{bmatrix}$

E 10.2. Show that the equations in E 10.1 are consistent, using both
methods discussed in Section 10.2c.

E 10.3. Use (10.5) to obtain from (10.2) the solutions given in Table
10.1.

E 10.4. Define

$$\mathbf{A} = \begin{bmatrix} 3 & 5 \\ 2 & 7 \\ 1 & 9 \end{bmatrix}.$$

(a) Find the rank of $\mathbf{A}$.

(b) Ascertain if equations $\mathbf{Ax} = \begin{bmatrix} 13 \\ 16 \\ 19 \end{bmatrix}$ are consistent.

(c) If the equations in (b) are consistent, find their solution using (10.7) and (10.9).

(d) Use (10.6) to show that (10.9) is algebraically identical to (10.7).

E 10.5. The third column in

$$M = \begin{bmatrix} 1 & 6 & 8 \\ 4 & 17 & 12 \\ 7 & 28 & 16 \\ 4 & 10 & -8 \end{bmatrix}$$

is a linear combination of the other two columns. How many LIN solutions are there to $\mathbf{Mx} = [19 \ \ 42 \ \ 65 \ \ 8]'$? Find such a set of solutions and from them derive one other solution.

E 10.6. Solve the two sets of equations

$$\begin{bmatrix} 2 & 4 & 3 & 5 \\ 1 & 9 & 7 & 0 \\ 4 & 22 & 17 & 5 \end{bmatrix} \mathbf{x} = \begin{bmatrix} 21 \\ 33 \\ 87 \end{bmatrix}$$

$$\begin{bmatrix} -1 & 4 & 3 & 2 \\ 2 & 9 & 7 & 1 \\ 5 & -3 & -2 & -5 \end{bmatrix} \mathbf{x} = \begin{bmatrix} 8 \\ 31 \\ 7 \end{bmatrix}$$

using the same generalized inverse in both cases, and find a linear transformation connecting the general solutions of the two sets of equations.

E 10.7. Find a set of LIN solutions to

$$\frac{1}{4} \begin{bmatrix} 1 & 1 & 1 & 1 \\ 1 & 1 & 1 & 1 \\ 1 & 1 & 1 & 1 \\ 1 & 1 & 1 & 1 \end{bmatrix} \mathbf{x} = \begin{bmatrix} 2 \\ 2 \\ 2 \\ 2 \end{bmatrix}.$$

E 10.8. Find a common solution to the following equations:

$$\begin{bmatrix} 4 & 5 & 1 \\ -5 & -6 & 2 \\ -1 & -1 & 3 \end{bmatrix} \mathbf{x} = \begin{bmatrix} 19 \\ -20 \\ -1 \end{bmatrix}$$

$$\begin{bmatrix} 4 & 5 & 7 \\ -5 & -6 & 3 \\ 9 & 11 & 4 \end{bmatrix} \mathbf{x} = \begin{bmatrix} 25 \\ -19 \\ 44 \end{bmatrix}.$$

E 10.9. (Manufacturing) Three different machines, A, B and C, are to be used eight hours a day in manufacturing four different products t, u, v and w. The number of hours that each machine must be used in making one unit of each product is shown in the accompanying table:

Machine	Product			
	t	u	v	w
A	1	2	1	2
B	7	0	2	0
C	1	0	0	4

i.e., to make one v requires one hour of machine A and two hours of machine B.

(a) Write equations that represent full utilization of all machines. Express these equations in terms of the numbers of each product made in an 8-hour day.

(b) Solve the equations derived in (a).

(c) Negative solutions to the equations are meaningless, but fractional solutions have meaning. What conditions therefore apply to the solution obtain in (b)?

(d) How many of each product are made if production of t is maximized while maintaining full utilization of all machines?

(e) What is the production for the four products if production of w is maximized while maintaining full utilization of all machines?

E 10.10. Repeat E 10.9 for full utilization of all machines over a 40-hour week rather than an 8-hour day. Solutions should be integers.

E 10.11. Repeat E 10.9 for full utilization of all machines over a regular week except that machine C is unavailable eight hours.

Chapter 11

EIGENROOTS AND EIGENVECTORS

The topic of this chapter is just the starting point of a large body of theoretical results, many of which are beyond the scope of this book. But for even the basic results, which are shown here and which are useful for work in economics, a number of the derivations are somewhat lengthy. These derivations are therefore not included here; they can be found in Searle (1982, Chapters 11 and 11a), for which appropriate page numbers are given throughout this chapter.

11.1 Basic Equation

Let us consider the equation

$$\mathbf{Au} = \lambda\mathbf{u} \qquad (11.1)$$

where $\mathbf{u}$ is a vector and λ is a scalar. Before going any further note that (11.1) exists only for $\mathbf{A}$ being square. Thus it is for only square matrices $\mathbf{A}$ that we consider (11.1); and in doing so we drop the word square for this chapter.

a. Non-null vectors

It is obvious that (11.1) is true when $\mathbf{u}$ is null. Aside from that uninteresting value for $\mathbf{u}$, let us ask "When else is (11.1) true?" The answer lies in writing (11.1) as

$$(\mathbf{A} - \lambda\mathbf{I})\mathbf{u} = \mathbf{0}, \qquad (11.2)$$

from which we know that if $\mathbf{A} - \lambda\mathbf{I}$ is non-singular the only solution is $\mathbf{u} = \mathbf{0}$. But if $\mathbf{A} - \lambda\mathbf{I}$ is singular, a non-null solution for $\mathbf{u}$ can be obtained from (10.15) of Section 10.4b as

$$\mathbf{u} = [(\mathbf{A} - \lambda\mathbf{I})^-(\mathbf{A} - \lambda\mathbf{I}) - \mathbf{I}]\mathbf{z}, \tag{11.3}$$

using a generalized inverse of $\mathbf{A} - \lambda\mathbf{I}$ and where $\mathbf{z}$ is arbitrary. We return to this element of arbitrariness in Section 11.3.

b.   Deriving roots

In the meantime observe that (11.3) exists so long as $\mathbf{A} - \lambda\mathbf{I}$ is singular, i.e., so long as

$$|\mathbf{A} - \lambda\mathbf{I}| = \mathbf{0}. \tag{11.4}$$

Therefore (11.4) is the condition for $\mathbf{u}$ and λ to exist such that (11.1) is true, i.e., pick λ so that the determinant of $\mathbf{A} - \lambda\mathbf{I}$ is zero.

c.   Definitions

Equation (11.4) is called the *characteristic equation* of $\mathbf{A}$. For $\mathbf{A}$ of order n, it is a polynomial equation in λ of order n, with n roots to be denoted by $\lambda_1, \lambda_2, \ldots, \lambda_n$, some of which may be zero. These roots are called *latent roots, characteristic roots, eigenvalues* or *eigenroots*. Corresponding to each root λ_i is a vector $\mathbf{u}_i$ satisfying (11.1):

$$\mathbf{A}\mathbf{u} = \lambda_i\mathbf{u} \quad \text{for} \quad i = 1, \ldots, n, \tag{11.5}$$

and these vectors $u_1, u_2, \ldots, u_n$ are correspondingly called *latent vectors, characteristic vectors* or *eigenvectors*. "Characteristic" or "eigen" are the usual names; "eigen" is used in this book. And, in preference to today's commonly used "eigenvalue", we use "eigenroot". It is, after all, a root of a polynomial equation. We give two examples.

Example 1

The matrix

$$\mathbf{A} = \begin{bmatrix} 1 & 4 \\ 9 & 1 \end{bmatrix}$$

has the characteristic equation

$$\left| \begin{bmatrix} 1 & 4 \\ 9 & 1 \end{bmatrix} - \begin{bmatrix} \lambda & 0 \\ 0 & \lambda \end{bmatrix} \right| = 0 \quad \text{i.e.,} \quad \left| \begin{matrix} 1 - \lambda & 4 \\ 9 & 1 - \lambda \end{matrix} \right| = 0. \tag{11.6}$$

Expanding the determinant in (11.6) gives

$$(1 - \lambda)^2 - 36 = 0; \quad \text{i.e., } \lambda = -5 \text{ or } 7.$$

Note that characteristic equations are, from (11.4), always of the form shown in (11.6): the determinant of **A**, amended by subtracting λ from each diagonal element, is equated to zero.

Methods for obtaining eigenvectors corresponding to solutions of a characteristic equation are discussed in Section 11.3, but meanwhile it can be seen here that

$$\begin{bmatrix} 1 & 4 \\ 9 & 1 \end{bmatrix} \begin{bmatrix} 2 \\ -3 \end{bmatrix} = -5 \begin{bmatrix} 2 \\ -3 \end{bmatrix} \quad \text{and} \quad \begin{bmatrix} 1 & 4 \\ 9 & 1 \end{bmatrix} \begin{bmatrix} 2 \\ 3 \end{bmatrix} = 7 \begin{bmatrix} 2 \\ 3 \end{bmatrix},$$

(11.7)

these being examples of (11.1). Thus $\begin{bmatrix} 2 \\ -3 \end{bmatrix}$ is an eigenvector corresponding to the eigenroot -5, and $\begin{bmatrix} 2 \\ 3 \end{bmatrix}$ is a vector for the root 7.

Example 2

The characteristic equation for

$$\mathbf{A} = \begin{bmatrix} 2 & 2 & 0 \\ 2 & 1 & 1 \\ -7 & 2 & -3 \end{bmatrix} \quad \text{is} \quad \begin{vmatrix} 2 - \lambda & 2 & 0 \\ 2 & 1 - \lambda & 1 \\ -7 & 2 & -3 - \lambda \end{vmatrix} = 0.$$

Diagonal expansion of the determinant (Section 4.5) gives

$$-\lambda^3 + \lambda^2(2+1-3) - \lambda \left\{ \begin{vmatrix} 2 & 2 \\ 2 & 1 \end{vmatrix} + \begin{vmatrix} 2 & 0 \\ -7 & -3 \end{vmatrix} + \begin{vmatrix} 1 & 1 \\ 2 & -3 \end{vmatrix} \right\} + |\mathbf{A}| = 0,$$

which, after simple arithmetic, reduces to

$$\lambda^3 - 13\lambda + 12 = 0, \quad \text{equivalent to} \quad (\lambda - 1)(\lambda^2 + \lambda - 12) = 0.$$

Solutions are $\lambda = 1$, 3 and -4, and these are the eigenroots of **A**.

d. Cayley-Hamilton theorem

A surprising feature of the characteristic equation is that it is satisfied by the matrix itself. This is known as the Cayley-Hamilton theorem. It means that if $p(\lambda) = 0$ is the characteristic equation, $p(\lambda)$

being the polynomial function of order n in λ coming from $|\mathbf{A} - \lambda\mathbf{I}|$, then $p(\mathbf{A}) = \mathbf{0}$, with the constant (non-λ) term in $p(\lambda)$ being multiplied by $\mathbf{I}$ in $p(\mathbf{A})$. For example, where the characteristic equation has just been derived as

$$\lambda^3 - 13\lambda + 12 = 0$$

it will be found that for $\mathbf{A}$ of that example

$$\mathbf{A}^3 - 13\mathbf{A} + 12\mathbf{I} = 0. \tag{11.8}$$

This kind of result is true for all matrices (Searle, 1982, p. 314).

11.2   Elementary Properties of Eigenroots

Four simple properties of eigenroots are stated and illustrated here in subsections a to d. In all four cases λ is taken as being an eigenroot of $\mathbf{A}$. Derivations of these properties are in Searle (1982, pp. 276-79). Subsection e presents eigenroots of some particular matrices.

a.   Eigenroots of powers of a matrix

$\mathbf{A}^n$ for positive integer n has λ^n as an eigenroot. And for non-singular $\mathbf{A}$ then λ^{-n} is an eigenroot of $(\mathbf{A}^{-1})^n$. For Example 1 it is easily seen that

$$\mathbf{A}^2 = \begin{bmatrix} 37 & 8 \\ 18 & 37 \end{bmatrix} \quad \text{and} \quad |\mathbf{A}^2 - \beta\mathbf{I}| = 0$$

for β being an eigenroot of $\mathbf{A}^2$ reduces to

$$(37 - \beta)^2 - 144 = 0, \quad \text{giving} \quad \beta^2 - 74\beta + 49(25) = 0,$$

so that $\beta = 25$ or 49; and these values are $\beta = \lambda^2$ of the eigenroots found in Example 1.

b.   Eigenroots of a scalar multiple of a matrix

An eigenroot of $c\mathbf{A}$ is $c\lambda$; and $c\mathbf{u}$ is an eigenvector of $c\mathbf{A}$; and $(\mathbf{A} + c\mathbf{I})$ has $\lambda + c$ as an eigenroot.

c.   Eigenroots of polynomials

If $f(\mathbf{A})$ is a polynomial in $\mathbf{A}$ (with the non-$\mathbf{A}$ term being a multiple of $\mathbf{I}$) then $f(\lambda)$ is an eigenroot of $f(\mathbf{A})$. For example, $\mathbf{A}^3 + 7\mathbf{A}^2 + \mathbf{A} + 5\mathbf{I}$ has $\lambda^3 + 7\lambda^2 + \lambda + 5$ as an eigenroot.

This result extends, of course, to infinite polynomials such as, for example, $e^{\mathbf{A}} = \sum_{i=0}^{\infty} \mathbf{A}^i / i!$ having eigenroot e^{λ}.

d.   Sum and product of eigenroots

These two results turn out to be very useful. Denoting the n eigenroots by λ_i for $i = 1, 2, \ldots, n$ we then have

$$\sum_{i=1}^{n} \lambda_i = \text{trace}(\mathbf{A}) \quad \text{and} \quad \prod_{i=1}^{n} \lambda_i = |\mathbf{A}_i|. \tag{11.9}$$

An important consequence of the second result in (11.9) is that whenever a matrix has an eigenroot of zero then that matrix is singular.

e.   Eigenroots of particular matrices

– i.   Transposed

The eigenroots of a transposed matrix are the same as those of the matrix itself. This is because

$$0 = |\mathbf{A} - \lambda\mathbf{I}| = |(\mathbf{A} - \lambda\mathbf{I})'| = |\mathbf{A}' - \lambda\mathbf{I}|.$$

– ii.   Orthogonal

For λ being an eigenroot of orthogonal $\mathbf{A}$, so is $1/\lambda$. Because $\mathbf{A}'\mathbf{A} = \mathbf{I}$ for square $\mathbf{A}$ means that $|\mathbf{A}| \neq 0$, Section 11.2d shows that all eigenroots are non-zero,

$$0 = |\mathbf{A}' - \lambda\mathbf{I}| = |\mathbf{A}||\mathbf{A}' - \lambda\mathbf{I}| = |\mathbf{A}\mathbf{A}' - \mathbf{A}\lambda\mathbf{I}| = |\mathbf{I} - \lambda\mathbf{A}| = |\mathbf{A} - \tfrac{1}{\lambda}\mathbf{I}|.$$

– iii.   Idempotent

Every eigenroot of an idempotent matrix is either 0 or 1. Using $\mathbf{A} = \mathbf{A}^2$ we derive this result from (11.1).

$$\lambda\mathbf{u} = \mathbf{A}\mathbf{u} = \mathbf{A}^2\mathbf{u} = \lambda^2\mathbf{u},$$

and because $\mathbf{u} \neq \mathbf{0}$ this means $\lambda = \lambda^2$ and so $\lambda = 0$ or 1.

The derivations of these simple results illustrate ways in which the equations $\mathbf{Au} = \lambda\mathbf{u}$ and $|\mathbf{A} - \lambda\mathbf{I})| = 0$ can be (and frequently are) used to obtain eigenroots of special forms of $\mathbf{A}$ or, indeed, of matrices related to $\mathbf{A}$.

11.3   Calculating Eigenvectors

Suppose λ_k is an eigenroot of $\mathbf{A}$. It is a solution of the characteristic equation $|\mathbf{A} - \lambda\mathbf{I}| = 0$. Calculating an eigenvector corresponding to λ_k requires finding a non-null $\mathbf{u}$ to satisfy $\mathbf{Au} = \lambda_k\mathbf{u}$, which is equivalent to solving

$$(\mathbf{A} - \lambda_k\mathbf{I})\mathbf{u} = \mathbf{0} \tag{11.10}$$

similar to (11.2). Since λ_k is such that $|\mathbf{A} - \lambda_k\mathbf{I}| = 0$, the matrix $\mathbf{A} - \lambda_k\mathbf{I}$ is singular and so at least one non-null solution (11.10) always exists (see the paragraph following the example in Section 10.4b). However many solutions there are, they can be found from

$$\mathbf{u}_k = [(\mathbf{A} - \lambda_k\mathbf{I})^-(\mathbf{A} - \lambda_k\mathbf{I}) - \mathbf{I}]\mathbf{z} \tag{11.11}$$

similar to (11.3), using a generalized inverse of $\mathbf{A} - \lambda_k\mathbf{I}$ and an arbitrary $\mathbf{z}$. Furthermore, (11.10) has

$$n - r(\mathbf{A} - \lambda_k\mathbf{I}) \text{ LIN solutions,} \tag{11.12}$$

all obtainable from (11.11).

Suppose in (11.11) we make the partitioning

$$\mathbf{A} - \lambda_k\mathbf{I} = \mathbf{B}_k = \begin{bmatrix} \mathbf{R}_k & \mathbf{C}_k \\ \mathbf{D} & \mathbf{E} \end{bmatrix} \tag{11.13}$$

for each eigenvalue λ_k, where $\mathbf{R}_k$ is non-singular with the same rank as $\mathbf{A} - \lambda_k\mathbf{I}$. Then using the algorithm of Section 9.3b we have

$$(\mathbf{A} - \lambda_k\mathbf{I})^- = \begin{bmatrix} \mathbf{R}_k^{-1} & \mathbf{0} \\ \mathbf{0} & \mathbf{0} \end{bmatrix}$$

and so (11.11), on partitioning the arbitrary $\mathbf{z}$ as $\mathbf{z}' = [-\mathbf{v}' \quad -\mathbf{w}']$, becomes

$$\mathbf{u}_k = \begin{bmatrix} -\mathbf{R}_k^{-1}\mathbf{C}_k\mathbf{w} \\ \mathbf{w} \end{bmatrix} \text{ for arbitrary } \mathbf{w} \text{ of order } n - r(\mathbf{A} - \lambda_k\mathbf{I}).$$

$$\tag{11.14}$$

Example 2 (*continued*)

Eigenroots of $\mathbf{A} = \begin{bmatrix} 2 & 2 & 0 \\ 2 & 1 & 1 \\ -7 & 2 & -3 \end{bmatrix}$ are $\lambda_1 = 1$, $\lambda_2 = 3$ and $\lambda_3 = -4$.

For $\lambda_1 = 1$, (11.13) and (11.14) give $\mathbf{A} - \lambda_1\mathbf{I} =$

$$\begin{bmatrix} 1 & 2 & 0 \\ 2 & 0 & 1 \\ -7 & 2 & -4 \end{bmatrix} \text{ and } \mathbf{u}_1 = \begin{bmatrix} \frac{1}{4} \begin{pmatrix} 0 & -2 \\ -2 & 1 \end{pmatrix} \begin{pmatrix} 0 \\ 1 \end{pmatrix} w \\ w \end{bmatrix} = \begin{bmatrix} -\frac{1}{2}w \\ \frac{1}{4}w \\ w \end{bmatrix}.$$
(11.15)

Similarly, with $\lambda_2 = 3$, (11.13) and (11.14) give $\mathbf{A} - \lambda_2\mathbf{I} =$

$$\begin{bmatrix} -1 & 2 & 0 \\ 2 & -2 & 1 \\ -7 & 2 & -6 \end{bmatrix} \text{ and } \mathbf{u}_2 = \begin{bmatrix} \frac{1}{2} \begin{pmatrix} -2 & -2 \\ -2 & -1 \end{pmatrix} \begin{pmatrix} 0 \\ 1 \end{pmatrix} w \\ w \end{bmatrix} = \begin{bmatrix} -w \\ -\frac{1}{2}w \\ w \end{bmatrix}.$$
(11.16)

and $\lambda_3 = -4$ yields $\mathbf{A} - \lambda_3\mathbf{I} =$

$$\begin{bmatrix} 6 & 2 & 0 \\ 2 & 5 & 1 \\ -7 & 2 & 1 \end{bmatrix} \text{ and } \mathbf{u}_3 = \begin{bmatrix} -\frac{1}{26} \begin{pmatrix} 5 & -2 \\ -2 & 6 \end{pmatrix} \begin{pmatrix} 0 \\ 1 \end{pmatrix} w \\ w \end{bmatrix} = \begin{bmatrix} w/13 \\ -3w/13 \\ w \end{bmatrix}.$$
(11.17)

In each of (11.15), (11.16) and (11.17) the w is arbitrary and can be chosen at will, e.g., to make elements of the **us** be integers. For example, using $w = 4$, 2 and 13 in (11.15), (11.16) and (11.17), respectively, gives one set of eigenvectors as

$$\mathbf{u}_1 = \begin{bmatrix} -2 \\ 1 \\ 4 \end{bmatrix}, \quad \mathbf{u}_2 = \begin{bmatrix} -2 \\ -1 \\ 2 \end{bmatrix} \text{ and } \mathbf{u}_3 = \begin{bmatrix} 1 \\ -3 \\ 13 \end{bmatrix}. \tag{11.18}$$

Verification that $\mathbf{A}\mathbf{u}_i = \lambda_i\mathbf{u}_i$ in each case is easy.

11.4 Similar Canonical Form

The two equations in (11.7), each of which is an example of $\mathbf{A}\mathbf{u} = \lambda\mathbf{u}$ of (11.1), can be written as a single matrix equation:

$$\begin{bmatrix} 1 & 4 \\ 9 & 1 \end{bmatrix} \begin{bmatrix} 2 & 2 \\ -3 & 3 \end{bmatrix} = \begin{bmatrix} 2 & 2 \\ -3 & 3 \end{bmatrix} \begin{bmatrix} -5 & 0 \\ 0 & 7 \end{bmatrix}. \tag{11.19}$$

We represent this as

$$\mathbf{AU} = \mathbf{UD}, \qquad (11.20)$$

where $\mathbf{U}$ is the matrix having the eigenvectors as columns and $\mathbf{D}$ is the diagonal matrix of the eigenroots where the sequence of those roots in $\mathbf{D}$ is the same as that of the vectors as columns of $\mathbf{U}$.

Equation (11.20) exists for every square matrix $\mathbf{A}$. This is so because for each eigenroot λ_i of $\mathbf{A}$ there is an eigenvector $\mathbf{u}_i$, and these satisfy the basic equation (11.1), namely $\mathbf{Au}_i = \lambda_i \mathbf{u}_i$ for $i = 1, 2, \ldots, n$. And these equations can be summarized in a single matrix equation

$$\mathbf{A}[\mathbf{u}_1 \quad \mathbf{u}_2 \quad \cdots \quad \mathbf{u}_n] = [\mathbf{u}_1 \quad \mathbf{u}_2 \quad \cdots \quad \mathbf{u}_n] \begin{bmatrix} \lambda_1 & 0 & \cdots & 0 \\ 0 & \lambda_2 & \cdots & 0 \\ \vdots & \vdots & \ddots & \vdots \\ 0 & 0 & \cdots & \lambda_n \end{bmatrix}.$$

We now define $\mathbf{U}$ as the matrix of n eigenvectors and $\mathbf{D}$ as the diagonal matrix of eigenroots, namely

$$\mathbf{U} = [\mathbf{u}_1 \quad \mathbf{u}_2 \quad \cdots \quad \mathbf{u}_n] \quad \text{and} \quad \mathbf{D} = \begin{bmatrix} \lambda_1 & 0 & \cdots & 0 \\ 0 & \lambda_2 & \cdots & 0 \\ \vdots & \vdots & \ddots & \vdots \\ 0 & 0 & \cdots & \lambda_n \end{bmatrix}, \qquad (11.21)$$

and get

$$\mathbf{AU} = \mathbf{UD}. \qquad (11.22)$$

The matrix $\mathbf{D}$ is known as the *canonical form under similarity*, or equivalently as the *similar canonical form*.

It is clear from (11.22) that if $\mathbf{U}^{-1}$ exists then

$$\mathbf{A} = \mathbf{UDU}^{-1}$$

which leads to the diagonal form

$$\mathbf{U}^{-1}\mathbf{AU} = \mathbf{D}$$

and to

$$\mathbf{A}^n = \mathbf{UD}^n\mathbf{U}, \qquad \text{and to} \qquad (\mathbf{A}^{-1})^n = \mathbf{U}(\mathbf{D}^{-1})^n\mathbf{U}^{-1}$$

providing $\mathbf{A}^{-1}$ exists. These last results are very useful for calculating powers, because $\mathbf{D}$ is diagonal and so its powers are easily obtained.

Unfortunately not every matrix $\mathbf{A}$ leads to a $\mathbf{U}$ for which $\mathbf{U}^{-1}$ exists. To ascertain if $\mathbf{U}^{-1}$ exists we need to distinguish between two different classes of matrices; the first are those for which a $\mathbf{U}^{-1}$ always exists. The second are those for which a theorem (the diagonability theorem; see Section 11.5) determines whether or not $\mathbf{U}^{-1}$ exists. In considering this situation we restrict attention, as usual, to real matrices.

When every eigenroot of a matrix occurs only once, the roots are said to be *simple roots* and $\mathbf{U}^{-1}$ exists. This is so because for every pair of different roots the corresponding eigenvectors are LIN. That is established (E 11.8) by showing that the assumption of two such vectors being independent leads to a contradiction.

Example 2 illustrates this: its eigenroots 1, 3 and -4 are all different, and from (11.18) its $\mathbf{U}$ is

$$\mathbf{U} = \begin{bmatrix} -2 & -2 & 1 \\ 1 & -1 & -3 \\ 4 & 2 & 13 \end{bmatrix}, \qquad \text{which is non-singular.}$$

11.5   Asymmetric Matrices, Multiple Roots

These are the difficult cases, for which one must resort to a theorem in order to know whether $\mathbf{U}^{-1}$ exists or not.

a.   Multiple roots

Eigenroots are solutions of the characteristic equation, which is a polynomial equation of order n, the order of $\mathbf{A}$. Therefore there are n roots. But they may not be all different. For example, if the characteristic equation simplifies down to being $(\lambda + 1)(\lambda - 1)^2 = 0$, then the three roots are $\lambda_1 = -1$, $\lambda_2 = 1$ and $\lambda_3 = 1$. Because the root $\lambda = 1$ occurs more than once it is said to be a *multiple root*, and the number of times it occurs, 2, is said to be its *multiplicity*. In general, we formulate $\mathbf{A}$ of order n as having s distinctly different eigenroots $\lambda_1, \ldots, \lambda_s$ with λ_k having multiplicity m_k for $k = 1, 2, \ldots, s$ and, of course, $\sum_{k=1}^{s} m_k = n$. Note two features of this formulation. If zero is an eigenroot (as it can be), it is one of the λ_ks; and simple eigenroots are also included, their multiplicities each being 1. These multiplicities play an important role in the theorem which determines the existence of $\mathbf{U}^{-1}$.

b.   Vectors for multiple roots

Corresponding to each simple eigenroot there is a single eigenvector. Corresponding to a root λ_k having multiplicity m_k one can always use an eigenvector for m_k columns of $\mathbf{U}$. And $\mathbf{AU} = \mathbf{UD}$ will exist. But with two or more (m_k) columns of $\mathbf{U}$ being the same, $\mathbf{U}^{-1}$ will not exist. Nevertheless, for some matrices it will be possible to have m_k different eigenvectors for each multiple root λ_k; then $\mathbf{U}^{-1}$ will exist.

c.   Diagonability theorem

The theorem which tells us when this is possible, is known as the *diagonability theorem*. It specifies the conditions under which all $\sum_{k=1}^{s} m_k = n$ eigenvectors will be linearly independent and so yield a non-singular $\mathbf{U}$.

> **Diagonability Theorem.** A matrix $\mathbf{A}_{n \times n}$, having eigenroots λ_k with multiplicity m_k for $k = 1, 2, \ldots, s$ and $\sum_{k=1}^{s} m_k = n$, has eigenvectors that are LIN if and only if
>
> $$r(\mathbf{A} - \lambda_k \mathbf{I}) = n - m_k \qquad (11.23)$$
>
> for all $k = 1, 2, \ldots, s$, whereupon $\mathbf{U}^{-1}$ exists and $\mathbf{A}$ is diagonable as $\mathbf{U}^{-1}\mathbf{AU} = \mathbf{D}$.

Each eigenroot satisfying (11.23) is called a *regular root* and when $\mathbf{U}^{-1}$ exists $\mathbf{A}$ is said to be a *regular matrix*. When $\mathbf{U}^{-1}$ does not exist $\mathbf{A}$ is said to be *deficient*, or *defective*. As indicated in Section 11.4a we need not worry about this theorem for simple matrices, nor need we do so for simple roots of deficient matrices.

Proof of the diagonability theorem is available in Searle (1982, pp. 305-8) , along with a number of lemmas needed for the proof. It is to be noted that the proof not only establishes the condition for a root of multiplicity m to have m different vectors, but also that those vectors are LIN.

Examples

The matrix of Example 2 is simple, because the three eigenroots are all different; and so its $\mathbf{U}^{-1}$ exists. On the other hand, the characteristic equation (11.4) for

$$\mathbf{A}_2 = \begin{bmatrix} -1 & -2 & -2 \\ 1 & 2 & 1 \\ -1 & -1 & 0 \end{bmatrix} \text{ reduces to } (\lambda + 1)(\lambda - 1)^2 = 0.$$

And for the multiple root $\lambda = 1$ with $m = 2$,

$$\mathbf{A}_2 - 1\mathbf{I} = \begin{bmatrix} -2 & -2 & -2 \\ 1 & 1 & 1 \\ -1 & -1 & -1 \end{bmatrix} \text{ has rank } 1 = 3 - 2 = n - m,$$

which satisfies (11.23) of the diagonability theorem. Therefore $\mathbf{A}_2$ has $\mathbf{U}^{-1}$ existing. But (11.4) for

$$\mathbf{A}_3 = \begin{bmatrix} 0 & 1 & 2 \\ 2 & 3 & 0 \\ 0 & 4 & 5 \end{bmatrix} \text{ is } (\lambda - 6)(\lambda - 1)^2 = 0,$$

also with multiple of $\lambda = 1$ with $m = 2$. But here

$$\mathbf{A}_3 - 1\mathbf{I} = \begin{bmatrix} -1 & 1 & 2 \\ 2 & 2 & 0 \\ 0 & 4 & 4 \end{bmatrix} \cong \begin{bmatrix} -1 & 0 & 0 \\ 2 & 4 & 4 \\ 0 & 4 & 4 \end{bmatrix} \cong \begin{bmatrix} -1 & 0 & 0 \\ 2 & 0 & 0 \\ 0 & 4 & 4 \end{bmatrix}$$

which has

$$\text{rank}(\mathbf{A}_3 - 1\mathbf{I}) = 2 \neq n - m = 3 - 2.$$

Therefore $\mathbf{A}_3$ does not satisfy (11.23) of the diagonability theorem and so has a singular $\mathbf{U}$, and thus $\mathbf{U}^{-1}$ does not exist.

11.6 Symmetric Matrices

Symmetric matrices have sufficient notable properties in regard to eigenroots and eigenvectors as to warrant special attention. Furthermore, the widespread use of symmetric matrices through their involvement in least squares equations and sums of squares make it worthwhile to discuss these properties in detail. Only real, symmetric matrices are considered.

a. Eigenroots all real

The eigenroots of a matrix of order n are roots of a polynomial equation of degree n and so are not necessarily real numbers. Some may be complex numbers, occurring in pairs as $a + ib$ and its complex conjugate $a - ib$, where $i = \sqrt{-1}$ and a and b are real. However, when $\mathbf{A}$ is symmetric and real then all its eigenroots are real.

b.   Symmetric matrices are diagonable

For every eigenroot of a symmetric matrix $\mathbf{A}$, condition (11.23) is satisfied: $r(\mathbf{A} - \lambda_k \mathbf{I}) = n - m_k$ for every λ_k. (Proof is given in Searle, 1982, pp. 307-8.) Therefore $\mathbf{A} = \mathbf{A}'$ is diagonable. Hence for any symmetric matrix $\mathbf{A}$, we have $\mathbf{U}^{-1}$ of $\mathbf{AU} = \mathbf{UD}$ existing, and $\mathbf{A} = \mathbf{UDU}^{-1}$ and $\mathbf{A}^p = \mathbf{UD}^p\mathbf{U}^{-1}$.

c.   Eigenvectors are orthogonal

Symmetric matrices have eigenvectors that are orthogonal to each other. We establish this for eigenvectors corresponding to different eigenroots and refer the reader to Searle (1982, p. 291) for those corresponding to a multiple eigenroot. Suppose two different eigenroots are $\lambda_1 \neq \lambda_2$ with corresponding eigenvectors $\mathbf{u}_1$ and $\mathbf{u}_2$. Then $\mathbf{A} = \mathbf{A}'$ and $\mathbf{Au}_k = \lambda_k \mathbf{u}_k$ give

$$\lambda_1 \mathbf{u}_2' \mathbf{u}_1 = \mathbf{u}_2' \lambda_1 \mathbf{u}_1 = \mathbf{u}_2' \mathbf{Au}_1 = (\mathbf{Au}_2)' \mathbf{u}_1 = (\lambda_2 \mathbf{u}_2)' \mathbf{u}_1 = \lambda_2 \mathbf{u}_2' \mathbf{u}_1.$$

But $\lambda_1 \neq \lambda_2$. Therefore $\mathbf{u}_2' \mathbf{u}_1 = 0$; i.e., $\mathbf{u}_1$ and $\mathbf{u}_2$ are orthogonal. Hence eigenvectors corresponding to different eigenroots are orthogonal.

d.   Canonical form under orthogonal similarity

We now have that $\mathbf{A} = \mathbf{A}'$ is diagonable, that eigenvectors corresponding to different eigenroots are orthogonal, and that m_k LIN eigenvectors corresponding to any eigenroot λ_k of multiplicity m_k can be obtained such that they are orthogonal. Those m_k eigenvectors are also orthogonal to eigenvectors corresponding to each other eigenroot. Hence eigenvectors for a symmetric matrix are *all* orthogonal to each other. On normalizing each vector (see Section 3.5a) by changing $\mathbf{u}$ to $(1/\sqrt{\mathbf{u}'\mathbf{u}})\mathbf{u}$ and arraying the normalized vectors in a matrix $\mathbf{U}$, we then have $\mathbf{U}$ as an orthogonal matrix and hence

$$\mathbf{U}'\mathbf{AU} = \mathbf{D} \qquad \text{with} \qquad \mathbf{UU}' = \mathbf{I}. \tag{11.24}$$

This is the *canonical form under orthogonal similarity* and it exists for all symmetric matrices.

Example

$$\mathbf{A} = \begin{bmatrix} 1 & 2 & 2 \\ 2 & 1 & 2 \\ 2 & 2 & 1 \end{bmatrix}$$

has characteristic equation which reduces to $(\lambda+1)^2(\lambda-5) = 0$. Hence $\lambda_1 = 5$ with $m_1 = 1$, and $\lambda_2 = -1$ with $m_2 = 2$. Using (11.14) we get

$$\mathbf{u}_1 = \left[\frac{1}{12} \begin{pmatrix} -4 & -2 \\ -2 & -4 \end{pmatrix} \begin{pmatrix} 2 \\ 2 \end{pmatrix} w \atop w \right] = \left[\begin{matrix} w \\ w \\ w \end{matrix} \right]$$

and

$$\mathbf{u}_2 = \left[\begin{matrix} -\frac{1}{2}[2 \; 2]\mathbf{w} \\ \mathbf{w} = \begin{bmatrix} w_1 \\ w_2 \end{bmatrix} \end{matrix} \right] = \left[\begin{matrix} -(w_1 + w_2) \\ w_1 \\ w_2 \end{matrix} \right].$$

Clearly one value for $\mathbf{u}_2$ is $[-2 \; 1 \; 1]' = \mathbf{v}_2$, say. Because $m_2 = 2$ there will be two possible values of $\mathbf{u}_2$ and they can be orthogonal. Hence a second value for $\mathbf{u}_2$ can be found by solving $\mathbf{v}_2'\mathbf{u}_2 = 0$, which is $2(w_1 + w_2) + w_1 + w_2 = 0$. Since $w_1 = -1$ and $w_2 = 1$ is a solution, a second value for $\mathbf{u}_2$ is $\mathbf{v}_3 = [0 \; -1 \; 1]'$. We therefore have three LIN orthogonal eigenvectors

$$\mathbf{u}_1 = \begin{bmatrix} 1 \\ 1 \\ 1 \end{bmatrix}, \qquad \mathbf{v}_2 = \begin{bmatrix} -2 \\ 1 \\ 1 \end{bmatrix} \quad \text{and} \qquad \mathbf{v}_3 = \begin{bmatrix} 0 \\ -1 \\ 1 \end{bmatrix}.$$

Arraying the normalized forms of these vectors as a matrix gives

$$\mathbf{U} = \frac{1}{\sqrt{6}} \begin{bmatrix} \sqrt{2} & -2 & 0 \\ \sqrt{2} & 1 & -\sqrt{3} \\ \sqrt{2} & 1 & \sqrt{3} \end{bmatrix}.$$

The reader should verify that $\mathbf{U}'\mathbf{A}\mathbf{U} = \mathbf{D} = \text{diag}\{5, \; -1, \; -1\}$ and $\mathbf{U}\mathbf{U}' = \mathbf{I}$.

e. Rank equals number of non-zero eigenroots

Define $z_\mathbf{A}$ as the number of zero eigenroots of the matrix $\mathbf{A}$; then $n - z_\mathbf{A}$ is the number of non-zero eigenroots. Since, for $\mathbf{A}$ being symmetric, $\mathbf{A} = \mathbf{U}\mathbf{D}\mathbf{U}'$ for non-singular (orthogonal) $\mathbf{U}$, the ranks of $\mathbf{A}$ and $\mathbf{D}$ are equal, $r_\mathbf{A} = r_\mathbf{D}$ (Theorem 2, Section 7.9). But the only non-zero elements in the diagonal matrix $\mathbf{D}$ are the non-zero eigenroots, and so its rank is the number of such eigenroots, $n - z_\mathbf{A}$. Hence

$$r_\mathbf{A} = n - z_\mathbf{A}, \qquad \text{for} \qquad \mathbf{A} = \mathbf{A}'; \tag{11.25}$$

i.e., for symmetric matrices rank equals the number of non-zero eigenroots.

This result is true not only for all symmetric matrices (because they are diagonable), but also for all diagonable matrices. Other than diagonability there is nothing inherent in the development of (11.25) that uses the symmetry of $\mathbf{A}$. Nevertheless, (11.25) is of importance because it applies to all symmetric matrices.

f. Spectral decomposition

Section 3.2e shows that $\mathbf{AA'} = \sum_j \mathbf{a}_j \mathbf{a}'_j$, where $\mathbf{a}_j$ is the jth column of $\mathbf{A}$. Applied to columns of $\mathbf{U}$ where $\mathbf{U}$ is orthogonal, $\mathbf{I} = \mathbf{UU'}$, this gives $\mathbf{I} = \sum_{i=1}^{n} \mathbf{u}_i \mathbf{u}'_i$. If the $\mathbf{u}_i$s are the eigenvectors of symmetric $\mathbf{A}$, pre-multiplication by $\mathbf{A}$ gives

$$\mathbf{A} = \mathbf{A} \sum \mathbf{u}_i \mathbf{u}'_i = \sum \mathbf{A} \mathbf{u}_i \mathbf{u}'_i = \sum \lambda_i \mathbf{u}_i \mathbf{u}'_i. \qquad (11.26)$$

This is the *spectral decomposition* of $\mathbf{A}$. It also applies to powers of $\mathbf{A}$:

$$\mathbf{A}^k = \sum \lambda_i^k \mathbf{u}_i \mathbf{u}'_i$$

for any positive integer k (and any negative integer if $\mathbf{A}$ is non-singular).

Illustration   (Transition probabilities)

For a transition probability matrix $\mathbf{P}$ and a known vector $\mathbf{t}$ we often want to evaluate $\mathbf{P}^k \mathbf{t}$ for some k. If $\mathbf{P}$ is symmetric,

$$\mathbf{P}^k \mathbf{t} = \sum \lambda_i^k \mathbf{u}_i \mathbf{u}'_i \mathbf{t}, \qquad (11.27)$$

and if k is sufficiently large that the largest eigenroot λ_L is such that λ_L^k greatly exceeds λ_i^k for any other eigenvalue, then an asymptotic value for $\mathbf{P}^k \mathbf{t}$ is $(\lambda_L^k \mathbf{u}'_L \mathbf{t}) \mathbf{u}_L$ where $\mathbf{u}_L$ is the normalized eigenvector corresponding to λ_L.

g. Non-negative definite (n.n.d.) matrices

Definition of n.n.d. matrices is given in Section 3.6f-ii and properties are discussed in Section 8.6. Since the definition is in terms of quadratic forms, n.n.d. matrices are usually taken as being symmetric, and so also have the following properties:

(i) All eigenroots are real (Section 11.5a).

(ii) They are diagonable (Section 11.5b).

(iii) Rank equals the number of non-zero eigenroots (Section 11.6e).

These lead to the following theorem and corollary. Proofs are to be found in Searle (1982, p. 310).

Theorem. The eigenroots of a symmetric matrix are all non-negative if and only if the matrix is n.n.d.

Corollary. The eigenroots of a symmetric matrix are all positive if and only if the matrix is positive definite (p.d.).

The consequences of this theorem are that when a symmetric matrix has every $\lambda \geq 0$, it is n.n.d.; if every $\lambda > 0$, it is p.d.; and if some λs are greater than zero and others (at least one) equal zero, then it is positive semi-definite (p.s.d.).

11.7 Dominant Eigenroots

Eigenroots play a key role in dealing with age distribution vectors, transition probability matrices and analysis of market stability. Their importance in the study of markets is that it is the dominant eigenroot which determines if markets will converge to an equilibrium. It also determines the steady-state solution in problems incorporating transition probability matrices.

Several methods are available for calculating a numerically dominant eigenroot of a matrix (e.g., Stewart, 1973, Chapter 7). They provide opportunity for finding the largest eigenroot of a matrix without having to solve the characteristic equation, dealing, *inter alia*, with matters such as multiple roots, multiple dominant roots and roots involving complex numbers, as well as considering the problem of the actual existence of a dominant root. We discuss only one of these topics here: an easy method for calculating the largest eigenroot *when it exists*. The existence of $\mathbf{A} = \mathbf{U}\mathbf{D}\mathbf{U}^{-1}$ and of a real-valued dominant root are assumed throughout.

From $\mathbf{A} = \mathbf{U}\mathbf{D}\mathbf{U}^{-1}$ we get

$$\mathbf{A}^k = \mathbf{U}\mathbf{D}^k\mathbf{U}^{-1} = \left\{_r \lambda_j^k \mathbf{u}_j\right\}_{j=1}^n \mathbf{U}^{-1} = \sum_{j=1}^n \lambda_j^k \mathbf{u}_j \mathbf{v}_j' \qquad (11.28)$$

on using (3.8) of Section 3.2e, where $\mathbf{v}_j'$ is the jth row of $\mathbf{U}^{-1}$.

Suppose that λ_1 is the dominant root, such that λ_1^k for some sufficiently large value of k is numerically so much greater than any of the values $\lambda_2^k, \ldots, \lambda_n^k$ that the latter may be taken as zero. Then (11.28) becomes

$$\mathbf{A}^k = \lambda_1^k \mathbf{u}_1 \mathbf{v}_1'. \tag{11.29}$$

Post-multiplying (11.29) by a non-null vector $\mathbf{x}$ gives, on defining $\theta = \mathbf{v}_1' \mathbf{x}$,

$$\mathbf{w}_k = \mathbf{A}^k \mathbf{x} = \lambda_1^k \mathbf{u}_1 \mathbf{v}_1' \mathbf{x} = \theta \lambda_1^k \mathbf{u}_1,$$

so defining $\mathbf{w}_k$. Then the ratio of the ith element of $\mathbf{w}_k$ to that of $\mathbf{w}_{k-1}$ is

$$\frac{w_{i,k}}{w_{i,k-1}} = \frac{\theta \lambda_1^k u_{i1}}{\theta \lambda_1^{k-1} u_{i1}} = \lambda_1. \tag{11.30}$$

This, λ_1 of (11.30), is the dominant eigenroot, and it has been developed by taking the ratio of elements of the two vectors $\mathbf{A}^k \mathbf{x}$ and $\mathbf{A}^{k-1} \mathbf{x}$ for any vector $\mathbf{x}$. Fortunately, computing routines are readily available for this kind of arithmetic. An elementary illustration of the arithmetic for a 2×2 matrix is shown in Searle (1982, pp. 294-6).

11.8 Singular-Value Decomposition

An analogue of the similar canonical form of square matrices is the singular-value decomposition of any matrix, square or rectangular. For a matrix $\mathbf{A}$, it is based on the reductions to canonical form under orthogonal similarity of the symmetric matrices $\mathbf{AA}'$ and $\mathbf{A}'\mathbf{A}$, as follows.

The singular-value decomposition of $\mathbf{A}_{p \times q}$ of rank r is

$$\mathbf{A} = \mathbf{L} \begin{bmatrix} \boldsymbol{\Delta} & 0 \\ 0 & 0 \end{bmatrix} \mathbf{M}', \tag{11.31}$$

where $\mathbf{L}$ and $\mathbf{M}$ are orthogonal and where $\boldsymbol{\Delta}^2$ is the diagonal matrix of the non-zero (positive) eigenroots of $\mathbf{AA}'$ and of $\mathbf{A}'\mathbf{A}$; and

$$\mathbf{L}'\mathbf{AA}'\mathbf{L} = \begin{bmatrix} \boldsymbol{\Delta}^2 & 0 \\ 0 & 0 \end{bmatrix}_{p \times p} \quad \text{and} \quad \mathbf{M}'\mathbf{A}'\mathbf{AM} = \begin{bmatrix} \boldsymbol{\Delta}^2 & 0 \\ 0 & 0 \end{bmatrix}_{q \times q}. \tag{11.32}$$

Comments

(i) Both $\mathbf{AA}'$ and $\mathbf{A}'\mathbf{A}$ are n.n.d. and they have the same non-zero eigenroots, which are all positive.

(ii) The matrices in (11.32) have different orders: $\mathbf{A}$ is $p \times q$, $\mathbf{\Delta}^2$ is $r \times r$, $\mathbf{L}$ is $p \times p$ and $\mathbf{M}$ is $q \times q$, and the matrices with leading submatrix $\mathbf{\Delta}^2$ are $p \times p$ in one case and $q \times q$ in the other.

(iii) Matrices $\mathbf{L}$ and $\mathbf{M}$ of (11.32) always exist but they are not unique, and not all pairs of them yield (11.31). However, for a particular choice of $\mathbf{L}$, one is always able to construct an $\mathbf{M}$ such that (11.31) and (11.32) are satisfied.

11.9 Exercises

E 11.1. Find the eigenroots and eigenvectors of the following matrices. In each case combine the eigenvectors into a matrix $\mathbf{U}$ and, where possible, verify that $\mathbf{U}^{-1}\mathbf{A}\mathbf{U} = \mathbf{D}$ where $\mathbf{D}$ is the diagonal matrix of the eigenroots.

$$\mathbf{B} = \begin{bmatrix} 1 & 4 & 1 \\ 2 & 1 & 0 \\ -1 & 3 & 1 \end{bmatrix} \qquad \mathbf{C} = \begin{bmatrix} 2 & -2 & 3 \\ 10 & -4 & 5 \\ 5 & -4 & 6 \end{bmatrix}$$

$$\mathbf{E} = \begin{bmatrix} 7 & 4 & -1 \\ 4 & 7 & -1 \\ -4 & -4 & 4 \end{bmatrix} \qquad \mathbf{F} = \begin{bmatrix} 9 & 15 & 3 \\ 6 & 10 & 2 \\ 3 & 5 & 1 \end{bmatrix}$$

$$\mathbf{G} = \begin{bmatrix} 4 & -2 & 0 \\ -2 & 3 & -2 \\ 0 & -2 & 2 \end{bmatrix} \qquad \mathbf{H} = \begin{bmatrix} 1 & 2 & 1 \\ 2 & 1 & -1 \\ 1 & -1 & 1 \end{bmatrix}$$

$$\mathbf{K} = \begin{bmatrix} -1 & -2 & 1 \\ -2 & 2 & -2 \\ 1 & -2 & -1 \end{bmatrix} \qquad \mathbf{L} = \begin{bmatrix} 1 & 4 & 2 \\ 4 & 1 & 2 \\ 8 & 8 & 1 \end{bmatrix}$$

$$\mathbf{M} = \begin{bmatrix} -9 & 2 & 6 \\ 2 & -9 & 6 \\ 6 & 6 & 7 \end{bmatrix} \qquad \mathbf{N} = \begin{bmatrix} 2 & 7 & 1 \\ 2 & 3 & 8 \\ 9 & 4 & 4 \end{bmatrix}$$

E 11.2. Of the matrices in E 11.1, show that $\mathbf{F} = \mathbf{B}^2 + \mathbf{B}$, and that if λ is an eigenroot of $\mathbf{B}$ then $\lambda^2 + \lambda$ is an eigenroot of $\mathbf{F}$.

E 11.3. Find the eigenroots of

$$\mathbf{T} = \begin{bmatrix} \frac{1}{2} & \frac{1}{2} & 0 \\ \frac{1}{4} & \frac{1}{2} & \frac{1}{4} \\ 0 & \frac{1}{2} & \frac{1}{2} \end{bmatrix}$$

and verify that

$$\mathbf{T}^k = \begin{bmatrix} \frac{1}{4} + (\frac{1}{2})^{k+1} & \frac{1}{2} & \frac{1}{4} - (\frac{1}{2})^{k+1} \\ \frac{1}{4} & \frac{1}{2} & \frac{1}{4} \\ \frac{1}{4} - (\frac{1}{2})^{k+1} & \frac{1}{2} & \frac{1}{4} + (\frac{1}{2})^{k+1} \end{bmatrix}.$$

E 11.4. Show that the eigenvectors of

$$\begin{bmatrix} a & b \\ c & d \end{bmatrix} \quad \text{are of the form} \quad \begin{bmatrix} -b \\ a - \lambda_1 \end{bmatrix} \quad \text{and} \quad \begin{bmatrix} -b \\ a - \lambda_2 \end{bmatrix}$$

where λ_1 and λ_2 are the eigenroots. Verify that $2d = \lambda_1 + \lambda_2$ when $a = d$.

E 11.5. Express $\mathbf{A} = \begin{bmatrix} 1 & 0 & 0 \\ u & v & 0 \\ x & y & z \end{bmatrix}$ in the form $\mathbf{A} = \mathbf{U}\mathbf{D}\mathbf{U}^{-1}$. In doing

so, notice that $\begin{bmatrix} 1 & 0 & 0 \\ a & 1 & 0 \\ b & c & 1 \end{bmatrix}^{-1} = \begin{bmatrix} 1 & 0 & 0 \\ -a & 1 & 0 \\ ac - b & -c & 1 \end{bmatrix}.$

E 11.6. Use the result in E 11.5 to show that the kth power of the transition probability matrix

$$\mathbf{P} = \begin{bmatrix} 1 & 0 & 0 \\ \frac{1}{2} & \frac{1}{2} & 0 \\ \frac{1}{2} - c(1 - c) & 2c(1 - c) & \frac{1}{2} - c(1 - c) \end{bmatrix}$$

is

$$\mathbf{P}^k = \begin{bmatrix} 1 & 0 & 0 \\ 1-(1/2)^k & (1/2)^k & 0 \\ 1+z^k-(1/2)^{k-1} & (1/2)^{k-1}-2z^k & z^k \end{bmatrix}$$

for

$$z = \tfrac{1}{2} + c(c-1).$$

E 11.7. For the transpose of the general 2×2 transition probability matrix

$$\mathbf{P}' = \begin{bmatrix} p_{11} & 1-p_{22} \\ 1-p_{11} & p_{22} \end{bmatrix}$$

find the two eigenroots of $\mathbf{P}'$ and verify that neither root can ever exceed 1 in absolute value.

E 11.8. Show that the eigenvectors corresponding to two different eigenroots are LIN.

E 11.9. Explain why a zero eigenroot means that its matrix has no inverse.

E 11.10. Show that if λ is an eigenroot of an orthogonal matix, then so is $1/\lambda$. What is the numerical value of one root when the matrix is of odd order?

E 11.11. Suppose $\mathbf{A}$ is idempotent and symmetric. Prove the following:
(a) Eigenvalues are 0 or 1.
(b) Trace equals rank.
(c) Diagonability theorem is satisfied.
(d) $\mathbf{A} = \mathbf{BB}'$ where $\mathbf{B}'\mathbf{B} = \mathbf{I}$.

E 11.12. Prove that idempotent matrices are diagonable.

E 11.13. Find the singular-value decomposition of $\begin{bmatrix} 10 & -5 \\ 2 & -11 \\ 6 & -8 \end{bmatrix}$.

E 11.14. (a) Find the rank of $\mathbf{M} = \begin{bmatrix} 1 & -3 & 5 \\ 0 & 1 & 7 \\ 2 & -4 & 28 \\ 3 & -13 & 0 \\ 4 & -14 & -3 \\ 5 & -9 & 30 \\ 6 & 2 & 173 \end{bmatrix}$.

Without any further arithmetic, for $\mathbf{M}$ given in part a):

(b) What feature of $\mathbf{S}$ makes $\mathbf{M'MSM'M} = \mathbf{M'M}$?

(c) Find the rank of $\mathbf{MSM'}$.

(d) Give the numerical values of the eigenroots of $\mathbf{MSM'}$.

(e) If $\mathbf{T}$ is a generalized inverse of $\mathbf{MM'}$, prove that $\mathbf{M'TM}$ is an identity matrix.

(f) Suggest two different generalized inverses of $\mathbf{M'}$.

E 11.15. Suppose

$$\mathbf{K} = \mathbf{K'}, \quad \mathbf{K} = \mathbf{K}^3, \quad \mathbf{K1} = \mathbf{0} \text{ and } \mathbf{K} \begin{bmatrix} 1 \\ 2 \\ -3 \end{bmatrix} = \begin{bmatrix} 1 \\ 2 \\ -3 \end{bmatrix}.$$

Calculate the following values, in each case giving reasons as to why the value can be calculated *without calculating* $\mathbf{K}$.

(a) The order of $\mathbf{K}$.   (b) The rank of $\mathbf{K}$.

(c) The trace of $\mathbf{K}$.   (d) The determinant of $\mathbf{K}$.

(e) The trace of $\mathbf{K}^{26}$.   (f) The determinant of $\mathbf{K}^{26}$.

(g) The trace of $6\mathbf{K}^{60} - 7\mathbf{K}^{37} + 3\mathbf{I}$.

(h) The determinant of $6\mathbf{K}^{60} - 7\mathbf{K}^{37} + 3\mathbf{I}$.

E 11.16. (Capital demand) A simplified version of a theoretical model
for the demand for capital goods uses $\mathbf{d}_{2\times1}$ to represent desired
levels of capital and labor and $\mathbf{x}_n$, the actual levels at time n,
with the relationship between the two being

$$\mathbf{x}_n = (\mathbf{I} - \mathbf{B})\mathbf{x}_{n-1} + \mathbf{Bd}.$$

For

$$\mathbf{B} = \begin{bmatrix} 0.1 & -0.5 \\ 0.3 & 0.9 \end{bmatrix},$$

demonstrate that as $n \to \infty$ the actual levels of capital and
labor do tend to the desired levels.

Chapter 12

MISCELLANEA

This chapter contains miscellaneous topics useful in applications of matrix algebra. Some sections summarize topics scattered throughout earlier chapters; other sections introduce new topics. Numbers in square braces refer to sections or, when preceded by E, to exercises.

12.1   Orthogonal Matrices: A Summary

a. The norm of $\mathbf{x}$ is $\sqrt{\mathbf{x}'\mathbf{x}}$. [3.5]

b. $\mathbf{x}$ is a unit vector when $\mathbf{x}'\mathbf{x} = 1$. [3.5]

c. Vectors $\mathbf{x}$ and $\mathbf{y}$ are orthogonal if $\mathbf{x}'\mathbf{y} = 0$. They are orthonormal when $\mathbf{x}'\mathbf{x} = 1 = \mathbf{y}'\mathbf{y}$ and $\mathbf{x}'\mathbf{y} = 0$. [3.5]

d. $\mathbf{A}$ is orthogonal when any two of (i) $\mathbf{A}$ is square, (ii) $\mathbf{A}\mathbf{A}' = \mathbf{I}$ and (iii) $\mathbf{A}'\mathbf{A} = \mathbf{I}$ are true; and any two imply the third. [5.7c]

e. Rows (columns) of an orthogonal matrix are orthonormal. [E 3.5]

f. Products of orthogonal matrices are orthogonal. [E 3.16]

g. Determinants of orthogonal matrices are $+1$ or -1. [4.3e-vi] Hence orthogonal matrices are non-singular; and $\mathbf{A}' = \mathbf{A}^{-1}$. [5.7c]

h. If λ is an eigenvalue of an orthogonal matrix, so is $1/\lambda$. [11.2e-ii]

i. When $\mathbf{A}$ is skew-symmetric [3.2g], $(\mathbf{I} - \mathbf{A})(\mathbf{I} + \mathbf{A})^{-1}$ is orthogonal. [E 5.22]

249

12.2 Idempotent Matrices: A Summary

The following properties pertain to idempotent matrices.

a. The definition of idempotency is $\mathbf{A}^2 = \mathbf{A}$. [3.4]

b. All idempotent matrices (except $\mathbf{I}$) are singular. [E 5.20]

c. The product of two idempotent matrices is idempotent if the matrices commute in multiplication. [E 3.17]

d. Eigenroots of an idempotent matrix are all 0 or 1. [11.2e-iii]

e. The converse of property d is partially true: all eigenroots being 0 or 1 define a matrix as being idempotent if it is diagonable [11.4d-iii] but not otherwise. For example, $\begin{bmatrix} 6 & 2 \\ -18 & -6 \end{bmatrix}$ has zero for both eigenroots, but is not idempotent.

f. For diagonable idempotent matrices $\mathbf{A}$ (which includes symmetric $\mathbf{A} = \mathbf{A}' = \mathbf{A}^2$), rank equals trace; i.e., $r_\mathbf{A} = \mathrm{tr}(\mathbf{A})$, and $\mathbf{I}_n - \mathbf{A}$ is idempotent, with $r(\mathbf{I}_n - \mathbf{A}) = n - r_\mathbf{A}$.

g. All symmetric, idempotent matrices (except $\mathbf{I}$) are p.s.d. (because of the corollary to the theorem in Section 11.5g).

h. A symmetric idempotent matrix $\mathbf{A}$ can be expressed as $\mathbf{A} = \mathbf{L}\mathbf{L}'$ for $\mathbf{L}'\mathbf{L} = \mathbf{I}_{r_\mathbf{A}}$. [E 11.11]

i. A special form of symmetric idempotent matrix which has many useful properties is $\mathbf{P} = \mathbf{I} - \mathbf{X}(\mathbf{X}'\mathbf{X})^-\mathbf{X}'$. [E 9.10]

j. Matrices $\mathbf{A}$ satisfying $\mathbf{A}^2 = \mathbf{A}$ are idempotent; similarly, those satisfying $\mathbf{A}^2 = \mathbf{0}$ are called *nilpotent* matrices, and those for which $\mathbf{A}^2 = \mathbf{I}$ could be called *unipotent*. [3.4]

12.3 Matrix $a\mathbf{I} + b\mathbf{J}$: A Summary

The summing vector $\mathbf{1}_n$ and the matrix $\mathbf{J}_n = \mathbf{1}_n\mathbf{1}_n'$ have all elements unity. [3.3] A useful combination of $\mathbf{I}$ and $\mathbf{J}$ is $a\mathbf{I} + b\mathbf{J}$ for scalars a and b; e.g.,

$$a\mathbf{I}_3 + b\mathbf{J}_3 = \begin{bmatrix} a+b & b & b \\ b & a+b & b \\ b & b & a+b \end{bmatrix}.$$

Useful properties are as follows.

a. Products of $(a\mathbf{I}_n + b\mathbf{J}_n)$-matrices have the same form:

$$(a\mathbf{I} + b\mathbf{J})(\alpha\mathbf{I} + \beta\mathbf{J}) = a\alpha\mathbf{I} + (\alpha\beta + \alpha b + nb\beta)\mathbf{J}.$$

b. $(a\mathbf{I}_n + b\mathbf{J}_n)^{-1} = \frac{1}{a}\left(\mathbf{I}_n - \frac{b}{a+nb}\mathbf{J}_n\right)$ for $a \neq 0$ and $a \neq -nb$.

c. $|a\mathbf{I}_n + b\mathbf{J}_n| = a^{n-1}(a + nb)$.

d. Eigenroots of $a\mathbf{I}_n + b\mathbf{J}_n$ are a, $n - 1$ times, and $a + nb$ once.

e. $a\mathbf{I}_n + b\mathbf{J}_n$ is diagonable.

f. A special case is for $a = 1$ and $b = -1/n$. This gives the centering matrix [3.3]

$$\mathbf{C}_n \equiv \mathbf{I} - \bar{\mathbf{J}}_n \quad \text{with} \quad \mathbf{x}'\mathbf{C}_n\mathbf{x} = \sum_{i=1}^{n}(x_i - \bar{x})^2$$

for $\bar{\mathbf{J}}_n = \mathbf{J}_n/n$. It and $\mathbf{C}_n$ are symmetric and idempotent.

12.4 Non-negative Definite Matrices

A n.n.d. matrix is $\mathbf{X}'\mathbf{X}$ for any $\mathbf{X}$; p.d. for $\mathbf{X}$ of full column rank. For n.n.d. $\mathbf{A}$ (real and symmetric):

a. $\mathbf{A}$ is diagonable (since it is symmetric). [11.5b]

b. $\mathbf{A} = \mathbf{B}'\mathbf{B}$ for some $\mathbf{B}$ of full column rank $r_{\mathbf{A}}$.

$$\mathbf{B}\mathbf{B}' \text{ is p.d.} \quad \text{and} \quad \mathbf{A}^+ = \mathbf{B}(\mathbf{B}'\mathbf{B})^{-2}\mathbf{B}'.$$

c. Every leading principal minor ≥ 0. [8.6a]

d. Every diagonal element ≥ 0. [8.6a]

e. $|\mathbf{A}| \geq 0$. [8.6a]

f. Every eigenroot ≥ 0. [11.5g]

g. If $\mathbf{A}$ is p.d., the four preceding cases of ≥ 0 become > 0.

12.5 Canonical Forms and Other Decompositions

We summarize here the four canonical forms and three decompositions of a matrix.

a. Equivalent canonical form

The equivalent canonical form of a matrix $\mathbf{A}$ of rank r is

$$\mathbf{PAQ} = \begin{bmatrix} \mathbf{I}_r & \mathbf{0} \\ \mathbf{0} & \mathbf{0} \end{bmatrix}$$

for $\mathbf{P}$ and $\mathbf{Q}$ non-singular (Section 8.2b).

b. Congruent canonical form

The diagonal form of a symmetric matrix of rank r is

$$\mathbf{PAP}' = \begin{bmatrix} \mathbf{D}_r & \mathbf{0} \\ \mathbf{0} & \mathbf{0} \end{bmatrix} \quad \text{for } \mathbf{A} = \mathbf{A}'$$

with $\mathbf{P}$ being non-singular and with $\mathbf{D}_r$ diagonal, of order r. The congruent canonical form (Section 8.3a) is

$$\mathbf{RAR}' = \begin{bmatrix} \mathbf{I}_r & \mathbf{0} \\ \mathbf{0} & \mathbf{0} \end{bmatrix} \quad \text{for } \mathbf{A} = \mathbf{A}'.$$

$\mathbf{P}$ is always real, but $\mathbf{R}$ will not be real when any element of $\mathbf{D}_r$ is negative. ($\mathbf{R}$ is always real for n.n.d. matrices.)

c. Similar canonical form

The similar canonical form [11.4] of any matrix $\mathbf{A}$ is $\mathbf{D}\{\lambda\}$, a diagonal matrix of the eigenroots of $\mathbf{A}$, given by

$$\mathbf{AU} = \mathbf{UD}\{\lambda\},$$

where $\mathbf{U}$ is the matrix of eigenvectors corresponding to the eigenroots of $\mathbf{D}\{\lambda\}$. When the diagonability theorem [11.4c-iii] is satisfied, $\mathbf{U}$ is non-singular and

$$\mathbf{U}^{-1}\mathbf{AU} = \mathbf{D}\{\lambda\}.$$

d. Orthogonal similar canonical form

With the same notation as in the preceding paragraph, the canonical form under orthogonal similarity for a symmetric matrix $\mathbf{A}$ is [11.5d]

$$\mathbf{U'AU} = \mathbf{D}\{\lambda\}, \quad \text{with} \quad \mathbf{A} = \mathbf{A'} \text{ and orthogonal } \mathbf{U}.$$

$\mathbf{U}^{-1}$ always exists. The eigenroots are always real and for n.n.d. matrices they are either positive or zero (all positive for p.d. matrices).

e. Singular-value decomposition

The singular-value decomposition of any matrix $\mathbf{A}$ is [11.7]

$$\mathbf{A} = \mathbf{L} \begin{bmatrix} \boldsymbol{\Delta}_r & \mathbf{0} \\ \mathbf{0} & \mathbf{0} \end{bmatrix} \mathbf{M'},$$

where $\mathbf{L}$ and $\mathbf{M}$ are orthogonal, and $\boldsymbol{\Delta}_r$ is the diagonal matrix of positive square roots of the (positive) eigenvalues of $\mathbf{A'A}$ (or equivalently of $\mathbf{AA'}$): $\boldsymbol{\Delta}_r = \sqrt{\boldsymbol{\Delta}^2}$ where

$$\mathbf{L'AA'L} = \begin{bmatrix} \boldsymbol{\Delta}^2 & \mathbf{0} \\ \mathbf{0} & \mathbf{0} \end{bmatrix} \quad \text{and} \quad \mathbf{M'A'AM} = \begin{bmatrix} \boldsymbol{\Delta}^2 & \mathbf{0} \\ \mathbf{0} & \mathbf{0} \end{bmatrix}.$$

f. Spectral decomposition

The spectral decomposition of a symmetric matrix $\mathbf{A}$ is [11.5f]

$$\mathbf{A} = \sum \lambda_i \mathbf{u}_i \mathbf{u}_i' \text{ for } \mathbf{A} = \mathbf{A'} \text{ and } \mathbf{A}\mathbf{u}_i = \lambda_i \mathbf{u}_i,$$

i.e., for λ_i and $\mathbf{u}_i$ being an eigenroot and corresponding eigenvector.

g. The LDU and LU decompositions

Any square matrix $\mathbf{A}$ can be factored as $\mathbf{A} = \mathbf{LDU}$, where $\mathbf{L}$ and $\mathbf{U}$ are, respectively, lower and upper unit triangular matrices and $\mathbf{D}$ is diagonal. Redefining $\mathbf{DU}$ as $\mathbf{U}$, still upper (but not unit) triangular, gives $\mathbf{A} = \mathbf{LU}$. When $\mathbf{A}$ is symmetric, the $\mathbf{LDU}$ decomposition becomes $\mathbf{UDU'}$; then expressing $\mathbf{D}$ as $\mathbf{D} = \mathbf{T}^2$ and redefining $\mathbf{TU'}$ as $\mathbf{U'}$ gives $\mathbf{A} = \mathbf{A'} = \mathbf{UU'}$. [3.9]

12.6   Matrix Functions

a.   Functions of matrices

Some functions of matrices which seem at first sight to have no meaning can, in fact, be defined. For example, $e^{\mathbf{A}}$ is defined by adapting the power series for a scalar,

$$e^x = 1 + x + x^2/2! + x^3/3! + \cdots,$$

to define $e^{\mathbf{A}}$ as

$$e^{\mathbf{A}} = \mathbf{I} + \mathbf{A} + \mathbf{A}^2/2! + \mathbf{A}^3/3! + \cdots = \sum_{i=0}^{\infty} \mathbf{A}^i/i!,$$

provided the series is convergent. This definition requires $\mathbf{A}$ to be square, in which case $e^{\mathbf{A}}$ has the same order as $\mathbf{A}$.

b.   Matrices of functions

This book deals mostly with matrices having elements which are real numbers. But matrices can have elements which are functions of one variable x, or of several variables $x_1, x_2, \ldots x_r$, represented as $\mathbf{x}$. Using $\mathbf{F}$ to represent matrices of such elements, their functional relationship to x or $\mathbf{x}$ can be represented by $\mathbf{F}(x)$ and $\mathbf{F}(\mathbf{x})$, respectively. Simple examples are

$$\mathbf{F}(x) = \begin{bmatrix} x^3 & e^x \\ 2^x & 3 + 4x^2 \end{bmatrix}$$

and for

$$\mathbf{x} = \begin{bmatrix} x_1 \\ x_2 \end{bmatrix}, \quad \mathbf{F}(\mathbf{x}) = \begin{bmatrix} x_1^2 & x_1 x_2 & x_2^2 \\ x_1 x_2 & x_2^3 & x_1 + x_2 \end{bmatrix}.$$

12.7   Direct Sums and Products

Two functions of matrices that have many practical uses are direct sums and direct products. Their definitions and properties are summarized here.

a.   Direct sums

The direct sum of three matrices $\mathbf{A}$, $\mathbf{B}$ and $\mathbf{C}$ is

$$\mathbf{A} \oplus \mathbf{B} \oplus \mathbf{C} = \begin{bmatrix} \mathbf{A} & 0 & 0 \\ 0 & \mathbf{B} & 0 \\ 0 & 0 & \mathbf{C} \end{bmatrix}. \tag{12.1}$$

$\mathbf{A}$, $\mathbf{B}$ and $\mathbf{C}$ can be of any order; and the null matrices in (12.1) are of appropriate orders, determined by their positions in (12.1) and the orders of $\mathbf{A}$, $\mathbf{B}$ and $\mathbf{C}$. A general expression for the direct sum of matrices $\mathbf{A}_i$ for $i = 1, 2, \ldots, k$ is

$$\bigoplus_{i=1}^{k} \mathbf{A}_i = \left\{ {}_d \mathbf{A}_i \right\}_{i=1}^{k},$$

a block diagonal matrix of the $\mathbf{A}_i$-matrices.

Providing the appropriate existence conditions are satisfied (e.g., conformability, non-singularity, and so on), the following properties of direct sums apply.

(i) $\mathbf{A} \oplus (-\mathbf{A}) \ne \mathbf{0}.$

(ii) $(\mathbf{A} \oplus \mathbf{B}) + (\mathbf{C} \oplus \mathbf{D}) = (\mathbf{A} + \mathbf{C}) \oplus (\mathbf{B} + \mathbf{D}).$

(iii) $(\mathbf{A} \oplus \mathbf{B})(\mathbf{C} \oplus \mathbf{D}) = \mathbf{AC} \oplus \mathbf{BD}.$

(iv) $(\mathbf{A} \oplus \mathbf{B})^{-1} = \mathbf{A}^{-1} \oplus \mathbf{B}^{-1}.$

(v) $|\mathbf{A} \oplus \mathbf{B}| = |\mathbf{A}|\,|\mathbf{B}|,$ if $\mathbf{A}$ and $\mathbf{B}$ are both square

$$= 0 \text{ otherwise.}$$

b.   Direct products

The direct product (often called the Kronecker product) of matrices $\mathbf{A}_{p \times q}$ and $\mathbf{B}_{m \times n}$ is defined as

$$\mathbf{A}_{p \times q} \otimes \mathbf{B}_{m \times n} = \left\{ {}_m\, a_{ij} \mathbf{B} \right\}_{i=1 \; j=1}^{p \quad q}. \tag{12.2}$$

It has order $pm \times qn$, its elements being every possible product of an element of $\mathbf{A}$ with an element of $\mathbf{B}$. A small example is

$$\begin{bmatrix} a_{11} & a_{12} \\ a_{21} & a_{22} \end{bmatrix} \otimes \mathbf{B} = \begin{bmatrix} a_{11}\mathbf{B} & a_{12}\mathbf{B} \\ a_{21}\mathbf{B} & a_{22}\mathbf{B} \end{bmatrix}.$$

Useful properties (assuming existence conditions) are as follows.

(i) $\qquad\qquad \mathbf{x}' \otimes \mathbf{y} = \mathbf{y}\mathbf{x}' = \mathbf{y} \otimes \mathbf{x}', \quad$ for vectors $\mathbf{x}$ and $\mathbf{y}$.

(ii) $\qquad\qquad \lambda \otimes \mathbf{A} = \lambda\mathbf{A} = \mathbf{A} \otimes \lambda = \mathbf{A}\lambda, \quad$ for scalar λ.

(iii) $\qquad\qquad [\mathbf{A}_1 \quad \mathbf{A}_2] \otimes \mathbf{B} = [\mathbf{A}_1 \otimes \mathbf{B} \quad \mathbf{A}_2 \otimes \mathbf{B}]$

$\qquad$ but

$\qquad\qquad\qquad \mathbf{A} \otimes [\mathbf{B}_1 \quad \mathbf{B}_2] \neq [\mathbf{A} \otimes \mathbf{B}_1 \quad \mathbf{A} \otimes \mathbf{B}_2].$

(iv) $\qquad\qquad (\mathbf{A} \otimes \mathbf{B})(\mathbf{X} \otimes \mathbf{Y}) = \mathbf{A}\mathbf{X} \otimes \mathbf{B}\mathbf{Y}.$

(v) $\qquad\qquad (\mathbf{A} \otimes \mathbf{B})' = \mathbf{A}' \otimes \mathbf{B}'$, in contrast to $(\mathbf{A}\mathbf{B})' = \mathbf{B}'\mathbf{A}'.$

(vi) $\qquad\qquad\qquad (\mathbf{A} \otimes \mathbf{B})^{-1} = \mathbf{A}^{-1} \otimes \mathbf{B}^{-1},$

$\qquad$ in contrast to

$\qquad\qquad\qquad (\mathbf{A}\mathbf{B})^{-1} = \mathbf{B}^{-1}\mathbf{A}^{-1}.$

(vii) $\qquad\qquad\qquad r(\mathbf{A} \otimes \mathbf{B}) = r(\mathbf{A})r(\mathbf{B}).$

(viii) $\qquad\qquad\qquad \mathrm{tr}(\mathbf{A} \otimes \mathbf{B}) = \mathrm{tr}(\mathbf{A})\mathrm{tr}(\mathbf{B}).$

(ix) $\qquad\qquad\qquad |\mathbf{A}_{a \times a} \otimes \mathbf{B}_{b \times b}| = |\mathbf{A}|^b |\mathbf{B}|^a.$

(x) $\mathbf{A} \otimes \mathbf{B}$ has eigenroots which are products of those of $\mathbf{A}$ and $\mathbf{B}$, and eigenvectors which are direct products of those of $\mathbf{A}$ and $\mathbf{B}$.

Sometimes the definition in (12.2) is called the *right direct product* to distinguish it from $\mathbf{B} \otimes \mathbf{A}$, which is then called the *left direct product*, and on rare occasions the right-hand side of (12.2) will be found defined as $\mathbf{B} \otimes \mathbf{A}$.

12.8 Vec and Vech Operators

A matrix operation dating back nearly a century is that of stacking the columns of a matrix one under the other to form a single column.

It has had a variety of names, and is currently known as "vec". Thus for

$$\mathbf{X} = \begin{bmatrix} 1 & 2 & 3 \\ 4 & 5 & 6 \end{bmatrix}, \quad \mathrm{vec}(\mathbf{X}) = \begin{bmatrix} 1 \\ 4 \\ 2 \\ 5 \\ 3 \\ 6 \end{bmatrix}.$$

An extension of $\mathrm{vec}(\mathbf{X})$ is $\mathrm{vech}(\mathbf{X})$ for square (primarily symmetric) $\mathbf{X}$. It is defined just as is $\mathrm{vec}(\mathbf{X})$ except that for each column of $\mathbf{X}$ only that part of it which is on and below the diagonal is put into $\mathrm{vech}(\mathbf{X})$— i.e., vector-half of $\mathbf{X}$. In this way, for symmetric $\mathbf{X}$, $\mathrm{vech}(\mathbf{X})$ contains only the distinct elements of $\mathbf{X}$; e.g., for

$$\mathbf{X} = \begin{bmatrix} 1 & 7 & 6 \\ 7 & 3 & 8 \\ 6 & 8 & 2 \end{bmatrix} = \mathbf{X}', \quad \mathrm{vech}(\mathbf{X}) = \begin{bmatrix} 1 \\ 7 \\ 6 \\ 3 \\ 8 \\ 2 \end{bmatrix}.$$

Henderson and Searle (1979) give history, properties and many applications of these operators, including proofs (as does Searle, 1982, Section 12.9) of the following major properties.

(i) $\mathrm{vec}(\mathbf{ABC}) = (\mathbf{C}' \otimes \mathbf{A})\mathrm{vec}(\mathbf{B})$

(ii) $\mathrm{tr}(\mathbf{AB}) = [\mathrm{vec}(\mathbf{A}')]'\mathrm{vec}(\mathbf{B})$

(iii) $\mathrm{vec}(\mathbf{AZ'BZC}) = [\mathrm{vec}(\mathbf{Z})]'(\mathbf{CA} \otimes \mathbf{B})'\mathrm{vec}(\mathbf{Z})$

(iv) $\mathrm{vec}(\mathbf{A}_{m \times n}) = \mathbf{I}_{m,n}\,\mathrm{vec}[(\mathbf{A}')_{n \times m}],$

where $\mathbf{I}_{m,n}$ is a particular kind of permutation matrix called a *vec-permutation matrix*. It occurs in the relationship of $\mathbf{B} \otimes \mathbf{A}$ to $\mathbf{A} \otimes \mathbf{B}$, each of which have as elements all possible products $a_{ij}b_{rs}$:

$$\mathbf{B}_{p \times q} \otimes \mathbf{A}_{m \times n} = \mathbf{I}_{m,p}\,(\mathbf{A}_{m \times n} \otimes \mathbf{B}_{p \times q})\,\mathbf{I}_{q,n}.$$

(v) For symmetric $\mathbf{X}$, elements of $\mathrm{vec}(\mathbf{X})$ are those of $\mathrm{vech}(\mathbf{X})$ with some repetitions. Therefore there are matrices $\mathbf{H}$ and $\mathbf{G}$ such that

for $\mathbf{X} = \mathbf{X}'$, $\mathrm{vech}\,\mathbf{X} = \mathbf{H}\,\mathrm{vec}\,\mathbf{X}$ and $\mathrm{vec}\,\mathbf{X} = \mathbf{G}\mathrm{vech}\,\mathbf{X}.$

These matrices have the properties that $\mathbf{HG} = \mathbf{I}$, $\mathbf{G}$ is unique of full column rank, and $\mathbf{H}$ is not unique, but is of full row rank; and for $\mathbf{X}$ of order n, $|\mathbf{H}(\mathbf{X} \otimes \mathbf{X})\mathbf{G}| = |\mathbf{X}|^{n+1}$.

Full discussion of the preceding items (iv) and (v) is given in Henderson and Searle (1981b).

12.9 Differential Calculus

This section makes simple use of, and should be omitted by readers not familiar with, the differential calculus.

a. Scalars

When differentiating a scalar function of several variables with respect to each variable, it is convenient to write the derivatives as a column vector. As a simple example, for

$$\lambda = 3x_1 + 4x_2 + 9x_3,$$

where λ, x_1, x_2 and x_3 are scalars, we write the derivatives of λ with respect to x_1, x_2 and x_3 as

$$\frac{\partial \lambda}{\partial \mathbf{x}} = \begin{bmatrix} \dfrac{\partial \lambda}{\partial x_1} \\ \dfrac{\partial \lambda}{\partial x_2} \\ \dfrac{\partial \lambda}{\partial x_3} \end{bmatrix} = \begin{bmatrix} \dfrac{\partial}{\partial x_1} \\ \dfrac{\partial}{\partial x_2} \\ \dfrac{\partial}{\partial x_3} \end{bmatrix} \lambda = \begin{bmatrix} 3 \\ 4 \\ 9 \end{bmatrix}.$$

The second vector shows how the symbol $\dfrac{\partial}{\partial \mathbf{x}}$ represents a whole vector of differential operators.

Noting that

$$\lambda = 3x_1 + 4x_2 + 9x_3 = \begin{bmatrix} 3 & 4 & 9 \end{bmatrix} \begin{bmatrix} x_1 \\ x_2 \\ x_3 \end{bmatrix} = \mathbf{a}'\mathbf{x},$$

so defining $\mathbf{a}$ and $\mathbf{x}$, we see that differentiating the scalar $\mathbf{a}'\mathbf{x} = \mathbf{x}'\mathbf{a}$ gives

$$\frac{\partial}{\partial \mathbf{x}}(\mathbf{a}'\mathbf{x}) = \frac{\partial}{\partial \mathbf{x}}(\mathbf{x}'\mathbf{a}) = \mathbf{a}. \qquad (12.3)$$

b. Vectors

Suppose (12.3) is applied to each element of $\mathbf{y}' = \mathbf{x}'\mathbf{A}$ expressed as

$$\mathbf{y}' = [y_1 \ y_2 \ \cdots \ y_i \ \cdots \ y_n] = [\mathbf{x}'\mathbf{a}_1 \ \mathbf{x}'\mathbf{a}_2 \ \cdots \ \mathbf{x}'\mathbf{a}_i \ \cdots \ \mathbf{x}'\mathbf{a}_n]$$
(12.4)

with $\mathbf{a}_i$ being the ith column of $\mathbf{A}$. Applying (12.3) to each element of (12.4) gives

$$
\frac{\partial \mathbf{y}'}{\partial \mathbf{x}} = \left[\frac{\partial y_1}{\partial \mathbf{x}} \ \frac{\partial y_2}{\partial \mathbf{x}} \ \cdots \ \frac{\partial y_n}{\partial \mathbf{x}} \right]
$$

$$
= \left[\frac{\partial \mathbf{x}'\mathbf{a}_1}{\partial \mathbf{x}} \ \frac{\partial \mathbf{x}'\mathbf{a}_2}{\partial \mathbf{x}} \ \cdots \ \frac{\partial \mathbf{x}'\mathbf{a}_n}{\partial \mathbf{x}} \right] = [\mathbf{a}_1 \ \mathbf{a}_2 \ \cdots \ \mathbf{a}_n] = \mathbf{A}.
$$

Thus

$$\frac{\partial}{\partial \mathbf{x}}(\mathbf{x}'\mathbf{A}) = \mathbf{A}. \tag{12.5}$$

Result (12.5) is for differentiating the row vector $\mathbf{y}' = \mathbf{x}'\mathbf{A}$ with respect to elements of $\mathbf{x}$. It is attained by arraying the $\partial y_i/\partial \mathbf{x}$ terms in a row—motivated by $\mathbf{y}'$ being a row. The extension of this to differentiating the column vector $\mathbf{y}$ would therefore seem to be to array the terms $\partial y_i/\partial \mathbf{x}$ as a very long column: for $\mathbf{y}_{p \times 1}$ and $\mathbf{x}_{q \times 1}$ it would have pq elements. As such it would not be consistent with $\partial \mathbf{y}'/\partial \mathbf{x} = \mathbf{A}'$ being a matrix of order $q \times p$. In fact, $\partial \mathbf{y}/\partial \mathbf{x}$ would be $\mathrm{vec}(\partial \mathbf{y}'/\partial \mathbf{x})$. To avoid this inconsistency, the convention is adopted that differentiating $\mathbf{y} = \mathbf{A}\mathbf{x}$ is the same as differentiating $\mathbf{y}' = \mathbf{x}'\mathbf{A}'$; i.e.,

$$\frac{\partial}{\partial \mathbf{x}}(\mathbf{A}\mathbf{x}) \equiv \frac{\partial}{\partial \mathbf{x}}(\mathbf{A}\mathbf{x})' \quad \text{which is} \quad \frac{\partial}{\partial \mathbf{x}}(\mathbf{x}'\mathbf{A}') = \mathbf{A}'. \tag{12.6}$$

For example, for $\mathbf{y} = \mathbf{A}\mathbf{x}$ with $\mathbf{A}$ of order 3×2,

$$\frac{\partial \mathbf{y}}{\partial \mathbf{x}} = \frac{\partial}{\partial \mathbf{x}}(\mathbf{A}\mathbf{x}) = \begin{bmatrix} \dfrac{\partial y_1}{\partial x_1} & \dfrac{\partial y_2}{\partial x_1} & \dfrac{\partial y_3}{\partial x_1} \\ \dfrac{\partial y_1}{\partial x_2} & \dfrac{\partial y_2}{\partial x_2} & \dfrac{\partial y_3}{\partial x_2} \end{bmatrix} = \mathbf{A}'. \tag{12.7}$$

Example

Applying (12.7) to

$$\mathbf{A}\mathbf{x} = \begin{bmatrix} 2 & 6 \\ 3 & -2 \\ 3 & 4 \end{bmatrix} \begin{bmatrix} x_1 \\ x_2 \end{bmatrix} = \begin{bmatrix} 2x_1 + 6x_2 \\ 3x_1 - 2x_2 \\ 3x_1 + 4x_2 \end{bmatrix}$$

gives

$$\frac{\partial}{\partial \mathbf{x}}(\mathbf{Ax}) = \left[\frac{\partial}{\partial \mathbf{x}}(2x_1 + 6x_2) \quad \frac{\partial}{\partial \mathbf{x}}(3x_1 - 2x_2) \quad \frac{\partial}{\partial \mathbf{x}}(3x_1 + 4x_2)\right]$$

$$= \begin{bmatrix} 2 & 3 & 3 \\ 6 & -2 & 4 \end{bmatrix} = \mathbf{A}'.$$

c.   Quadratic forms

Adapting the principle of differentiation by parts, differentiating $\mathbf{x}'\mathbf{Ax}$ with respect to $\mathbf{x}$ gives

$$\begin{aligned}
\frac{\partial}{\partial \mathbf{x}}(\mathbf{x}'\mathbf{Ax}) &= \frac{\partial}{\partial \mathbf{x}}(\mathbf{x}'\mathbf{P}) + \frac{\partial}{\partial \mathbf{x}}(\mathbf{Qx}) \quad \text{for} \quad \mathbf{P} = \mathbf{Ax} \quad \text{and} \quad \mathbf{Q} = \mathbf{x}'\mathbf{A} \\
&= \mathbf{P} + \mathbf{Q}', \qquad \text{from (12.5) and (12.6)} \\
&= \mathbf{Ax} + \mathbf{A}'\mathbf{x}.
\end{aligned}$$

Usually $\mathbf{A}$ is taken as symmetric, $\mathbf{A} = \mathbf{A}'$, in which case

$$\frac{\partial}{\partial \mathbf{x}}(\mathbf{x}'\mathbf{Ax}) = 2\mathbf{Ax}. \tag{12.8}$$

d.   Inverses

Suppose elements of $\mathbf{A}_{r \times c}$ are functions of a scalar $\mathbf{x}$. The differential of $\mathbf{A}$ with respect to x is defined as the matrix of each element of $\mathbf{A}$ differentiated with respect to x:

$$\frac{\partial \mathbf{A}}{\partial x} = \left\{ \frac{\partial a_{ij}}{\partial x} \right\} \quad \text{for} \quad i = 1, \ldots, r \quad \text{and} \quad j = 1, \ldots, c. \tag{12.9}$$

Applying this to the product $\mathbf{A}\mathbf{A}^{-1} = \mathbf{I}$ for non-singular $\mathbf{A}$ gives

$$\frac{\partial \mathbf{A}^{-1}}{\partial x} = -\mathbf{A}^{-1}\frac{\partial \mathbf{A}}{\partial x}\mathbf{A}^{-1}; \tag{12.10}$$

and (12.9) applied to $\mathbf{A}\mathbf{A}^{-}\mathbf{A} = \mathbf{A}$, followed by pre-multiplication by $\mathbf{A}\mathbf{A}^{-}$ and post-multiplication by $\mathbf{A}^{-}\mathbf{A}$ gives

$$\mathbf{A}\frac{\partial \mathbf{A}^{-}}{\partial x}\mathbf{A} = -\mathbf{A}\mathbf{A}^{-}\frac{\partial \mathbf{A}}{\partial x}\mathbf{A}^{-}\mathbf{A},$$

which does, of course, reduce to (12.10) for non-singular $\mathbf{A}$.

e. Traces

Differentiating a scalar θ that is a function of elements of $\mathbf{X}_{p \times q}$, is defined by

$$\frac{\partial \theta}{\partial \mathbf{X}} = \left\{ \frac{\partial \theta}{\partial x_{ij}} \right\} \quad \text{for} \quad i = 1, \ldots, p \text{ and } j = 1, \ldots, q.$$

Trace functions involving $\mathbf{X}$ are special cases of this. For $\mathbf{X}$ having functionally independent elements

$$\frac{\partial}{\partial x_{ij}} \text{tr}(\mathbf{X}\mathbf{A}) = \frac{\partial}{\partial x_{ij}} \sum_r \sum_s x_{rs} a_{sr} = a_{ji}$$

and the matrix of all such results is

$$\frac{\partial}{\partial \mathbf{X}} \text{tr}(\mathbf{X}\mathbf{A}) = \mathbf{A}'. \tag{12.11}$$

But (12.11) does not hold when elements of $\mathbf{X}$ are functionally related as, for example, when $\mathbf{X}$ is symmetric. Then, as is shown in Henderson and Searle [1981a, equation (58)],

$$\frac{\partial}{\partial \mathbf{X}} \text{tr}(\mathbf{X}\mathbf{A}) = \mathbf{A} + \mathbf{A}' - \left\{ {}_d \, a_{ii} \right\} \quad \text{for symmetric } \mathbf{X}. \tag{12.12}$$

f. Determinants

For having all elements functionally independent, it is clear that

$$\frac{\partial \mathbf{X}}{\partial x_{ij}} = \mathbf{E}_{ij},$$

a null matrix except for unity as the (i, j)th element, the matrix $\mathbf{E}_{ij} = \mathbf{e}_i \mathbf{e}_j'$ being as defined in Section 3.2f. Hence

$$\frac{\partial (\mathbf{P}\mathbf{X})}{\partial x_{ij}} = \mathbf{P}\mathbf{E}_{ij} \quad \text{and so} \quad \frac{\partial (\mathbf{P}\mathbf{X})}{\partial \mathbf{X}} = \mathbf{0} \quad \text{implies} \quad \mathbf{P} = \mathbf{0}.$$

Similar results apply for symmetric $\mathbf{X}$, for which

$$\frac{\partial \mathbf{X}}{\partial x_{ij}} = \mathbf{E}_{ij} + \mathbf{E}_{ji} - \delta_{ij} \mathbf{E}_{ij} \tag{12.13}$$

for $\delta_{ij} = 0$ when $i \neq j$ and $\delta_{ij} = 1$ when $i = j$.

Recalling from equation (5.6) of Section 5.4 that $|\mathbf{X}| = \sum_j x_{ij}|\mathbf{X}_{ij}|$ for each i, where $\mathbf{X} = \{x_{ij}\}$ and $|\mathbf{X}_{ij}|$ is the cofactor of x_{ij}, we have

$$\frac{\partial|\mathbf{X}|}{\partial x_{ij}} = |\mathbf{X}_{ij}| \quad \text{for } \mathbf{X} \text{ having functionally independent elements.}$$

$$(12.14)$$

But, similar to (12.14),

$$
\begin{aligned}
\frac{\partial|\mathbf{X}|}{\partial x_{ij}} &= |\mathbf{X}_{ij}| + |\mathbf{X}_{ji}| - \delta_{ij}|\mathbf{X}_{ij}| \quad \text{for symmetric } \mathbf{X}, \\
&= (2 - \delta_{ij})|\mathbf{X}_{ij}| \quad \text{for symmetric } \mathbf{X}. \quad (12.15)
\end{aligned}
$$

Therefore, on denoting the adjoint (or adjugate) matrix of $\mathbf{X}$ by adj $\mathbf{X}$ (as in Section 5.5c), and on arraying results (12.14) and (12.15) as matrices, we have, for $\mathbf{X}^{-1} = \text{adj}(\mathbf{X})/|\mathbf{X}|$,

$$\frac{\partial|\mathbf{X}|}{\partial\mathbf{X}} = (\text{adj}\,\mathbf{X})' = |\mathbf{X}|(\mathbf{X}^{-1})' \quad \text{for } \mathbf{X} \text{ having functionally unrelated elements} \quad (12.16)$$

and

$$
\begin{aligned}
\frac{\partial|\mathbf{X}|}{\partial\mathbf{X}} &= 2(\text{adj}\,\mathbf{X}) - \left\{_d\, |\mathbf{X}_{ii}|\right\} \\
&= |\mathbf{X}|\left[2\mathbf{X}^{-1} - \left\{_d\, x^{ii}\right\}\right] \quad \text{for symmetric } \mathbf{X}, \quad (12.17)
\end{aligned}
$$

where x^{ii} is the ith diagonal element of $\mathbf{X}^{-1}$. Furthermore, for nonsingular $\mathbf{X}$

$$
\begin{aligned}
\frac{\partial}{\partial\mathbf{X}}\log|\mathbf{X}| &= \frac{1}{|\mathbf{X}|}\frac{\partial|\mathbf{X}|}{\partial\mathbf{X}} \\
&= \mathbf{X}^{-1'}, \text{ from (12.16), for } \mathbf{X} \text{ with functionally independent elements} \quad (12.18) \\
&= 2\mathbf{X}^{-1} - \text{diag}(\mathbf{X}^{-1}), \text{ from (12.17), for symmetric } \mathbf{X}. \quad (12.19)
\end{aligned}
$$

Results (12.11) and (12.18) are well known; (12.12) and (12.19) come from Henderson and Searle (1981a).

Another useful derivative is $\partial \log |\mathbf{X}|/\partial w$. For $\mathbf{X}$ having functionally independent elements, it is

$$\frac{\partial}{\partial w}\log|\mathbf{X}| = \frac{1}{|\mathbf{X}|}\frac{\partial|\mathbf{X}|}{\partial w} = \frac{1}{|\mathbf{X}|}\sum_i\sum_j\frac{\partial|\mathbf{X}|}{\partial x_{ij}}\frac{\partial x_{ij}}{\partial w}. \quad (12.20)$$

Observing that, in general, $\text{tr}(\mathbf{K}'\mathbf{M}) = \sum_i \sum_j k_{ij} m_{ij}$, gives

$$\frac{\partial}{\partial w} \log |\mathbf{X}| = \frac{1}{|\mathbf{X}|} \text{tr} \left[\left(\frac{\partial |\mathbf{X}|}{\partial \mathbf{X}} \right)' \frac{\partial \mathbf{X}}{\partial w} \right] = \text{tr} \left[\left(\frac{1}{|\mathbf{X}|} \frac{\partial |\mathbf{X}|}{\partial \mathbf{X}} \right)' \frac{\partial \mathbf{X}}{\partial w} \right]$$

$$= \text{tr} \left(\mathbf{X}^{-1} \frac{\partial \mathbf{X}}{\partial w} \right), \quad \text{from (12.18).} \quad (12.21)$$

When $\mathbf{X}$ is symmetric, differentiation is needed only with respect to the distinct elements of $\mathbf{X}$ so that, in contrast to (12.20),

$$\frac{\partial}{\partial w} \log |\mathbf{X}| = \frac{1}{|\mathbf{X}|} \sum_{i \leq j} \sum \frac{\partial |\mathbf{X}|}{\partial x_{ij}} \frac{\partial x_{ij}}{\partial w} \quad \text{for symmetric } \mathbf{X}.$$

But this, in combination with (12.17), gives

$$\frac{\partial}{\partial w} \log |\mathbf{X}| = \frac{1}{|\mathbf{X}|} \sum_{i \leq j} \sum (2 - \delta_{ij}) |\mathbf{X}_{ij}| \frac{\partial x_{ij}}{\partial w} \quad \text{for } \mathbf{X} = \mathbf{X}'$$

$$= \frac{1}{|\mathbf{X}|} \sum_i \sum_j |\mathbf{X}_{ij}| \frac{\partial x_{ij}}{\partial w} = \text{tr} \left(\mathbf{X}^{-1} \frac{\partial \mathbf{X}}{\partial w} \right),$$

which is (12.21); i.e., (12.21) applies for symmetric and asymmetric $\mathbf{X}$.

g. Jacobians

When $\mathbf{y}$ is a vector of n differentiable functions of the n elements of $\mathbf{x}$, of such a nature that the transformation $\mathbf{x} \to \mathbf{y}$ is 1-to-1, then the matrix

$$\mathbf{J}_{\mathbf{x} \to \mathbf{y}} = \left(\frac{\partial \mathbf{x}}{\partial \mathbf{y}} \right)' = \left\{ \frac{\partial x_i}{\partial y_j} \right\} \quad \text{for} \quad i, j = 1, \ldots, n \quad (12.22)$$

is the *Jacobian matrix* of the transformation from $\mathbf{x}$ to $\mathbf{y}$. For example, for the linear transformation $\mathbf{y} = \mathbf{Ax}$, the 1-to-1 property ensures existence of $\mathbf{A}^{-1}$ and so $\mathbf{x} = \mathbf{A}^{-1}\mathbf{y}$. Hence, using (12.7), the Jacobian matrix (12.22) for

$$\mathbf{y} = \mathbf{Ax}$$

is

$$\mathbf{J}_{\mathbf{x} \to \mathbf{y}} = \left[\frac{\partial (\mathbf{A}^{-1}\mathbf{y})}{\partial \mathbf{y}} \right]' = (\mathbf{A}^{-1'})' = \mathbf{A}^{-1}.$$

Determinants of Jacobian matrices are required when making a change of variables in integral calculus. Suppose in an integral involving xs, a 1-to-1 transformation to ys is to be made. The integral is transformed by substituting for the xs in terms of the ys and by replacing $dx_1 dx_2 dx_3 \cdots dx_n$ (denoted by $d\mathbf{x}$) by $\|\mathbf{J}_{\mathbf{x}\to\mathbf{y}}\| dy_1 dy_2 dy_3 \cdots dy_n$, where $\|\mathbf{J}_{\mathbf{x}\to\mathbf{y}}\|$ is the absolute value of the determinant of $\mathbf{J}_{\mathbf{x}\to\mathbf{y}}$. In brief, for the transformation

$$\mathbf{x} \to \mathbf{y} \text{ make the replacement } d\mathbf{x} \to \|\mathbf{J}_{\mathbf{x}\to\mathbf{y}}\| d\mathbf{y}. \qquad (12.23)$$

Although $\mathbf{J}_{\mathbf{x}\to\mathbf{y}}$ is the Jacobian matrix, $\|\mathbf{J}_{\mathbf{x}\to\mathbf{y}}\|$ is known as the *Jacobian* of the transformation.

Because the transformation in (12.23) is from $\mathbf{x}$ to $\mathbf{y}$, elements of $\mathbf{y}$ will usually be expressed as functions of those of $\mathbf{x}$ (rather than vice versa). Hence derivation of $\mathbf{J}_{\mathbf{y}\to\mathbf{x}}$ with elements $\partial y_i/\partial x_j$ will be easier to obtain than $\mathbf{J}_{\mathbf{x}\to\mathbf{y}}$ with elements $\partial x_i/\partial y_j$. This presents no problem, though, because it can be shown that $\mathbf{J}_{\mathbf{x}\to\mathbf{y}}\mathbf{J}_{\mathbf{y}\to\mathbf{x}} = \mathbf{I}$ and so we always have

$$\|\mathbf{J}_{\mathbf{x}\to\mathbf{y}}\| = 1/\|\mathbf{J}_{\mathbf{y}\to\mathbf{x}}\|. \qquad (12.24)$$

Hence, whichever of $\mathbf{J}_{\mathbf{x}\to\mathbf{y}}$ and $\mathbf{J}_{\mathbf{y}\to\mathbf{x}}$ is most easily obtainable from the transformation can be used in (12.23), either $\|\mathbf{J}_{\mathbf{x}\to\mathbf{y}}\|$ or $1/\|\mathbf{J}_{\mathbf{y}\to\mathbf{x}}\|$ of (12.24). This being so, relying upon memory for using (12.23) can raise a doubt: "Is it $\mathbf{J}_{\mathbf{x}\to\mathbf{y}}$ or $\mathbf{J}_{\mathbf{y}\to\mathbf{x}}$?" This is especially important when symbols other than $\mathbf{x}$ and $\mathbf{y}$ are being used. Fortunately a useful mnemomic[1] is available for those who like such an aid. Denote the variables by "o" for "old" and "n" for "new", where the transformation is thought of as being from "old" variables to "new" variables. Then in (12.23) the Jacobian matrix is $\mathbf{J}_{o\to n}$ where the subscripts are "on"; in contrast, the wrong matrix $\mathbf{J}_{n\to o}$ has subscripts "no"—and it is *not* "no"; i.e., it is not $\mathbf{J}_{\mathbf{y}\to\mathbf{x}}$.

Example

For the transformation from x_1, x_2 to y_1, y_2 represented by

$$\left.\begin{array}{l} y_1 = e^{x_1} - x_2 \\ y_2 = x_2 \end{array}\right\} \text{ we have } \|\mathbf{J}_{\mathbf{y}\to\mathbf{x}}\| = \left\| \begin{array}{cc} e^{x_1} & -1 \\ 0 & 1 \end{array} \right\| = e^{x_1} = y_1 + y_2.$$

The inverse relationship is, for $r = 1/(y_1 + y_2)$,

$$\left.\begin{array}{l} x_1 = \log(y_1 + y_2) \\ x_2 = y_2 \end{array}\right\} \text{ with } \|\mathbf{J}_{\mathbf{x}\to\mathbf{y}}\| = \left\| \begin{array}{cc} r & r \\ 0 & 1 \end{array} \right\| = 1/(y_1 + y_2).$$

---

[1] We are indebted to Daniel L. Solomon for this.

Sometimes variables are arrayed not as vectors but as matrices, in which case, putting the matrices in vector form enables (12.22) and (12.23) to be used for deriving Jacobians. In general

$$\mathbf{J_{X \to Y}} = \frac{\partial \text{vec}(\mathbf{X})}{\partial \text{vec}(\mathbf{Y})}$$

and

$$\mathbf{J_{X \to Y}} = \frac{\partial \text{vech}(\mathbf{X})}{\partial \text{vech}(\mathbf{Y})} \quad \text{for } \mathbf{X} \text{ and } \mathbf{Y} \text{ symmetric.}$$

These techniques provide straightforward derivation of results such as those shown in Table 12.1. Henderson and Searle (1981a) give details and discuss earlier derivations.

Table 12.1. Jacobians for Certain Matrix Transformations

	Jacobian: $\|\mathbf{J_{X \to Y}}\|$	
	Matrices Having	
Transformation:	Functionally	Symmetric Matrices
$\mathbf{X} \to \mathbf{Y}$	Independent Elements	$\mathbf{X} = \mathbf{X}'$ and $\mathbf{Y} = \mathbf{Y}'$
$\mathbf{Y}_{p \times q} = \mathbf{A} \mathbf{X}_{p \times q} \mathbf{B}$	$\|\mathbf{A}\|^{-q} \|\mathbf{B}\|^{-p}$	Not applicable
$\mathbf{Y}_{p \times p} = \mathbf{A} \mathbf{X}_{p \times p} \mathbf{A}'$	$\|\mathbf{A}\|^{-2p}$	$\|\mathbf{A}\|^{-(p+1)}$
$\mathbf{Y}_{n \times n} = \mathbf{X}^{-1}$	$\|\mathbf{X}\|^{2n}$	$\|\mathbf{X}\|^{n+1}$
$\mathbf{Y}_{n \times n} = \mathbf{X}^k$	$\left(\prod\limits_{i=1}^{n} \prod\limits_{j=1}^{n} \sum\limits_{r=1}^{k} \lambda_i^{k-r} \lambda_j^{r-1} \right)^{-1}$	$\left(\prod\limits_{i=1}^{n} \prod\limits_{j \geq i}^{n} \sum\limits_{r=1}^{k} \lambda_i^{k-r} \lambda_j^{r-1} \right)^{-1}$

Note: All determinants are taken positively, and in the last line $\lambda_1, \lambda_2, \ldots, \lambda_n$ are eigenvalues of $\mathbf{X}$; symmetric $\mathbf{X}$ differs from nonsymmetric only in having $j \geq i$.

h. Hessians

Another matrix of partial differentials is the Hessian. If θ is a function of n variables $x_1, x_2, \ldots, x_n$, the matrix of second-order partial derivatives of θ with respect to the xs is called a *Hessian*: for $i, j = 1, 2, \ldots, n$,

$$\mathbf{H} = \left\{ \frac{\partial^2 \theta}{\partial x_i \partial x_j} \right\} = \frac{\partial^2 \theta}{\partial \mathbf{x} \partial \mathbf{x}'}.$$

Illustration (Maximum likelihood)

One of the properties of maximum likelihood in statistics is that, under certain conditions (see, for example, Casella and Berger, 1990,

Section 7.4.1), the variance-covariance matrix of a set of simultaneously estimated parameters is the inverse of minus the expected value of the Hessian of the likelihood with respect to those parameters.

12.10   Exercises

E 12.1. From result i in Section 12.1, calculate orthogonal matrices from

$$
\begin{bmatrix} 0 & -1 & 1 \\ 1 & 0 & 2 \\ -1 & -2 & 0 \end{bmatrix}
\quad \text{and} \quad
\begin{bmatrix} 0 & 1 & -2 \\ -1 & 0 & 5 \\ 2 & -5 & 0 \end{bmatrix},
$$

and use them to verify results e through h in Section 12.1.

E 12.2. (a) Calculate $\mathbf{P}$ of result i in Section 12.2, using

$$
\mathbf{X} = \begin{bmatrix} 1 & 2 \\ 0 & -1 \\ -1 & 0 \end{bmatrix}.
$$

(b) Verify appropriate results in Section 12.2 for $\mathbf{P}$ of E 12.2a.
(c) Repeat E 12.2(b) for the algebraic form of $\mathbf{P}$.

E 12.3. Using an orthogonal matrix calculated in E 12.1 as $\mathbf{Q}$, and $\mathbf{P}$ from E 12.2, show that $\mathbf{Q'PQ}$ is idempotent.

E 12.4. Calculate $\mathbf{B}^2$ and eigenvalues for

$$
\mathbf{B} = \tfrac{1}{8} \begin{bmatrix} 11 & 3 & 3 \\ 1 & 1 & 1 \\ -12 & -4 & -4 \end{bmatrix}.
$$

E 12.5. For $\mathbf{Q} = \mathbf{X(YX)}^{-1}\mathbf{Y}$, prove the following:
(a) $\mathbf{Q}$ is idempotent.
(b) $\mathbf{Q} = \mathbf{I}$ when $\mathbf{X}$ and $\mathbf{Y}$ are non-singular.
(c) $\text{tr}(\mathbf{Q})$ = number of rows in $\mathbf{Y}$.

E 12.6. For $\mathbf{D} = \mathbf{X(X'V}^{-1}\mathbf{X)}^{-1}\mathbf{X'V}^{-1}$ with $\mathbf{X}$ rectangular and $\mathbf{V}$ symmetric, prove the following:

(a) $\mathbf{D}$ is idempotent.

(b) $\mathbf{D}' = \mathbf{V}^{-1}\mathbf{D}\mathbf{V}$.

(c) $(\mathbf{I} - \mathbf{D}')\mathbf{V}^{-1}(\mathbf{I} - \mathbf{D}) = (\mathbf{I} - \mathbf{D}')\mathbf{V}^{-1} = \mathbf{V}^{-1}(\mathbf{I} - \mathbf{D})$.

E 12.7. For $\mathbf{A}$ being idempotent of order n, prove $r_{\mathbf{I}-\mathbf{A}} = n - r_{\mathbf{A}}$ in two ways: (a) using the trace-equal-to-rank property of $\mathbf{A}$, and (b) considering LIN solutions of $(\mathbf{I} - \mathbf{A})\mathbf{x} = \mathbf{0}$.

E 12.8. Prove that (a) a symmetric, and (b) a diagonable, matrix is idempotent if and only if its eigenvalues are all zero or unity.

E 12.9. Suppose $\mathbf{A}$ and $\mathbf{V}$ are symmetric, and $\mathbf{V}$ is p.d. If $\mathbf{A}\mathbf{V}$ has eigenvalues that are all zero or unity, prove that $\mathbf{A}\mathbf{V}$ is idempotent.

E 12.10. Prove that

$$\begin{vmatrix} a & c\mathbf{1}' \\ c\mathbf{1} & (a-b)\mathbf{I}_r + b\mathbf{J}_r \end{vmatrix} = (a-b)^{r-1}[a^2 + (r-1)ab - rc^2].$$

E 12.11. Prove that

$$e^{a\mathbf{I}_r + b\mathbf{J}_r} = e^a\mathbf{I}_r + \tfrac{1}{r}(e^{a+rb} - e^a)\mathbf{J}_r.$$

E 12.12. (a) Show that $a\mathbf{I} + b\mathbf{J}$ of order n is diagonable.

(b) For $n = 4$ show that

$$\frac{1}{\sqrt{4}}\begin{bmatrix} 1 \\ 1 \\ 1 \\ 1 \end{bmatrix}, \quad \frac{1}{\sqrt{2}}\begin{bmatrix} 1 \\ -1 \\ 0 \\ 0 \end{bmatrix}, \quad \frac{1}{\sqrt{6}}\begin{bmatrix} 1 \\ 1 \\ -2 \\ 0 \end{bmatrix} \quad \text{and} \quad \frac{1}{\sqrt{12}}\begin{bmatrix} 1 \\ 1 \\ 1 \\ -3 \end{bmatrix}$$

are orthogonal vectors.

E 12.13. Under what conditions is $a\mathbf{I} + b\mathbf{J}$ of order n (a) idempotent and (b) orthogonal?

E 12.14. When $\mathbf{V}$ is p.d., prove that $(\mathbf{X}'\mathbf{V}^{-1}\mathbf{X})^{-}\mathbf{X}'\mathbf{V}^{-1}$ is a generalized inverse of $\mathbf{X}$.

E 12.15. Prove that $\mathbf{A}(\mathbf{A}'\mathbf{A})^{-2}\mathbf{A}'$ is the Moore-Penrose inverse of the n.n.d. matrix $\mathbf{A}\mathbf{A}'$.

E 12.16. If $\mathbf{K}$ is square of order n, what values of $\mathbf{K}$ make $e^{\mathbf{K}}$ equal to the following?
(a) $e\mathbf{I}$ (b) $\mathbf{I} - \mathbf{K} + e\mathbf{K}$
(c) $\mathbf{I}$ (d) $e^{\lambda}\mathbf{I}$
(e) $\mathbf{I} + \mathbf{K}$ (f) $\mathbf{I} + \mathbf{K}(e^{n} - 1)/n$

E 12.17. For two simple arithmetic matrices of your own creation (other than $\mathbf{I}$, $\mathbf{J}$, $\mathbf{0}$ and diagonal matrices) confirm items (i) through (x) of Section 12.7.

E 12.18. Prove: $\text{vec}(\mathbf{PQ}) = (\mathbf{Q}' \otimes \mathbf{I})\text{vec}(\mathbf{P}) = (\mathbf{Q}' \otimes \mathbf{P})\text{vec}(\mathbf{I}) = (\mathbf{I} \otimes \mathbf{P})\text{vec}(\mathbf{Q})$.

E 12.19. For $\mathbf{X} = \begin{bmatrix} 1 & 2 \\ 2 & 5 \end{bmatrix}$, derive $\mathbf{G}$ and $\mathbf{H}$ of Section 12.8 and illustrate their properties stated there.

E 12.20. Show that with respect to x_1 and x_2, the Jacobian of

$$y_1 = 6x_1^2 x^2 + 2x_1 x_2 + x_2^2 \quad \text{and} \quad y_2 = 2x_1^3 + x_1^2 + 2x_1 x_2$$

is the same as the Hessian of

$$y = 2x_1^3 x_2 + x_1^2 x_2 + x_1 x_2^2.$$

E 12.21. Prove: $\partial(\mathbf{PX} + \mathbf{X}'\mathbf{Q})/\partial\mathbf{X} = \mathbf{0}$ implies $\mathbf{P} = -\mathbf{Q}'$.

Part III

WORKING WITH MATRICES

Chapter 13

APPLYING LINEAR EQUATIONS TO ECONOMICS

Economics has several opportunities to use matrix algebra in understanding the behavior of firms, consumers and the market. Some of these applications take the form of linear equations $\mathbf{Ax} = \mathbf{y}$, where $\mathbf{A}$ is a matrix of coefficients and $\mathbf{x}$ and $\mathbf{y}$ are vectors of unknown and known values, respectively.

This chapter deals with three applications of matrices in economics, each using linear equations. First, we present the cost minimization approach to the theory of the firm and the determination of cross-price effects. Second, the consumer's demand, specifically the effects of price changes on demand, are presented. For a complete description of both the theory of the firm and of consumer theory, see Varian (1992) and Silberberg (1990). Finally, we present input-output analysis of an economy poised to satisfy the demand.

13.1 Cost Minimization in a Firm

Economists assume that a firm wishes to maximize its profits. Alternatively, a firm could minimize its costs when purchasing its factors of production. Assuming a firm operates in perfectly competitive output and input markets where the firm has no effect on output price or

271

input prices, one can write the firm's cost minimization requirement as

$$\text{Minimize}\quad \mathbf{w}'\mathbf{x},$$

provided the firm operates on its production function, which is often expressed as $F(\mathbf{x}) = y$. In the firm's problem $\mathbf{w}'$ is a $1 \times n$ vector of factor prices, $\mathbf{x}$ is an $n \times 1$ vector of inputs, with $\mathbf{x}$ being a function of $\mathbf{w}'$ and the output y, and y, a scalar, is a measure of the firm's output.

Ultimately the demand function for inputs, $\mathbf{x}$, must satisfy necessary first-order conditions for cost minimization. These conditions are

$$\begin{aligned}
\mathbf{w} - \lambda\mathbf{k} &= \mathbf{0} \qquad\qquad (13.1)\\
F(\mathbf{x}) - y &= 0
\end{aligned}$$

where λ is a Lagrangian multiplier and $\mathbf{k}$ is the vector of derivatives of $F(\mathbf{x}) = y$ with respect to each x_i;

$$\mathbf{k} = \begin{bmatrix} \partial y/\partial x_1 \\ \partial y/\partial x_2 \\ \vdots \\ \partial y/\partial x_n \end{bmatrix}.$$

The first-order conditions suggest that the *economic rate of substitution* (the ratio of input prices to each other) must equal the *marginal rate of technical substitution* (the physical trade-off of goods) while maintaining a constant cost. If this equality does not hold, then arbitrage will occur and one factor of production will be substituted for another factor until costs are minimized. These first-order conditions are considered to be identities, true for all values of $\mathbf{w}'$ and y. We can differentiate the first-order conditions with respect to $\mathbf{w}'$ to obtain

$$\mathbf{I}_n - \lambda\mathbf{RS}\ -\ \mathbf{k}\boldsymbol{\ell}' = \mathbf{0} \qquad\qquad (13.2)$$

and

$$\mathbf{k}'\mathbf{S} = \mathbf{0} \qquad\qquad (13.3)$$

where

$$\mathbf{R} = \frac{\partial\mathbf{k}}{\partial\mathbf{x}'} = \frac{\partial^2 y}{\partial\mathbf{x}\partial\mathbf{x}'},$$

since

$$\mathbf{k} = \frac{\partial y}{\partial\mathbf{x}} \qquad\qquad (13.4)$$

as described following (13.1). And

$$\mathbf{S} = \frac{\partial \mathbf{x}}{\partial \mathbf{w}'};$$

both $\mathbf{R}$ and $\mathbf{S}$ are square, of order n, and $\mathbf{R}$ is symmetric. Finally for (13.2) and (13.3) which, in many cases, is a null vector,

$$\boldsymbol{\ell}' = \frac{\partial \lambda}{\partial \mathbf{w}'}.$$

Rearranging and combining (13.2) and (13.3) gives the equations

$$\begin{bmatrix} \lambda \mathbf{R} & \mathbf{k} \\ \mathbf{k}' & 0 \end{bmatrix} \begin{bmatrix} \mathbf{S} \\ \boldsymbol{\ell}' \end{bmatrix} = \begin{bmatrix} \mathbf{I} \\ \mathbf{0} \end{bmatrix}. \tag{13.5}$$

The matrix $\begin{bmatrix} \mathbf{S} \\ \boldsymbol{\ell}' \end{bmatrix}$, known as the *substitution matrix*, can, if $\mathbf{R}^{-1}$ exists, be obtained from (13.5) as

$$\begin{bmatrix} \mathbf{S} \\ \boldsymbol{\ell}' \end{bmatrix} = \left\{ \begin{bmatrix} \frac{1}{\lambda}\mathbf{R}^{-1} & 0 \\ 0 & 0 \end{bmatrix} + \begin{bmatrix} -\frac{1}{\lambda}\mathbf{R}^{-1}\mathbf{k} \\ \mathbf{I} \end{bmatrix} (-\theta) \begin{bmatrix} -\frac{1}{\lambda}\mathbf{k}'\mathbf{R}^{-1} & \mathbf{I} \end{bmatrix} \right\} \begin{bmatrix} \mathbf{I} \\ \mathbf{0} \end{bmatrix}$$

$$= \begin{bmatrix} \frac{1}{\lambda}\mathbf{R}^{-1} \\ 0 \end{bmatrix} + \begin{bmatrix} \frac{-\theta}{\lambda^2}\mathbf{R}^{-1}\mathbf{k}\mathbf{k}'\mathbf{R}^{-1} \\ \frac{\theta}{\lambda}\mathbf{k}'\mathbf{R}^{-1} \end{bmatrix} \quad \text{for} \quad \theta = \frac{\lambda}{\mathbf{k}'\mathbf{R}^{-1}\mathbf{k}}, \tag{13.6}$$

after using the first partitioned inverse at the end of Section 5.10.

This is an important result in the analysis of the firm because it captures the cross-price effects for the firm. These effects can be described simply as $\partial x_i/\partial w_j$ and $\partial x_j/\partial w_i$, for all i, j. Notice that the partitioned matrix in (13.5) is symmetric because $\mathbf{R}$ is. Hence its inverse is symmetric. Thus the cross-price effects, captured by $\mathbf{S}$, are symmetric. In other words $\partial x_i/\partial w_j = \partial x_j/\partial w_i$ for all i and j.

Illustration (Production)

Suppose you owned a firm which uses two inputs, x_1 and x_2, in the production of y. You wish to minimize costs where w_1 and w_2 are the prices of the inputs. Your production function can be described as $F(\mathbf{x}) = x_1 x_2 = y$. Then for equation (13.5)

$$\mathbf{k} = \begin{bmatrix} x_2 \\ x_1 \end{bmatrix}, \qquad \mathbf{R} = \begin{bmatrix} 0 & 1 \\ 1 & 0 \end{bmatrix} \quad \text{and we take } \boldsymbol{\ell}' \text{ null};$$

and so that equation is

$$\begin{bmatrix} 0 & \lambda & x_2 \\ \lambda & 0 & x_1 \\ x_2 & x_1 & 0 \end{bmatrix} \begin{bmatrix} \dfrac{\partial x_1}{\partial w_1} & \dfrac{\partial x_2}{\partial w_1} \\ \dfrac{\partial x_1}{\partial w_2} & \dfrac{\partial x_2}{\partial w_2} \\ 0 & 0 \end{bmatrix} = \begin{bmatrix} 1 & 0 \\ 0 & 1 \\ 0 & 0 \end{bmatrix}. \tag{13.7}$$

Thus from (13.6) we can get the solution for $\begin{bmatrix} \mathbf{S} \\ \boldsymbol{\ell}' \end{bmatrix}$. See E 13.3. The result identifies the impact of a change in the cost of an input on the change in your factor demand.

13.2   Consumer's Utility Maximization and Expenditure Minimization

Basic economic theory assumes that consumers maximize utility subject to a budget constraint.   This leads to a characterization of consumer behavior such that the consumer marginal rate of substitution between two goods is equal to the economic rate of substitution. If not, a consumer would give up a unit of one good, purchase a unit of another good and increase utility. The resulting demand functions for the consumer, the *Marshallian demand functions*, identify the amount of each good a consumer desires for a given level of prices and income.

Alternatively, a consumer can be assumed to minimize the cost of achieving a fixed level of utility. The functions describing the cost-minimizing bundle of goods necessary to achieve utility at fixed prices are known as the *Hicksian* or *compensated demand functions*. These are not observable since they depend on utility. But Marshallian demand functions are simply the market demand functions we observe.

Economists are interested in the effect on demand of a change in the price of a good or goods. The *Slutsky equation* is used to describe this effect and captures the fundamental relationship between the Marshallian demand curve derived from utility maximization and the Hicksian demand curve derived from expenditure minimization. The Slutsky equation says the effect of a price change of a good on demand can be split into the *substitution effect* and the *income effect*. The substitution effect describes the substitution of one good for another when a price changes, holding income constant. The income effect of a price change captures the change in demand when a consumer's real income changes. When a single price changes, the Slutsky equation can be written as

$$\frac{\partial x_j(\mathbf{p}, y)}{\partial p_i} = \frac{\partial h_j(\mathbf{p}, u)}{\partial p_i} - \frac{\partial x_j(\mathbf{p}, y)}{\partial y} x_i \tag{13.8}$$

where

$x_j(\mathbf{p}, y)$ is the Marshallian demand for good j

p_i is the price of good i

$h_j(\mathbf{p}, u)$ is the Hicksian demand for good j

$\mathbf{p}$ is a vector of prices for all goods in the consumer's bundle

y is income.

If one were to examine the effects of changes in all prices one can generalize the Slutsky equation. In the two-good case the Slutsky equation is

$$
\begin{bmatrix}
\dfrac{\partial x_1(\mathbf{p}, y)}{\partial p_1} & \dfrac{\partial x_1(\mathbf{p}, y)}{\partial p_2} \\[2ex]
\dfrac{\partial x_2(\mathbf{p}, y)}{\partial p_1} & \dfrac{\partial x_2(\mathbf{p}, y)}{\partial p_2}
\end{bmatrix}
=
\begin{bmatrix}
\dfrac{\partial h_1(\mathbf{p}, u)}{\partial p_1} & \dfrac{\partial h_1(\mathbf{p}, u)}{\partial p_2} \\[2ex]
\dfrac{\partial h_2(\mathbf{p}, u)}{\partial p_1} & \dfrac{\partial h_2(\mathbf{p}, u)}{\partial p_2}
\end{bmatrix}
$$
$$
-
\begin{bmatrix}
\dfrac{\partial x_1(\mathbf{p}, y)}{\partial y} \\[2ex]
\dfrac{\partial x_2(\mathbf{p}, y)}{\partial y}
\end{bmatrix}
\begin{bmatrix} x_1 & x_2 \end{bmatrix} \qquad (13.9)
$$

or

$$
\mathbf{M} = \mathbf{H} - \mathbf{Q}\mathbf{x}'
$$

where $\mathbf{M}$ is a $n \times n$ matrix of derivatives of the Marshallian demand curves with respect to prices, $\mathbf{H}$ is a $n \times n$ matrix of derivatives of the Hicksian demand curve with respect to prices, $\mathbf{Q}$ is a $n \times 1$ vector of the derivatives of the Marshallian demand curves with respect to income, and $\mathbf{x}'$ is a $1 \times n$ vector of goods, represented by the derivative of income with respect to prices.

Recall that the Marshallian demand curves are observable but the Hicksian demand curves are not. Hence $\mathbf{H}$ is often an unknown but of interest to economists since it isolates the substitution effect of a price change.

Illustration   (Utility function)

Suppose a consumer has

$$
u(x_1, x_2) = x_1^a x_2^{1-a}
$$

as a Cobb-Douglas utility function. A monotonic transformation of this utility function has the same preferences, so we can define this utility function as

$$
u(x_1, x_2) = a \, \ell n \, x_1 + (1 - a) \, \ell n \, x_2.
$$

The Marshallian demand functions can be derived and found to be [see Varian (1992, pp. 121-122) and Silberberg (1990, pp. 323-338)]

$$
x_1(p_1, p_2, y) = \frac{ay}{p_1} \quad \text{and} \quad x_2(p_1, p_2, y) = \frac{(1 - a)y}{p_2}.
$$

The Hicksian demand functions could be derived and are

$$h_1(p_1, p_2, u) = ap_1^{a-1}p_2^{1-a}u$$
$$h_2(p_1, p_2, u) = (1-a)p_1^a p_2^{-a}u.$$

The Slutsky equation becomes

$$
\begin{bmatrix} \dfrac{-ay}{p_1^2} & 0 \\[2ex] 0 & \dfrac{-(1-a)y}{p_2^2} \end{bmatrix} = \begin{bmatrix} \dfrac{a(a-1)y}{p_1^2} & \dfrac{a(1-a)y}{p_1 p_2} \\[2ex] \dfrac{a(1-a)y}{p_1 p_2} & \dfrac{-a(1-a)y}{p_2^2} \end{bmatrix}
$$
$$
- \begin{bmatrix} \dfrac{a}{p_1} \\[2ex] \dfrac{(1-a)}{p_2} \end{bmatrix} \begin{bmatrix} \dfrac{ay}{p_1} & \dfrac{(1-a)y}{p_2} \end{bmatrix}.
$$

Notice that when there are n prices and goods, these matrices are of order n. An economist could get data on demand and observe $\mathbf{M}$, $\mathbf{Q}$ and $\mathbf{x}$ and determine $\mathbf{H}$; alternatively, if the form of a consumer's utility function is known, we can use the operations of matrix algebra to quantify and isolate the substitution and income effects.

13.3 Input-Output Analysis

Input-output analysis deals with the interdependencies of various sectors of an economy be it that of a nation, a state, county or a corporation. Consider an economy with n sectors each producing a single commodity. Let

$$x_i = \text{gross output of commodity } i,$$
$$a_{ij} = \text{amount of commodity } i \text{ used in producing one unit of } j,$$
$$y_i = \text{final demand for commodity } i.$$

Now, gross output for each commodity must equal the sum of the amounts of that commodity used in producing all the commodities, plus the final demand for the commodity itself. Therefore with

$$\mathbf{x} = \{x_i\}, \quad \mathbf{y} = \{y_i\} \quad \text{and} \quad \mathbf{A} = \{a_{ij}\} \quad \text{for} \quad i, j = 1, 2, \ldots, n$$

we must have

$$\mathbf{x} = \mathbf{A}\mathbf{x} + \mathbf{y}. \tag{13.10}$$

The matrix **A** is called the *input-output coefficient matrix*: all its elements are, by definition, non-negative, as are elements of **x** if those of **y** are.

Assuming **A** remains constant over some period of time, we can determine **x**, the gross output of all sectors, required for any level of **y**, the final demand, by solving the above equation for **x**. Writing it as

$$(\mathbf{I} - \mathbf{A})\mathbf{x} = \mathbf{y}$$

and pre-multiplying by $(\mathbf{I} - \mathbf{A})^{-1}$, provided $(\mathbf{I} - \mathbf{A})$ is non-singular gives

$$\mathbf{x} = (\mathbf{I} - \mathbf{A})^{-1}\mathbf{y}. \tag{13.11}$$

Illustration (Input-output analysis)

Consider a four-sector economy in which there are two industries (agriculture and manufacturing), one primary factor of production (labor) and a government sector that consumes output from both industries, and utilizes labor as well. The government produces nothing for the economy and its consumption from the industry and labor sectors represents the final demand for the commodities produced by these sectors. Each industry uses some output of the other industry as well as labor in its production processes. The labor force requires output from both industries as well as labor services to sustain itself. There is no unemployment or excess demand for labor, and capital stock and inventories are held constant through time. The input-output table for the economy is shown in Table 13.1. Adding across each row gives the total output of each industry and the total labor force employed.

Each column shows the inputs a given sector requires in order to produce its total output, i.e., the production function of that sector. For example, the second column lists the output required from each production sector in order to meet the needs of the manufacturing process. To produce 4000 machines, the manufacturing industry requires 400 tons of agricultural output, 800 machines and 4800 employees.

Table 13.1. Input-Output Table

Production Sector	Consumption Sector				Total Output
	Agriculture	Manufacturing	Labor	Final Demand by Government	
Agriculture (tons)	600	400	1400	600	3000
Manufacturing (machines)	1500	800	700	1000	4000
Labor (employees)	900	4800	700	600	7000

Recall that the element a_{ij} of **A** is defined as the amount of commodity i used in producing one unit of commodity j. If we divide the (i, j)th element of the input-output table by the sum of row j, the result is a_{ij}. For example, $a_{12} = \frac{400}{4000} = 0.1$: it takes 0.1 ton of agricultural output to produce one machine. In this way the input-output coefficient matrix is obtained:

$$\mathbf{A} = \begin{bmatrix} \dfrac{600}{3000} & \dfrac{400}{4000} & \dfrac{1400}{7000} \\ \dfrac{1500}{3000} & \dfrac{800}{4000} & \dfrac{700}{7000} \\ \dfrac{900}{3000} & \dfrac{4800}{4000} & \dfrac{700}{7000} \end{bmatrix} = \begin{bmatrix} 0.2 & 0.1 & 0.2 \\ 0.5 & 0.2 & 0.1 \\ 0.3 & 1.2 & 0.1 \end{bmatrix}$$

and

$$(\mathbf{I} - \mathbf{A})^{-1} = \frac{1}{0.264} \begin{bmatrix} 0.60 & 0.33 & 0.17 \\ 0.48 & 0.66 & 0.18 \\ 0.84 & 0.99 & 0.59 \end{bmatrix}.$$

Suppose prices are relatively stable, technology changes slowly and the government sector now wishes to use (a) 1100 tons of agricultural products, (b) 1200 machines and (c) 792 employees in the next time period. We must determine the size of the labor force, and the levels of gross output by each of the agricultural and manufacturing sectors, **x**, given the government's final demand represented by the new **y** vector.

$$\mathbf{y} = \begin{bmatrix} 1100 \\ 1200 \\ 792 \end{bmatrix}.$$

To determine the level of agricultural and manufacturing activity and the labor force required under the new demand levels, we calculate **x** for the new value of **y**:

$$\mathbf{x} = (\mathbf{I} - \mathbf{A})^{-1}\mathbf{y} = \frac{1}{0.264}\begin{bmatrix} 0.60 & 0.33 & 0.17 \\ 0.48 & 0.66 & 0.18 \\ 0.84 & 0.99 & 0.59 \end{bmatrix}\begin{bmatrix} 1100 \\ 1200 \\ 792 \end{bmatrix} = \begin{bmatrix} 4510 \\ 5540 \\ 9770 \end{bmatrix}.$$

Thus, in order to meet the new government demands, 4510 tons of agricultural output and 5540 machines will have to be produced and 9770 employees will be required.

Three general comments can be made concerning input-output analysis. First, most input-output tables are expressed in dollar transactions rather than physical units. This avoids the problem of defining units and making units comparable. Second, input-output tables consider household demand as a component of final demand (**y**) rather than intermediate demand (**Ax**), and avoid the implication that there is a "production function" for the labor force. Third, the elements of $(\mathbf{I} - \mathbf{A})^{-1}$ contain economic meaning. Its (i, j)th element is the amount of commodity i that the economy is required to produce in order to deliver one unit of commodity j as final output.

Equation (13.11) may also apply to incremental changes in final demand rather than to absolute values. Thus, if $\mathbf{y}^*$ is the vector of such changes, $\mathbf{x}^* = (\mathbf{I} - \mathbf{A})^{-1}\mathbf{y}^*$ is the vector of changes in gross output that are needed to satisfy the changes in demand, $\mathbf{y}^*$.

13.4 Exercises

E 13.1. (Cobb-Douglas production function) Suppose your firm produced a single output with two inputs using a Cobb-Douglas production function, $dx_1^a x_2^b = y$. Your cost function is $w_1 x_1 + w_2 x_2$. Determine the cross-price effects when you minimize costs.

E 13.2. (Utility maximization) As an economic analyst for the Department of Commerce you determine that the utility of consumers of computers can be described by a Cobb-Douglas functional form, $u(x_1, x_2) = x_1^{1/2} x_2^{1/2}$ where x_1 and x_2 are hardware and software, respectively. Income of consumers is $1 million; price of hardware is $2000 and price of software is $400.

Determine if the Slutsky equation holds for the consumers. What are the substitution effects of changes in all prices?

E 13.3. Using (13.6), derive the 3×2 matrix in (13.7) and show that your result does indeed satisfy (13.7).

Note: E 13.4 through E 13.10 deal with input-output analysis.

E 13.4. Consider aggregating all industrial and service functions of an economy into two sectors: durable goods production and non-durable goods production. Each sector produces goods for the other, as well as for final consumption. Suppose we have the following sales data (in millions of dollars):

Sales from:	Sales to:		
	Durable	Non-durable	Consumers
Durable	24	90	6
Non-durable	12	45	93

(a) What is the total sales volume for each industry?

(b) What is the Gross National Product (GNP), defined as the sum of the values of final goods produced for consumption purposes? What is the "value added" by each industry? (Recall that the sum of values added over all sectors equal GNP.)

(c) Derive the economy's input-output coefficient matrix.

(d) What would be the dollar value of each industry's sales if consumers purchased $10 million of durables and $90 million of non-durables?

(e) What value of output must the durable goods industry produce in order for the economy to deliver one dollar's worth of non-durables to the consumers?

E 13.5. Consider a vertically integrated steel firm with coal mines, ore mines, blast furnaces that produce iron ingots, and rolling mills that produce steel sheets. Assume that the following input-output coefficient matrix accurately reflects the production processes involved:

Inputs	Outputs			
	Coal	Ore	Ingots	Sheets
Coal (tons)	0.1	0.1	0.5	0.1
Ore (tons)	0.0	0.0	1.0	0.0
Ingots	0.0	0.0	0.0	1.0
Sheets	0.0	0.0	0.0	0.0

Suppose the firm wishes to sell 3000 tons of coal, no ore, 6000 ingots and 12,000 rolled sheets to the rest of the economy in which it operates, with no purchases of any inputs from outside the company.

(a) How much of each product will have to be produced?

(b) Where is the coal utilized?

(c) Interpret the $(2, 4)$ element in $(\mathbf{I} - \mathbf{A})^{-1}$ and relate this to the original production function for steel sheets implied by the input-output coefficient matrix.

E 13.6. In Table 13.1 take the value of agricultural output to be $1000/ton, that of a machine to be $2000 and that of an employee to be $4000. Rework Table 13.1 in terms of dollar values, check it using (13.11) and find the output required to satisfy government demands of $600,000 worth of agricultural produce, $2,000,000 worth of machines and $2,400,000 for labor.

E 13.7. Suppose in E 13.6 the three production sectors charge each other and the government variable prices as in the following table.

Production Sector	Price Charged			
	Agriculture	Manuf.	Labor	Govt.
Agriculture (per ton)	$ 700	$ 900	$ 800	$1000
Manufacturing (per machine)	2000	$1537\frac{1}{2}$	1800	2010
Labor (per employee)	3840	3700	2530	3355

(a) Using these prices, rework and check Table 13.1.

(b) What agricultural and manufacturing output is needed (tons of produce and numbers of machines) and how many employees are needed to satisfy a government demand of \$800,000, \$3,200,000 and \$2,400,000 worth of agricultural produce, machines and employees, respectively?

E 13.8. Suppose an input-output table for n production sectors has the form of Table 13.1. Define the $n \times n$ matrix for the production sectors as $\mathbf{M}$, the final demand vector as $\mathbf{y}$ and the vector of production totals as $\mathbf{x}$. Then the row sums of the partitioned matrix $[\mathbf{M} \ \mathbf{y}]$ are the elements of $\mathbf{x}$.

(a) State this relationship between $\mathbf{M}$, $\mathbf{y}$ and $\mathbf{x}$ as an equation.

(b) Define a diagonal matrix $\mathbf{D}$ such that $\mathbf{MD}^{-1}$ is the input-output coefficient matrix in (13.11).

(c) Show that (13.11) always holds for such a table.

E 13.9. Three different machines, A, B and C, are to be used eight hours a day in manufacturing four different products t, u, v and w. The number of hours that each machine must be used in making one unit of each product is shown in the accompanying table:

Machine	Product			
	t	u	v	w
A	1	2	1	2
B	7	0	2	0
C	1	0	0	4

i.e., to make one v requires 1 hour of machine A and 2 hours of machine B.

(a) In terms of the numbers of each product made in an 8-hour day, what equations represent full utilization of all machines?

(b) Solve the equations derived in (a).

(c) Negative solutions to the equations are meaningless, but fractional solutions have meaning. What conditions therefore apply to the solution obtained in (a)?

(d) How many of each product are made if production of t is maximized whilst maintaining full production of all machines?

(e) What production of the four products maximizes production of w, subject to maintaining full utilization of all machines?

E 13.10. (a) Repeat Exercise 13.9 for full utilization of all machines over a 40-hour week rather than an 8-hour day. Confine solutions to being integers.

(b) Repeat (a) for machine C unavailable 8 hours a week.

Chapter 14

REGRESSION ANALYSIS

Regression analysis is a statistical procedure which is frequently used in the study of economic data. Only a modest acquaintance with statistics is assumed for reading this chapter and the next, for it is hoped that these two chapters will spark in the reader an interest both in the statistical procedures described, and in the merits of using matrix algebra for presenting them. Many books provide more extensive presentations then are given here. We name but a few: Kmenta (1986), Judge et al. (1988), Griffiths et al. (1993) and Greene (1997).

14.1 Simple Regression Model

Regression deals with situations where a relationship is thought to exist among variables. To begin, we deal with the variable y being related to just one variable, x. This is called *simple regression*. An easy example (which often gets newspaper publicity) is the relationship between lifetime income (y_i for person i) and number of years (x_i for person i) of formal education at high school and beyond.

a. Model specification

In customarily talking of regression, we usually mean linear regression. This is where, for simple regression, a variable y is thought to be a linear function of some other variable x, usually measured on the same observational unit. From a series of such measurements the objective of regression analysis is to estimate the coefficients in that linear function. To be specific, suppose that on each of N units we measure variables y and x. Suppose further that for a given value of x we assume the average value of y in the population of units (of which we are assuming we have just a random sample) to be

$$\mathrm{E}(y) = \alpha + \beta x. \qquad (14.1)$$

In this equation $\mathrm{E}(y)$ represents the average or expected value of y corresponding to x: and α and β are some unknown constants. Regression analysis simply uses the observations on y and x to obtain estimates of the α and β in equation (14.1). The variable denoted by y is usually called the *dependent variable* (because it depends on x) and the x-variable is correspondingly referred to as an *independent variable* or nowadays often as a *regressor variable*.

Equation (14.1) is an attempt at describing how we think one variable is related to others, and as such is an example of *model building*. The model here is said to be a *linear model*, linear because we envisage $\mathrm{E}(y)$ as linear in the unknowns α and β, which are called *parameters*. As such, the analysis we are to describe, referred to so far as regression analysis, is more correctly called simple *linear regression*, or (for more than one x-variable) *multiple linear regression* analysis.

b.   Data

In the preceding example, the income of someone with x years of education will not be exactly $\alpha + \beta x$. Indeed, this is already recognized in writing the model equation (14.1) as $\mathrm{E}(y) = \alpha + \beta x$ rather than as $y = \alpha + \beta x$. More particularly, if y_i and x_i are the values of y and x for the person labeled i, we write

$$\mathrm{E}(y_i) = \alpha + \beta x_i \qquad (14.2)$$

where $\mathrm{E}(y_i)$ is not the same as y_i. The difference, $y_i - \mathrm{E}(y_i)$, represents the deviation of the observed y_i from its expected value $\mathrm{E}(y_i)$ and is written as

$$e_i = y_i - \mathrm{E}(y_i) = y_i - \alpha - \beta x_i. \qquad (14.3)$$

Hence

$$y_i = \alpha + \beta x_i + e_i. \qquad (14.4)$$

The deviation e_i defined in (14.3) represents the extent to which an observed y_i differs from its expected value $\mathrm{E}(y_i)$. Thus the es include all manner of discrepancies between observed ys and their expected values and, as such, are considered as random variables, usually called *random errors* or *random residuals*.

c. Estimation

The parameters α and β are deemed to be fixed, unknown constants. To estimate them we use the procedure known as *ordinary least squares*. For this purpose, α and β are treated as being mathematical variables in order to find the values of them which minimize

$$\Delta = \sum_{i=1}^{N} [y_i - \mathrm{E}(y_i)]^2 = \sum_{i=1}^{N} (y_i - \alpha - \beta x_i)^2. \tag{14.5}$$

The values of α and β which minimize (14.5) are then taken as the estimators (the least squares estimators) of α and β, to be denoted $\hat{\alpha}$ and $\hat{\beta}$.

To do all this, first note that

$$\Delta = N\alpha^2 + \beta^2 \sum x_i^2 + \sum y_i^2 - 2\alpha \sum y_i - 2\beta \sum x_i y_i + 2\alpha\beta \sum x_i,$$

where every summation is over $i = 1, 2, \ldots, N$. Then, on writing $\bar{x}. = \sum x_i/N$ and $\bar{y}. = \sum y_i/N$ and rearranging terms

$$\Delta = \sum y_i^2 + N\alpha^2 + \beta^2 \sum x_i^2 - 2N\alpha\bar{y}. - 2\beta \sum x_i y_i + 2\alpha\beta N\bar{x}..$$

To achieve the minimization we need to differentiate Δ with respect to each of α and β:

$$\frac{\partial \Delta}{\partial \alpha} = 2N\alpha - 2N\bar{y}. + 2\beta N\bar{x}.$$

$$\frac{\partial \Delta}{\partial \beta} = 2\beta \sum x_i^2 - 2 \sum x_i y_i + 2\alpha N\bar{x}..$$

Equating these to zero, canceling the 2s and changing α and β to $\hat{\alpha}$ and $\hat{\beta}$, respectively, gives

$$N\hat{\alpha} + \hat{\beta}N\bar{x}. = N\bar{y}. \tag{14.6}$$

and

$$\hat{\alpha}N\bar{x}. + \hat{\beta} \sum x_i^2 = \sum x_i y_i. \tag{14.7}$$

Hence from (14.6)

$$\hat{\alpha} = \bar{y}. - \hat{\beta}\bar{x}. \tag{14.8}$$

and on substituting this into (14.7)

$$\hat{\beta} = \frac{\sum x_i y_i - N\bar{x}.\bar{y}.}{\sum x_i^2 - N\bar{x}.^2}. \tag{14.9}$$

These expressions for $\hat{\alpha}$ and $\hat{\beta}$ are estimators for α and β. They are not estimates. Estimates are the numerical values of $\hat{\alpha}$ and $\hat{\beta}$ obtained from using (numerical) data in $\hat{\alpha}$ and $\hat{\beta}$.

Illustration (Business start-ups)

Suppose we have five values of y_i which are the number of small business start-ups in five eastern seaboard states of the USA in 1998. Correspondingly, x_i is the average state unemployment rate (%) for those same states in 1998. Then calculation of the individual terms needed for $\hat{\alpha}$ and $\hat{\beta}$ of (14.8) and (14.9) are shown in Table 14.1.

Table 14.1. Data and Calculations for $\hat{\alpha}$ and $\hat{\beta}$

i	y_i	x_i	y_i^2	x_i^2	$x_i y_i$
1	62	2	3844	4	124
2	60	9	3600	81	540
3	57	6	3249	36	342
4	48	3	2304	9	144
5	23	5	529	25	115
Total	250	25	13526	155	1265
Mean	50	5			

With these data we get from (14.9)

$$\hat{\beta} = \frac{1265 - 5(5)50}{155 - 5(5^2)} = \frac{15}{30} = \frac{1}{2}$$

and then from (14.8)

$$\hat{\alpha} = 50 - \tfrac{1}{2}(5) = 47.5.$$

The meaning of these results is that from $\hat{\alpha} = 47.5$ we can say that on average even when there is no unemployment there will be an estimated 47.5 start-ups in a state; and from $\hat{\beta} = 0.5$ there will be an average increase of 0.5 start-ups for each 1% of unemployment.

In today's world, of course, one never sees the last three columns of Table 14.1, because all that arithmetic is done using computers. Those last three columns are shown here simply to emphasize precisely what is meant by the summations in $\hat{\alpha}$ and $\hat{\beta}$.

14.2 Multiple Linear Regression

a. The model

Having used Section 14.1 to introduce linear regression in its simplest form, of just one (independent) x-variable, we go straight to the case of having k such variables, $x_1, x_2, \ldots, x_k$. The remainder of this chapter deals with this case.

In place of (14.2) we have

$$\mathrm{E}(y_i) = \alpha + \beta_1 x_{i1} + \beta_2 x_{i2} + \cdots + \beta_k x_{ik} \qquad (14.10)$$

where x_{ij}, for $j = 1, 2, \ldots, k$, is the value of the jth x-variable observed on the ith observational unit. The term not involving the x-variables, namely α, is known as the *intercept*. It would be the value of the $\mathrm{E}(y_i)$ whenever the observed values of all the x-variables were zero.

For notational simplicity we now change α to be β_0. More than that, we change it to be $\beta_0 x_{i0}$ for every x_{i0} being unity, i.e., 1.0. This, together with defining

$$e_i = y_i - \mathrm{E}(y_i) \qquad (14.11)$$

just as in (14.3), leads to having

$$y_i = \beta_0 x_{i0} + \beta_1 x_{i1} + \beta_2 x_{i2} + \cdots + \beta_k x_{ik} + e_i. \qquad (14.12)$$

Now define

$$\mathbf{x}_i' = [x_{i0} \quad x_{i1} \quad x_{i2} \quad \cdots \quad x_{ik}]$$

and

$$\boldsymbol{\beta}' = [\beta_0 \quad \beta_1 \quad \beta_2 \quad \cdots \quad \beta_k]. \qquad (14.13)$$

Then (14.12) can be expressed as

$$y_i = \mathbf{x}_i' \boldsymbol{\beta} + e_i. \qquad (14.14)$$

Further, if we have N sets of observations, i.e., N equations (14.14) for $i = 1, 2, \ldots, N$, they can be written as

$$\mathbf{y} = \begin{bmatrix} y_1 \\ y_2 \\ \vdots \\ y_i \\ \vdots \\ y_N \end{bmatrix} = \begin{bmatrix} \mathbf{x}_1' \\ \mathbf{x}_2' \\ \vdots \\ \mathbf{x}_i' \\ \vdots \\ \mathbf{x}_N' \end{bmatrix} \boldsymbol{\beta} + \mathbf{e} \quad \text{for} \quad \mathbf{e} = \begin{bmatrix} e_1 \\ e_2 \\ \vdots \\ e_i \\ \vdots \\ e_N \end{bmatrix}. \qquad (14.15)$$

Thus

$$\mathbf{y} = \mathbf{X}\boldsymbol{\beta} + \mathbf{e} \qquad (14.16)$$

for

$$\mathbf{X} = \begin{bmatrix} \mathbf{x}_1' \\ \mathbf{x}_2' \\ \vdots \\ \mathbf{x}_i' \\ \vdots \\ \mathbf{x}_N' \end{bmatrix} = \begin{bmatrix} x_{10} & x_{11} & \cdots & x_{1j} & \cdots & x_{1k} \\ x_{20} & x_{21} & \cdots & x_{2j} & \cdots & x_{2k} \\ \vdots & \vdots & & \vdots & & \\ x_{i0} & x_{i1} & \cdots & x_{ij} & \cdots & x_{1k} \\ \vdots & \vdots & & & & \\ x_{N0} & x_{N1} & \cdots & x_{Nj} & \cdots & x_{Nk} \end{bmatrix} = \left\{ {}_m x_{ij} \right\}_{i=1,j=0}^{N\quad k}.$$

$$(14.17)$$

Equation (14.16) is possibly the most widely used equation in statistics. Here, it represents linear regression for as many x-variables, k, and as many sets of observations, N, as we care to have. But, as will be seen in Chapter 15, it can also represent linear models which themselves embrace analysis of variance and analysis of covariance.

b. The meaning of linear

In the context of regression it is important to note the meaning of "linear". It refers to the parameters, not to the x-variables. Thus in (14.12) every term is of the form β_x. From high-school geometry we know that the graph of $y = \beta x$ is a straight line. As a result we describe βx as being linear in β. So also are βx^2 and $\beta x_1 x_2$. But $\beta^2 x$ is not linear in β.

14.3   Estimation

a.   The general result

Just as $\mathbf{y} = \mathbf{X}\boldsymbol{\beta} + \mathbf{e}$ of (14.16) comes from y_i of (14.12), so can

$$E(\mathbf{y}) = \mathbf{X}\boldsymbol{\beta} \qquad (14.18)$$

be derived from $E(y_i)$ of (14.10). Then, just as in (14.5), we seek as the least squares estimator of $\boldsymbol{\beta}$ the value of $\boldsymbol{\beta}$ which minimizes

$$\Delta = \sum_{i=1}^{N} [y_i - E(y_i)]^2. \qquad (14.19)$$

Using $\mathbf{y}$, and $E(\mathbf{y})$ of (14.18), this is

$$\begin{aligned} \Delta &= [\mathbf{y} - E(\mathbf{y})]'\,[\mathbf{y} - E(\mathbf{y})] \\ &= (\mathbf{y}' - \boldsymbol{\beta}'\mathbf{X}')(\mathbf{y} - \mathbf{X}\boldsymbol{\beta}) \\ &= \mathbf{y}'\mathbf{y} - 2\boldsymbol{\beta}'\mathbf{X}'\mathbf{y} + \boldsymbol{\beta}'\mathbf{X}'\mathbf{X}\boldsymbol{\beta}. \end{aligned} \tag{14.20}$$

Applying the differentiation results in Section 12.9, equating $\partial\Delta/\partial\boldsymbol{\beta}$ to $\mathbf{0}$ and in doing so, rewriting $\boldsymbol{\beta}$ as $\hat{\boldsymbol{\beta}}$, leads to

$$\mathbf{X}'\mathbf{X}\hat{\boldsymbol{\beta}} = \mathbf{X}'\mathbf{y}. \tag{14.21}$$

These are known as the *normal equations*. Their solution as

$$\hat{\boldsymbol{\beta}} = (\mathbf{X}'\mathbf{X})^{-1}\mathbf{X}'\mathbf{y} \tag{14.22}$$

depends on $\mathbf{X}'\mathbf{X}$ being invertible. This will always be so when the columns of $\mathbf{X}$ are linearly independent; i.e., when $\mathbf{X}$ has full column rank. With regression data this is almost always the case: the columns of observed x-values are almost always linearly independent. Taking that as given we have (14.22) as the general expression for the least squares estimator of $\boldsymbol{\beta}$ in the regression model equation $\mathbf{y} = \mathbf{X}\boldsymbol{\beta} + \mathbf{e}$.

Rewriting equations (14.6) and (14.7) for simple regression in matrix-vector form gives

$$\begin{bmatrix} N & x_{\cdot} \\ x_{\cdot} & \Sigma x_i^2 \end{bmatrix} \begin{bmatrix} \hat{\alpha} \\ \hat{\beta} \end{bmatrix} = \begin{bmatrix} y_{\cdot} \\ \Sigma x_i y_i \end{bmatrix}. \tag{14.23}$$

For multiple regression, with $\mathbf{X}$ of (14.17) and $\mathbf{y}$ of (14.15), the equations $\mathbf{X}'\mathbf{X}\hat{\boldsymbol{\beta}} = \mathbf{X}'\mathbf{y}$ of (14.21) with $\boldsymbol{\beta}$ of (14.13) have precisely the same form as (14.23) but of order $k+1$ rather than 2 and with

$$\mathbf{X}'\mathbf{X} = \begin{bmatrix} N & x_{\cdot 1} & x_{\cdot 2} & \cdots & x_{\cdot k} \\ x_{\cdot 1} & \Sigma x_{i1}^2 & \Sigma x_{i1}x_{i2} & \cdots & \Sigma x_{i1}x_{ik} \\ x_{\cdot 2} & \Sigma x_{i1}x_{i2} & \Sigma x_{i2}^2 & \cdots & \Sigma x_{i2}x_{ik} \\ \vdots & \vdots & \vdots & \vdots & \vdots \\ x_{\cdot k} & \Sigma x_{i1}x_{ik} & \Sigma x_{i2}x_{ik} & \cdots & \Sigma x_{ik}^2 \end{bmatrix} \text{ and } \mathbf{X}'\mathbf{y} = \begin{bmatrix} y_{\cdot} \\ \Sigma x_{i1}y_i \\ \vdots \\ \Sigma x_{ik}y_i \end{bmatrix}. \tag{14.24}$$

Observe that the first row and column of $\mathbf{X}'\mathbf{X}$ and the first element of $\mathbf{X}'\mathbf{y}$ arise from the nature of x_{i0} in $\mathbf{X}$ of (14.8), namely that $x_{i0} = 1$ for all $i = 1, 2, \ldots, N$. Also observe that each off-diagonal element of $\mathbf{X}'\mathbf{X}$ is the inner product of one column of $\mathbf{X}$ with another, and thus is a sum of products; and each diagonal element of $\mathbf{X}'\mathbf{X}$ is the inner product of a column of $\mathbf{X}$ with itself, and thus is a sum of squares. These characteristics of $\mathbf{X}'\mathbf{X}$ are evident in (14.24).

b.   Using deviations from means

Despite the preceding descriptions of elements of $\mathbf{X'X}$, it is well known that the regression coefficients $\beta_1, \ldots, \beta_k$ can be estimated using a matrix and vector that are just like $\mathbf{X'X}$ and $\mathbf{X'y}$ only involving sums of squares and products corrected for their means. Indeed, this is the customary manner in which estimates are calculated. We now establish this formulation. First, in using the summing vector $\mathbf{1}_N$ of order N, defining

$$\mathbf{X}_* = \begin{bmatrix} x_{11} & x_{12} & \cdots & x_{1k} \\ x_{21} & x_{22} & \cdots & x_{2k} \\ \vdots & \vdots & & \vdots \\ x_{N1} & x_{N2} & \cdots & x_{Nk} \end{bmatrix}, \quad \text{gives} \quad \mathbf{X} = [\mathbf{1}_N \ \mathbf{X}_*], \quad (14.25)$$

with $\mathbf{X}_*$ simply being $\mathbf{X}$ without its first column of 1s. Similarly define

$$\boldsymbol{\beta}_* = [\beta_1 \ \beta_2 \ \cdots \ \beta_k]'$$

which is just $\boldsymbol{\beta}$ of (14.13) without β_0. Then E$(\mathbf{y})$ of (14.18) is

$$\text{E}(\mathbf{y}) = \mathbf{X}\boldsymbol{\beta} = [\mathbf{1}_N \ \mathbf{X}_*] \begin{bmatrix} \beta_0 \\ \boldsymbol{\beta}_* \end{bmatrix} = \beta_0 \mathbf{1}_N + \mathbf{X}_*\boldsymbol{\beta}_*. \quad (14.26)$$

We now introduce the $N \times N$ centering matrix

$$\mathbf{C} = \mathbf{I}_N - \bar{\mathbf{J}}_N$$

of (3.14), where $\bar{\mathbf{J}}$ is square of order N with every element being $1/N$. As a result

$$\mathbf{C}\mathbf{1}_N = \mathbf{0} \quad \text{and} \quad \mathbf{C}^2 = \mathbf{C} = \mathbf{C}'. \quad (14.27)$$

Also

$$\mathbf{C}\mathbf{y} = \left\{ _c \ y_i - \bar{y}. \right\}_{i=1}^N. \quad (14.28)$$

Consequently

$$\text{E}(\mathbf{C}\mathbf{y}) = \mathbf{C}\mathbf{1}_N \beta_0 + \mathbf{C}\mathbf{X}_*\boldsymbol{\beta}_* = \mathbf{C}\mathbf{X}_*\boldsymbol{\beta}_*. \quad (14.29)$$

From this, as a simple extension of E$(\mathbf{y}) = \mathbf{X}\boldsymbol{\beta}$ yielding $\hat{\boldsymbol{\beta}}$ of (14.22) we now estimate $\hat{\boldsymbol{\beta}}_*$ as

$$\hat{\boldsymbol{\beta}}_* = [(\mathbf{C}\mathbf{X}_*)'\mathbf{C}\mathbf{X}_*]^{-1}(\mathbf{C}\mathbf{X}_*)'\mathbf{C}\mathbf{y}$$
$$= (\mathbf{X}_*'\mathbf{C}\mathbf{X}_*)^{-1}\mathbf{X}_*'\mathbf{C}\mathbf{y}. \quad (14.30)$$

This, using the principle shown in (14.28), reduces to

$$
\hat{\boldsymbol{\beta}}_* = \left[\left\{ \frac{1}{m} \sum_{i=1}^{N} (x_{ij} - \bar{x}_{\cdot j})(x_{ij\cdot} - \bar{x}_{\cdot j'}) \right\}_{j,j'=1}^{k} \right]^{-1}
$$
$$
\times \left\{ \frac{1}{c} \sum_{i=1}^{N} (x_{ij} - \bar{x}_{\cdot j})(y_i - \bar{y}_{\cdot}) \right\}_{j=1}^{k} . \qquad (14.31)
$$

And this is the well-known result in terms of sums of squares and products. For example, if $k = 1$, (14.31) reduces to (14.9).

To estimate β_0 we use $\mathbf{X} = [\mathbf{1}_n \quad \mathbf{X}_*]$ and $\boldsymbol{\beta}' = [\beta_0 \quad \boldsymbol{\beta}'_*]$ in $\mathbf{X}'\mathbf{X}\hat{\boldsymbol{\beta}} = \mathbf{X}'\mathbf{y}$ and get

$$
\begin{bmatrix} N & N\bar{\mathbf{x}}' \\ N\bar{\mathbf{x}} & \mathbf{X}'_*\mathbf{X}_* \end{bmatrix} \begin{bmatrix} \hat{\beta}_0 \\ \hat{\beta}_* \end{bmatrix} = \begin{bmatrix} N\bar{y} \\ \mathbf{X}'_*\mathbf{y} \end{bmatrix} \qquad (14.32)
$$

where $\bar{\mathbf{x}}' = [\bar{x}_{\cdot 1} \ \bar{x}_{\cdot 2} \ \cdots \ \bar{x}_{\cdot j} \ \cdots \ \bar{x}_{\cdot k}]$. By inverting the matrix on the left-hand side of (14.32) it will be found (see E 14.3) that $\hat{\boldsymbol{\beta}}_*$ from (14.32) is the same as in (14.31). And then from the first equation in (14.32)

$$
\hat{\beta}_0 = \frac{1}{N} \left(N\bar{y}_{\cdot} - N\bar{\mathbf{x}}'\hat{\boldsymbol{\beta}}_* \right)
$$
$$
= \bar{y}_{\cdot} - \bar{\mathbf{x}}'\hat{\boldsymbol{\beta}}_*
$$
$$
= \bar{y}_{\cdot} - \sum_{j=1}^{k} \hat{\beta}_j \bar{x}_{\cdot j} . \qquad (14.33)
$$

Illustration (*continued*)

We continue the example of Section 14.1d. For five states of the USA y_i is the number of small business start-ups in state i in 1998, x_{i1} is the average 1998 unemployment rate and x_{i2} is financial incentives—measured in millions of dollars—offered by state i to start-up businesses. These data (including $x_{i0} = 1$ for $i = 1, \ldots, 5$) are shown in Table 14.2.

Table 14.2. Data for $E(y_i) = \beta_0 + \beta_1 x_{i1} + \beta_2 x_{i2}$

i	y_i	x_{i0}	x_{i1}	x_{i2}
1	62	1	2	6
2	60	1	9	10
3	57	1	6	4
4	48	1	3	13
5	23	1	5	2
$N = 5$	$y. = 250$	$x._0 = 5$	$x._1 = 25$	$x._2 = 35$
	$\bar{y}. = 50$		$\bar{x}._1 = 5$	$\bar{x}._2 = 7$

Then the estimated β is calculated as $\hat{\beta}$ from (14.22) using

$$\mathbf{y} = \begin{bmatrix} 62 \\ 60 \\ 57 \\ 48 \\ 23 \end{bmatrix}, \quad \mathbf{X} = \begin{bmatrix} 1 & 2 & 6 \\ 1 & 9 & 10 \\ 1 & 6 & 4 \\ 1 & 3 & 13 \\ 1 & 5 & 2 \end{bmatrix} \quad \text{and} \quad \mathbf{X'y} = \begin{bmatrix} 250 \\ 1265 \\ 1870 \end{bmatrix}, \quad (14.34)$$

with

$$(\mathbf{X'X})^{-1} = \begin{bmatrix} 5 & 25 & 35 \\ 25 & 155 & 175 \\ 35 & 175 & 325 \end{bmatrix}^{-1} = \tfrac{1}{480} \begin{bmatrix} 790 & -80 & -42 \\ -80 & 16 & 0 \\ -42 & 0 & 6 \end{bmatrix}. \quad (14.35)$$

Hence from (14.22)

$$\hat{\beta} = \begin{bmatrix} \hat{\beta}_0 \\ \hat{\beta}_1 \\ \hat{\beta}_2 \end{bmatrix} = \tfrac{1}{480} \begin{bmatrix} 790 & -80 & -42 \\ -80 & 16 & 0 \\ -42 & 0 & 6 \end{bmatrix} \begin{bmatrix} 250 \\ 1265 \\ 1870 \end{bmatrix} = \begin{bmatrix} 37 \\ \tfrac{1}{2} \\ 1\tfrac{1}{2} \end{bmatrix}, \quad (14.36)$$

and so the estimated regression line is

$$\widehat{E(y)} = 37 + \tfrac{1}{2}x_1 + 1\tfrac{1}{2}x_2.$$

Using the deviations from means calculations of (14.30) we need

$$\mathbf{CX_*} = (\mathbf{I} - \bar{\mathbf{J}}_5) \begin{bmatrix} 2 & 6 \\ 9 & 10 \\ 6 & 4 \\ 3 & 13 \\ 5 & 2 \end{bmatrix} \quad \text{and} \quad \mathbf{Cy} = (\mathbf{I} - \bar{\mathbf{J}}_5) \begin{bmatrix} 62 \\ 60 \\ 57 \\ 48 \\ 23 \end{bmatrix}$$

and with $\bar{x}_{.1} = 5$ and $\bar{x}_{.2} = 7$ we get

$$\mathbf{CX}_* = \begin{bmatrix} -3 & -1 \\ 4 & 3 \\ 1 & -3 \\ -2 & 6 \\ 0 & -5 \end{bmatrix}$$

with

$$\mathbf{X}'_*\mathbf{CX}_* = (\mathbf{CX}_*)'\mathbf{CX}_* = \begin{bmatrix} 30 & 0 \\ 0 & 80 \end{bmatrix}. \tag{14.37}$$

And

$$\mathbf{Cy} = \begin{bmatrix} 12 \\ 10 \\ 7 \\ -2 \\ -27 \end{bmatrix} \quad \text{with} \quad \mathbf{X}'_*\mathbf{Cy} = (\mathbf{CX}_*)'\mathbf{Cy} = \begin{bmatrix} 15 \\ 120 \end{bmatrix}$$

so that in (14.30)

$$\hat{\boldsymbol{\beta}}_* = \begin{bmatrix} \hat{\beta}_1 \\ \hat{\beta}_2 \end{bmatrix} = \begin{bmatrix} 30 & 0 \\ 0 & 80 \end{bmatrix}^{-1} \begin{bmatrix} 15 \\ 120 \end{bmatrix} = \begin{bmatrix} \frac{1}{2} \\ 1\frac{1}{2} \end{bmatrix}, \tag{14.38}$$

just as in (14.36). And from (14.33)

$$\begin{aligned} \hat{\beta}_0 &= \bar{y}. - \hat{\beta}_1\bar{x}_{.1} - \hat{\beta}_2\bar{x}_{.2} \\ &= 50 - \tfrac{1}{2}(5) - 1\tfrac{1}{2}(7) = 37. \end{aligned}$$

The meaning of these results is that even if unemployment and financial incentives are zero, we can estimate the average start-ups in a state to be 37 per year; with an additional $\frac{1}{2}$ start-up for every 1% of unemployment and a further $1\frac{1}{2}$ start-ups for each \$1,000,000 of financial incentive.

Note. The data used here are such that both $\hat{\boldsymbol{\beta}}$ from $\mathrm{E}(y_i) = \alpha + \beta x_i$ in Section 14.1 and $\hat{\beta}_1$ from $\mathrm{E}(y_i) = \alpha + \beta_1 x_{i1} + \beta_2 x_{i2}$ are the same, namely $\frac{1}{2}$. The algebraic reason for this is described in the Appendix to this chapter, Section 14.15. In economic terms, it suggests that there is no specification bias in the model. The coefficient $\hat{\boldsymbol{\beta}}$ captures the effect of x on y in Section 14.1 assuming there are no other variables in the model. For this specification to be unbiased there must be no additional variable appropriate for the model specification or the excluded model variables need not have an effect on the included model variables. In our example, the equality of $\hat{\beta}$ in Section 14.1 and $\hat{\beta}_1$ in (14.38) indicates that financial incentives do not have an impact on the unemployment rate. For a more detailed discussion of specification bias see Gujarati (1995), pages 204-207.

14.4 Statistical Model

Properties of $\hat{\beta} = (\mathbf{X'X})^{-1}\mathbf{X'y}$ come from properties of $\mathbf{e}$ in the model equation $\mathbf{y} = \mathbf{X\beta} + \mathbf{e}$ of (14.16). Based on the definition in (14.11) we have

$$\mathbf{e} = \mathbf{y} - \mathrm{E}(\mathbf{y}) \quad \text{and hence} \quad \mathrm{E}(\mathbf{e}) = \mathrm{E}(\mathbf{y}) - \mathrm{E}(\mathbf{y}) = \mathbf{0}.$$

This is a direct outcome of the definition of $\mathbf{e}$. Hence the expected outcome of all random disturbances is zero.

Since $\mathbf{e}$ is considered to be random, it has a variance. In contrast to $\mathrm{E}(\mathbf{e}) = 0$ coming from the definition of $\mathbf{e}$, we now attribute a variance-covariance to $\mathbf{e}$. We treat every e_i as having the same variance σ^2 (homoscedastic), and being uncorrelated (zero covariance) with all other e_is (no autocorrelation). Hence the variance-covariance matrix of $\mathbf{e}$ is

$$\mathrm{var}(\mathbf{e}) = \mathrm{E}\left\{[\mathbf{e} - \mathrm{E}(\mathbf{e})]\,[\mathbf{e} - \mathrm{E}(\mathbf{e})]'\right\} = \mathrm{E}(\mathbf{ee'}) = \sigma^2\mathbf{I}_N.$$

Then

$$\mathrm{var}(\mathbf{y}) = \mathrm{var}(\mathbf{X\beta} + \mathbf{e}) = \mathrm{var}(\mathbf{e}) = \sigma^2\mathbf{I}_N.$$

And we summarize the mean and variance of $\mathbf{y}$ in the notation

$$\mathbf{y} \sim (\mathbf{X\beta}, \quad \sigma^2\mathbf{I}_N).$$

Notice that up to this point we are not yet assuming normality.

14.5 Unbiasedness and Variances

The expected value of $\hat{\beta}$ is β. This is so because $\mathrm{E}(\mathbf{y}) = \mathbf{X\beta}$ and so

$$\mathrm{E}(\hat{\beta}) = \mathrm{E}(\mathbf{X'X})^{-1}\mathbf{X'y} = (\mathbf{X'X})^{-1}\mathbf{X'}\mathrm{E}(\mathbf{y}) = (\mathbf{X'X})^{-1}\mathbf{X'X\beta} = \beta,$$

i.e., $\hat{\beta}$ is an unbiased estimator of β; and the variance of $\hat{\beta}$ is

$$\begin{aligned}
\text{var}(\hat{\beta}) &= \text{var}\left[(\mathbf{X'X})^{-1}\mathbf{X'y}\right] \\
&= (\mathbf{X'X})^{-1}\mathbf{X'}\text{var}(\mathbf{y})\mathbf{X}(\mathbf{X'X})^{-1} \\
&= (\mathbf{X'X})^{-1}\mathbf{X'}\sigma^2\mathbf{IX}(\mathbf{X'X})^{-1} \\
&= (\mathbf{X'X})^{-1}\sigma^2.
\end{aligned} \tag{14.39}$$

A similar result holds for $\hat{\beta}_*$: from (14.30) we get

$$\text{var}(\hat{\beta}_*) = (\mathbf{X'_*CX_*})^{-1}\sigma^2 \tag{14.40}$$

where $\mathbf{X'_*CX_*}$ is the k-order square matrix of corrected sums of squares and products of the xs, the matrix which is inverted in (14.30). And then from (14.33) it is clear that

$$\text{cov}\left(\hat{\beta}_0, \hat{\beta}_*\right) = -\bar{\mathbf{x}}'\text{var}\left(\hat{\beta}_*\right). \tag{14.41}$$

Another useful result is

$$\text{cov}\left(\bar{y}., \hat{\beta}_*\right) = 0. \tag{14.42}$$

This is so because, as is evident in (14.31), the term involving ys in each element of $\hat{\beta}_*$ is $y_i - \bar{y}.$; and $\text{cov}(\bar{y}., y_i - \bar{y}.) = \sigma^2/N - \sigma^2/N = 0$. Then, using (14.33),

$$\begin{aligned}
\text{var}(\hat{\beta}_0) &= \text{var}\left(\bar{y}. - \bar{\mathbf{x}}'\hat{\beta}_*\right) \\
&= \left[1/N + \bar{\mathbf{x}}'(\mathbf{X'_*CX_*})^{-1}\bar{\mathbf{x}}\right]\sigma^2.
\end{aligned} \tag{14.43}$$

14.6 Estimating the Variance

Having estimated β as $\hat{\beta}$, the estimated value of $\text{E}(\mathbf{y}) = \mathbf{X}\beta$ is $\mathbf{X}\hat{\beta}$; and it is customarily called the *predicted* y, denoted by

$$\hat{\mathbf{y}} = \mathbf{X}\hat{\beta} = \mathbf{X}(\mathbf{X'X})^{-1}\mathbf{X'y} = \mathbf{Hy} \text{ on defining } \mathbf{H} = \mathbf{X}(\mathbf{X'X})^{-1}\mathbf{X'}. \tag{14.44}$$

$\mathbf{H}$ is often called the *"hat" matrix* : it converts $\mathbf{y}$ to $\hat{\mathbf{y}}$.

An important use for $\hat{\mathbf{y}}$ is that the sum of squares of the deviations of the observed ys from their predicted values is known as the *error sum of squares* or *residual sum of squares*:

$$\text{SSE} = \sum_{i=1}^{N}(y_i - \hat{y}_i)^2 = (\mathbf{y}-\hat{\mathbf{y}})'(\mathbf{y}-\hat{\mathbf{y}}) = \mathbf{y'My} \text{ for } \mathbf{M} = \mathbf{I} - \mathbf{H}. \tag{14.45}$$

The quadratic form $\mathbf{y}'\mathbf{M}\mathbf{y}$ comes from using $\hat{\mathbf{y}} = \mathbf{H}\mathbf{y}$, noting that $\mathbf{H}$ is symmetric and idempotent as is, therefore, $\mathbf{M}$.

An alternative form for SSE is

$$\text{SSE} = \mathbf{y}'(\mathbf{I} - \mathbf{H})\mathbf{y} = \mathbf{y}'\mathbf{y} - \hat{\boldsymbol{\beta}}'\mathbf{X}'\mathbf{y}. \tag{14.46}$$

This is sometimes useful for computing SSE: $\mathbf{y}'\mathbf{y}$ is the total sum of squares of the observations, and $\hat{\boldsymbol{\beta}}'\mathbf{X}'\mathbf{y}$ is the sum of products of the elements of the solution $\hat{\boldsymbol{\beta}}$ with their corresponding elements of the right-hand side, $\mathbf{X}'\mathbf{y}$, of the equations from which $\hat{\boldsymbol{\beta}}$ is derived, namely $\mathbf{X}'\mathbf{X}\hat{\boldsymbol{\beta}} = \mathbf{X}'\mathbf{y}$.

An expression for SSE based on corrected sums of squares and products can also be developed:

$$\text{SSE} = \mathbf{y}'\mathbf{C}\mathbf{y} - \hat{\boldsymbol{\beta}}_*'\mathbf{X}_*'\mathbf{C}\mathbf{y}. \tag{14.47}$$

Clearly, (14.47) is analogous to (14.46). $\mathbf{y}'\mathbf{C}\mathbf{y}$ is the corrected sum of squares of the ys, and $\hat{\boldsymbol{\beta}}_*'\mathbf{X}_*'\mathbf{C}\mathbf{y}$ is the sum of products of the elements of the solution vector $\hat{\boldsymbol{\beta}}_*$ (excluding $\hat{\beta}_0$) with the corresponding elements of the right-hand side $\mathbf{X}_*'\mathbf{C}\mathbf{y}$ of the equations from which $\hat{\boldsymbol{\beta}}_*'$ is derived, namely $\mathbf{X}_*'\mathbf{C}\mathbf{X}_*\hat{\boldsymbol{\beta}}_*' = \mathbf{X}_*'\mathbf{C}\mathbf{y}$, from which comes (14.30). Derivation of (14.47) from (14.46) is left to the reader as E 14.4.

In (14.46) we have SSE $= \mathbf{y}'(\mathbf{I} - \mathbf{H})\mathbf{y}$. And we have for $\mathbf{y} \sim (\boldsymbol{\mu}, \mathbf{V})$ that $E(\mathbf{y}'\mathbf{A}\mathbf{y}) = \text{tr}(\mathbf{A}\mathbf{V}) + \boldsymbol{\mu}'\mathbf{A}\boldsymbol{\mu}$ (see Section 15.14). Applying this to SSE with $\mathbf{y} \sim (\mathbf{X}\boldsymbol{\beta}, \sigma^2\mathbf{I}_N)$ gives

$$E[\text{SSE}] = \text{tr}\left[(\mathbf{I} - \mathbf{H})\sigma^2\mathbf{I}_N\right] + \boldsymbol{\beta}'\mathbf{X}'(\mathbf{I} - \mathbf{H})\mathbf{X}\boldsymbol{\beta}. \tag{14.48}$$

With $\mathbf{H}\mathbf{X}$ being null; and with $\mathbf{I} - \mathbf{H}$ being idempotent (and so having trace equal to rank) we have, after a little simplification (see E 14.4) that

$$E(\text{SSE}) = [N - (k+1)]\sigma^2, \tag{14.49}$$

where k is the number of x-variables, as in (14.10). Hence an unbiased estimator of σ^2 is

$$\hat{\sigma}^2 = \frac{\text{SSE}}{N - (k+1)}. \tag{14.50}$$

Illustration (*continued*)

Using (14.34) and (14.36) in (14.44) gives

$$\hat{\mathbf{y}} = \mathbf{X}\hat{\beta} = \begin{bmatrix} 1 & 2 & 6 \\ 1 & 9 & 10 \\ 1 & 6 & 4 \\ 1 & 3 & 13 \\ 1 & 5 & 2 \end{bmatrix} \begin{bmatrix} 37 \\ \frac{1}{2} \\ 1\frac{1}{2} \end{bmatrix} = \begin{bmatrix} 47 \\ 56\frac{1}{2} \\ 46 \\ 58 \\ 42\frac{1}{2} \end{bmatrix}.$$

Therefore, with **y** from (14.34)

$$\mathbf{y} - \hat{\mathbf{y}} = \begin{bmatrix} 62 \\ 60 \\ 57 \\ 48 \\ 23 \end{bmatrix} - \begin{bmatrix} 47 \\ 56\frac{1}{2} \\ 46 \\ 58 \\ 42\frac{1}{2} \end{bmatrix} = \begin{bmatrix} 15 \\ 3\frac{1}{2} \\ 11 \\ -10 \\ -19\frac{1}{2} \end{bmatrix},$$

and so from the first expression in (14.45)

$$\text{SSE} = 15^2 + (3\tfrac{1}{2})^2 + 11^2 + 10^2 + (19\tfrac{1}{2})^2 = 838\tfrac{1}{2}. \tag{14.51}$$

An alternative expression for SSE is (14.46). It requires $\mathbf{y}'\mathbf{y} = \Sigma y_i^2 = 13,526$ obtainable from (14.34). Then with $\hat{\beta}'$ from (14.36) and $\mathbf{X}'\mathbf{y}$ from (14.35), equation (14.46) is

$$\text{SSE} = 13,526 - [37 \quad \tfrac{1}{2} \quad 1\tfrac{1}{2}] \begin{bmatrix} 250 \\ 1265 \\ 1870 \end{bmatrix} = 13,526 - 12,687\tfrac{1}{2} = 838\tfrac{1}{2}, \text{ as before.}$$

Then (14.50) gives

$$\hat{\sigma}^2 = \frac{838\frac{1}{2}}{5-3} = 419\tfrac{1}{4}.$$

14.7 Partitioning the Total Sum of Squares

The total sum of squares, which we call SST, is

$$\text{SST} = \mathbf{y}'\mathbf{y} = \sum_{i=1}^{N} y_i^2.$$

And the sum of squares of deviations of the observed y_is from their predicted values is, from (14.46) and (14.47),

$$
\begin{aligned}
\text{SSE} &= \mathbf{y}'\mathbf{y} - \hat{\boldsymbol{\beta}}'\mathbf{X}'\mathbf{y} = \mathbf{y}'\mathbf{C}\mathbf{y} - \hat{\boldsymbol{\beta}}'_*\mathbf{X}'_*\mathbf{C}\mathbf{y} \\
&= \mathbf{y}'\mathbf{y} - N\bar{y}^2 - \hat{\boldsymbol{\beta}}'_*\mathbf{X}'_*\mathbf{C}\mathbf{y}
\end{aligned}
$$

The difference is

$$
\begin{aligned}
\text{SSR} &= \text{SST} - \text{SSE} \\
&= \hat{\boldsymbol{\beta}}'\mathbf{X}'\mathbf{y} = \mathbf{y}'\mathbf{X}(\mathbf{X}'\mathbf{X})^{-1}\mathbf{X}'\mathbf{y} \qquad (14.52) \\
&= N\bar{y}^2 + \hat{\boldsymbol{\beta}}'_*\mathbf{X}'_*\mathbf{C}\mathbf{y}. \qquad (14.53)
\end{aligned}
$$

It represents that portion of SST attributable to having fitted the regression and is called the *sum of squares due to regression*, SSR. It is also called the *reduction in sum of squares*. Clearly, SST = SSR + SSE, as now summarized for developing the traditional analysis of variance table:

$$
\begin{aligned}
\text{SSR} &= \hat{\boldsymbol{\beta}}'\mathbf{X}'\mathbf{y} &&= N\bar{y}^2 + \hat{\boldsymbol{\beta}}'_*\mathbf{X}'_*\mathbf{C}\mathbf{y} \\
\text{SSE} &= \mathbf{y}'\mathbf{y} - \hat{\boldsymbol{\beta}}'\mathbf{X}'\mathbf{y} = \mathbf{y}'\mathbf{y} - N\bar{y}^2 - \hat{\boldsymbol{\beta}}'_*\mathbf{X}'_*\mathbf{C}\mathbf{y} \qquad (14.54) \\
\hline
\text{SST} &= \mathbf{y}'\mathbf{y} &&= \mathbf{y}'\mathbf{y}
\end{aligned}
$$

Throughout all this note that $\hat{\boldsymbol{\beta}}'_*\mathbf{X}'_*\mathbf{C}\mathbf{y}$ is the sum of products of elements of $\hat{\boldsymbol{\beta}}'_*$ with those of $\mathbf{X}'_*\mathbf{C}\mathbf{y}$, namely

$$
\hat{\boldsymbol{\beta}}'_*\mathbf{X}'_*\mathbf{C}\mathbf{y} = \sum_{j=1}^{k} \hat{\beta}_j \left[\sum_{i=1}^{N} (x_{ij} - \bar{x}_{\cdot j})(y_i - \bar{y}_{\cdot}) \right]. \qquad (14.55)
$$

$N\bar{y}^2$ occurring in SSR of (14.53) is recognizable as the usual correction for the mean, to be denoted as SSM:

$$
\text{SSM} = N\bar{y}^2. \qquad (14.56)
$$

Subtracting it from SSR gives what we call

$$
\text{SSR}_m = \text{SSR} - \text{SSM} = \hat{\boldsymbol{\beta}}'_*\mathbf{X}'_*\mathbf{C}\mathbf{y}, \qquad (14.57)
$$

the regression sum of squares corrected for the mean. In this way (14.54) becomes

$$
\begin{array}{rcl}
\text{SSM} & = & N\bar{y}^2 \\
\text{SSR}_m & = & \hat{\boldsymbol{\beta}}'_*\mathbf{X}'_*\mathbf{C}\mathbf{y} \\
\text{SSE} & = & \mathbf{y}'\mathbf{y} - N\bar{y}^2 - \hat{\boldsymbol{\beta}}'_*\mathbf{X}'_*\mathbf{C}\mathbf{y} \\
\hline
\text{SST} & = & \mathbf{y}'\mathbf{y}
\end{array}
\tag{14.58}
$$

Similar to SSR_m we also have

$$
\text{SST}_m = \text{SST} - \text{SSM} = \mathbf{y}'\mathbf{y} - N\bar{y}^2 = \mathbf{y}'\mathbf{C}\mathbf{y}
\tag{14.59}
$$

as the corrected sum of squares of the ys. Then, with $\text{SST}_m = \text{SSR}_m + \text{SSE}$, we have the summary

$$
\begin{array}{rcl}
\text{SSR}_m & = & \hat{\boldsymbol{\beta}}'_*\mathbf{X}'_*\mathbf{C}\mathbf{y} \\
\text{SSE} & = & \mathbf{y}'\mathbf{y} - N\bar{y}^2 - \hat{\boldsymbol{\beta}}'_*\mathbf{X}'_*\mathbf{C}\mathbf{y} \\
\hline
\text{SST}_m & = & \mathbf{y}'\mathbf{y} - N\bar{y}^2
\end{array}
\tag{14.60}
$$

The format of (14.60) is identical to that of (14.54); in the one case, (14.54), uncorrected sums of squares are used with total SST, and in the other, (14.60), corrected sums of squares are used with total SST_m. The error term SSE is the same in both places.

The summary shown in (14.60) is the basis of the traditional analysis of variance table for fitting regression, considered in Section 14.11.

Example (*continued*)

From (14.51), $\text{SSE} = 838\frac{1}{2}$, and from the data themselves in (14.34), $\text{SST} = \Sigma y_i^2 = 13,526$ and $\text{SSM} = N\bar{y}^2 = 5(50^2) = 12,500$. Therefore the summaries of (14.54), (14.58) and (14.60) are as shown in Table 14.3.

Table 14.3. Partitioning of Sum of Squares

Equations (14.54)	Equations (14.58)	Equations (14.60)
	SSM $=12,500$	
SSR $=12,687\frac{1}{2}$	SSR$_m =$ $187\frac{1}{2}$	SSR$_m =$ $187\frac{1}{2}$
SSE $=$ $838\frac{1}{2}$	SSE $=$ $838\frac{1}{2}$	SSE $=$ $838\frac{1}{2}$
SST $=13,526$	SST $=13,526$	SST$_m =1,026$

14.8 Multiple Correlation

A measure of the goodness of fit of the regression is the *multiple correlation coefficient*, estimated as the product moment correlation between the observed y_is and the predicted $\hat{y}_i$s. Denoted by R, it can be calculated as

$$R^2 = \frac{\left[\sum(y_i - \bar{y}.)(\hat{y}_i - \bar{\hat{y}}.)\right]^2}{\sum(y_i - \bar{y}.)^2 \sum(\hat{y}_i - \bar{\hat{y}}.)^2}, \tag{14.61}$$

which is often referred to as the *coefficient of determination*. Fortunately there is an easier and equivalent formula based on the following equalities:

$$\sum(y_i - \bar{y}.)^2 = \mathbf{y}'\mathbf{C}\mathbf{y} = \text{SST}_m \tag{14.62}$$

and

$$\sum(y_i - \bar{y}.)(\hat{y}_i - \bar{\hat{y}}.) = \sum(\hat{y}_i - \bar{\hat{y}}.)^2 = \text{SSR}_m. \tag{14.63}$$

It is left to the reader (E 14.7) to derive (14.62) and (14.63).

Using those results in (14.61) gives

$$R^2 = (\text{SSR}_m)^2/\text{SST}_m(\text{SSR}_m) = \text{SSR}_m/\text{SST}_m. \tag{14.64}$$

In the example,

$$R^2 = 187\tfrac{1}{2}/1026 = .183.$$

Intuitively the ratio $\text{SSR}_m/\text{SST}_m$ of (14.64) has appeal, since it represents that fraction of the total (corrected) sum of squares which is accounted for by fitting the model—in this case fitting the regression. Thus although R has traditionally been thought of and used as a multiple correlation coefficient in some sense, its more frequent use nowadays is in the form (14.64), where it represents the fraction of the total sum of squares accounted for by fitting the model.

Care must be taken in using (14.64) for, although $\text{SSR}_m = \text{SSR} - N\bar{y}^2$ and $\text{SST}_m = \text{SST} - N\bar{y}^2$ in the intercept model, the ratio is SSR/SST in the no-intercept model, as presented in Section 14.13.

14.9 Testing Linear Hypotheses

a. Stating a hypothesis

Suppose in the example that we wanted to test the joint hypothesis

$$H:\left\{\begin{array}{rcl} \beta_1 & = & \frac{1}{2} \\ \beta_1 + 3\beta_2 & = & 3 \end{array}\right., \text{ equivalent to } H:\begin{bmatrix} 0 & 1 & 0 \\ 0 & 1 & 3 \end{bmatrix}\begin{bmatrix} \beta_0 \\ \beta_1 \\ \beta_2 \end{bmatrix} = \begin{bmatrix} \frac{1}{2} \\ 3 \end{bmatrix}.$$
$$(14.65)$$

Its second form is an example of

$$H:\mathbf{K}'\boldsymbol{\beta} = \mathbf{m} \qquad (14.66)$$

where $\boldsymbol{\beta}$ is the vector of parameters in the model $\text{E}(\mathbf{y}) = \mathbf{X}\boldsymbol{\beta}$, and where $\mathbf{K}'$ and $\mathbf{m}$ are a matrix and vector, respectively, of known values that define the hypothesis of interest. In (14.65)

$$\mathbf{K}' = \begin{bmatrix} 0 & 1 & 0 \\ 0 & 1 & 3 \end{bmatrix} \quad \text{and} \quad \mathbf{m} = \begin{bmatrix} \frac{1}{2} \\ 3 \end{bmatrix}.$$

In this way (14.66) is a quite general formulation of any linear hypothesis: every linear hypothesis can be expressed this way. For example, $H:\boldsymbol{\beta} = \mathbf{0}$ and $H:\boldsymbol{\beta}_* = \mathbf{0}$ can be expressed, respectively, as

$$H:\begin{bmatrix} 1 & 0 & 0 \\ 0 & 1 & 0 \\ 0 & 0 & 1 \end{bmatrix}\begin{bmatrix} \beta_0 \\ \beta_1 \\ \beta_2 \end{bmatrix} = \begin{bmatrix} 0 \\ 0 \\ 0 \end{bmatrix} \text{ and } H:\begin{bmatrix} 0 & 1 & 0 \\ 0 & 0 & 1 \end{bmatrix}\begin{bmatrix} \beta_0 \\ \beta_1 \\ \beta_2 \end{bmatrix} = \begin{bmatrix} 0 \\ 0 \end{bmatrix}.$$
$$(14.67)$$

Only one limitation is placed on the form $H:\mathbf{K}'\boldsymbol{\beta} = \mathbf{m}$, and it is a very practical one. $\mathbf{K}'\boldsymbol{\beta}$ must not contain any redundancies. For example, in

$$H:\left\{\begin{array}{rcl} \beta_1 & = & \frac{1}{2} \\ \beta_1 + 3\beta_2 & = & 3 \\ 2\beta_1 + 3\beta_2 & = & 3\frac{1}{2} \end{array}\right., \text{ i.e., } \begin{bmatrix} 0 & 1 & 0 \\ 0 & 1 & 3 \\ 0 & 2 & 3 \end{bmatrix}\begin{bmatrix} \beta_0 \\ \beta_1 \\ \beta_2 \end{bmatrix} = \begin{bmatrix} \frac{1}{2} \\ 3 \\ 3\frac{1}{2} \end{bmatrix},$$
$$(14.68)$$

the last equation is clearly the sum of the other two. If the statements $\beta_1 = \frac{1}{2}$ and $\beta_1 + 3\beta_2 = 3$ are part of the hypothesis, there is no need to also have $2\beta_1 + 3\beta_2 = 3\frac{1}{2}$. In this sense $2\beta_1 + 3\beta_2 = 3\frac{1}{2}$ is redundant and its presence along with $\beta_1 = \frac{1}{2}$ and $\beta_1 + 3\beta_2 = 3$ is not permitted in the hypothesis. In general, functions of $\boldsymbol{\beta}$ are not permitted in $\mathbf{K}'\boldsymbol{\beta}$ if they can be expressed as linear combinations of other such functions in $\mathbf{K}'\boldsymbol{\beta}$. This is tantamount to not permitting any rows of $\mathbf{K}'$ to be linearly dependent on other rows, as occurs in (14.68). Hence in stating any linear hypothesis as $H\colon \mathbf{K}'\boldsymbol{\beta} = \mathbf{m}$, the matrix $\mathbf{K}'$ must always be of full row rank. Not only is this limitation practical, as just explained, but it also materially affects development of the general expression for the F-statistic for testing $H\colon \mathbf{K}'\boldsymbol{\beta} = \mathbf{m}$ quoted in the next subsection.

b.   The F-statistic

In (14.43) we summarized the mean and variance of $\mathbf{y}$ by writing $\mathbf{y} \sim (\mathbf{X}\boldsymbol{\beta},\ \sigma^2\mathbf{I}_N)$. But no particular probability distribution was attributed to $\mathbf{y}$; in particular, not even the normal distribution was mentioned. But it is now, for it is the underpinning of the $\mathcal{F}$-distribution which is the basis for the F-statistic. Thus we assume $\mathbf{y}$ to have a multinormal distribution and write

$$\mathbf{y} \sim \mathcal{N}(\mathbf{X}\boldsymbol{\beta}\ \ \sigma^2\mathbf{I}). \tag{14.69}$$

Important consequences of this (see Section 14.11) lead to showing (as is done in many places, e.g., Searle, 1977, Section 3.6a) that the F-statistic for testing $H\colon \mathbf{K}'\boldsymbol{\beta} = \mathbf{m}$ is

$$F(H) = Q/\hat{\sigma}^2 r_{\mathbf{K}}, \text{ with } r_{\mathbf{K}} \text{ and } N - k - 1 \text{ degrees of freedom,} \tag{14.70}$$

$r_{\mathbf{K}}$ being the rank of $\mathbf{K}$, equal to the number of rows in $\mathbf{K}'$. The numerator sum of squares in $F(H)$ of (14.70) is

$$Q = (\mathbf{K}'\hat{\boldsymbol{\beta}} - \mathbf{m})' \left[\mathbf{K}'(\mathbf{X}'\mathbf{X})^{-1}\mathbf{K} \right]^{-1} (\mathbf{K}'\hat{\boldsymbol{\beta}} - \mathbf{m}); \tag{14.71}$$

and

$$\hat{\boldsymbol{\beta}} = (\mathbf{X}'\mathbf{X})^{-1}\mathbf{X}'\mathbf{y} \text{ and } \hat{\sigma}^2 = (\mathbf{y}'\mathbf{y} - \hat{\boldsymbol{\beta}}'\mathbf{X}'\mathbf{y})/(N - k - 1), \tag{14.72}$$

coming from (14.22), and (14.46) with (14.50).

Illustration (*continued*)

For the hypothesis of (14.65) we illustrate the calculation of Q of (14.71). First, with $\hat{\beta}$ from (14.36)

$$\mathbf{K}'\hat{\beta} - \mathbf{m} = \begin{bmatrix} 0 & 1 & 0 \\ 0 & 1 & 3 \end{bmatrix} \begin{bmatrix} 37 \\ \frac{1}{2} \\ 1\frac{1}{2} \end{bmatrix} - \begin{bmatrix} \frac{1}{2} \\ 3 \end{bmatrix} = \begin{bmatrix} 0 \\ 2 \end{bmatrix},$$

and with $(\mathbf{X}'\mathbf{X})^{-1}$ from (14.35)

$$[\mathbf{K}'(\mathbf{X}'\mathbf{X})^{-1}\mathbf{K}]^{-1} = \left\{ \begin{bmatrix} 0 & 1 & 0 \\ 0 & 1 & 3 \end{bmatrix} \frac{1}{480} \begin{bmatrix} 790 & -80 & -42 \\ -80 & 16 & 0 \\ -42 & 0 & 6 \end{bmatrix} \begin{bmatrix} 0 & 0 \\ 1 & 1 \\ 0 & 3 \end{bmatrix} \right\}^{-1}$$

$$= 480 \begin{bmatrix} 16 & 16 \\ 16 & 70 \end{bmatrix}^{-1} = \frac{5}{9} \begin{bmatrix} 70 & -16 \\ -16 & 16 \end{bmatrix}.$$

Therefore Q of (14.71) is

$$Q = \begin{bmatrix} 0 & 2 \end{bmatrix} \frac{5}{9} \begin{bmatrix} 70 & -16 \\ -16 & 16 \end{bmatrix} \begin{bmatrix} 0 \\ 2 \end{bmatrix} = 35\frac{5}{9}.$$

c. Equivalent statements of a hypothesis

A feature of testing hypotheses is, of course, that even when a hypothesis can be expressed in different equivalent ways, for each of those ways the F-statistic given by (14.70) will be the same. For example, the hypothesis (14.65),

$$H: \begin{cases} \beta_1 & = \frac{1}{2} \\ \beta_1 + 3\beta_2 & = 3 \end{cases}, \quad \text{is equivalent to} \quad H: \begin{cases} \beta_1 & = \frac{1}{2} \\ \beta_2 & = \frac{5}{6} \end{cases}.$$

For the latter

$$\mathbf{K}' = \begin{bmatrix} 0 & 1 & 0 \\ 0 & 0 & 1 \end{bmatrix}, \quad \mathbf{m} = \begin{bmatrix} \frac{1}{2} \\ \frac{5}{6} \end{bmatrix}, \quad \mathbf{K}'\hat{\beta} - \mathbf{m} = \begin{bmatrix} \frac{1}{2} \\ 1\frac{1}{2} \end{bmatrix} - \begin{bmatrix} \frac{1}{2} \\ \frac{5}{6} \end{bmatrix} = \begin{bmatrix} 0 \\ \frac{2}{3} \end{bmatrix},$$

$$[\mathbf{K}'(\mathbf{X}'\mathbf{X})^{-1}\mathbf{K}]^{-1} = \left\{ \begin{bmatrix} 0 & 1 & 0 \\ 0 & 0 & 1 \end{bmatrix} \frac{1}{480} \begin{bmatrix} 790 & -80 & -42 \\ -80 & 16 & 0 \\ -42 & 0 & 6 \end{bmatrix} \begin{bmatrix} 0 & 0 \\ 1 & 0 \\ 0 & 1 \end{bmatrix} \right\}^{-1}$$

$$= \begin{bmatrix} 30 & 0 \\ 0 & 80 \end{bmatrix}$$

and (14.70) gives

$$Q = \begin{bmatrix} 0 & \frac{2}{3} \end{bmatrix} \begin{bmatrix} 30 & 0 \\ 0 & 80 \end{bmatrix} \begin{bmatrix} 0 \\ \frac{2}{3} \end{bmatrix} = 35\frac{5}{9}, \quad \text{as before.}$$

d.   Hypotheses not involving the intercept

On many occasions hypotheses of interest do not involve β_0; for example, $H: \beta_1 = \beta_2$ which can be written as $H: [0 \ \ 1 \ \ -1][\beta_0 \ \ \beta_1 \ \ \beta_2]' = 0$, and F obtained from (14.70) and (14.71). But sometimes it is convenient to use not $(\mathbf{X}'\mathbf{X})^{-1}$ but $(\mathbf{X}'_*\mathbf{C}\mathbf{X}_*)^{-1}$, the inverse of the matrix of sums of squares and products of the xs, corrected for their means. To do this we write $H: \beta_1 = \beta_2$ as $H: [1 \ \ -1][\beta_1 \ \ \beta_2] = 0$ or, in general terms,

$$H: \mathbf{K}'_*\boldsymbol{\beta}_* = \mathbf{m}_*.$$

Then the appropriate F-statistic is

$$F_* = \frac{(\mathbf{K}'_*\hat{\boldsymbol{\beta}}_* - \mathbf{m}_*)'[\mathbf{K}'_*(\mathbf{X}'_*\mathbf{C}\mathbf{X}_*)^{-1}\mathbf{K}_*]^{-1}(\mathbf{K}'_*\hat{\boldsymbol{\beta}}_* - \mathbf{m}_*)}{\hat{\sigma}^2 r_{\mathbf{K}_*}} \tag{14.73}$$

with $r_{\mathbf{K}_*} = k$. This is readily derived from F of (14.70) by writing $\mathbf{K}' = [01 \ \ \mathbf{K}'_*]$ and $\mathbf{m} = [0 \ \ \mathbf{m}'_*]'$. See E 14.9b. In the example

$$\mathbf{K}'_* = \begin{bmatrix} 1 & 0 \\ 1 & 3 \end{bmatrix}, \quad \mathbf{m}_* = \begin{bmatrix} \frac{1}{2} \\ 3 \end{bmatrix}, \quad \hat{\boldsymbol{\beta}} = \begin{bmatrix} \frac{1}{2} \\ 1\frac{1}{2} \end{bmatrix}, \quad \text{and} \quad \mathbf{K}'_*\hat{\boldsymbol{\beta}}_* = \begin{bmatrix} \frac{1}{2} \\ 5 \end{bmatrix},$$

$$\mathbf{X}'_*\mathbf{C}\mathbf{X}_* = \begin{bmatrix} 30 & 0 \\ 0 & 80 \end{bmatrix}, \quad \text{from (14.37).}$$

Therefore for $F_* = Q_*/\hat{\sigma}^2 r_{\mathbf{K}_*}$,

$$\mathbf{K}'_*\hat{\boldsymbol{\beta}}_* - \mathbf{m}_* = \begin{bmatrix} \frac{1}{2} \\ 5 \end{bmatrix} - \begin{bmatrix} \frac{1}{2} \\ 3 \end{bmatrix} = \begin{bmatrix} 0 \\ 2 \end{bmatrix};$$

and

$$[\mathbf{K}'_*(\mathbf{X}'_*\mathbf{C}\mathbf{X}_*)\mathbf{K}_*]^{-1} = \left\{ \begin{bmatrix} 1 & 0 \\ 1 & 3 \end{bmatrix} \begin{bmatrix} 1/30 & 0 \\ 0 & 1/80 \end{bmatrix} \begin{bmatrix} 1 & 1 \\ 0 & 3 \end{bmatrix} \right\}^{-1}$$

$$= 30 \begin{bmatrix} 1 & 1 \\ 1 & 4\frac{3}{8} \end{bmatrix}^{-1}$$

so that

$$Q_* = [0 \ 2]30 \begin{bmatrix} 1 & 1 \\ 1 & 4\frac{3}{8} \end{bmatrix}^{-1} \begin{bmatrix} 0 \\ 2 \end{bmatrix} = \frac{4(30)1}{4\frac{3}{8} - 1} = 35\frac{5}{9},$$

as before.

e. Special cases

At least three special cases of the general linear hypothesis $H: \mathbf{K}'\boldsymbol{\beta} = \mathbf{m}$ are of interest: (i) that all elements of $\boldsymbol{\beta}$ are zero, i.e., $H: \boldsymbol{\beta} = \mathbf{0}$; (ii) that each element of $\boldsymbol{\beta}$ equals some specific value, i.e., $H: \boldsymbol{\beta} = \boldsymbol{\beta}_+$; and (iii) that some linear combination of the elements of $\boldsymbol{\beta}$ equals a specific value, e.g., for given $\boldsymbol{\lambda}$ and m, $H: \boldsymbol{\lambda}'\boldsymbol{\beta} = m$. For each of these the F-statistic can be derived using (14.70), as is now shown. A fourth case, of wide applicability, is dealt with in Section 14.14.

(i) $H: \boldsymbol{\beta} = \mathbf{0}$. The F-statistic for testing this hypothesis is derived from (14.70) expressing the equations $\boldsymbol{\beta} = \mathbf{0}$ as $\mathbf{K}'\boldsymbol{\beta} = \mathbf{m}$. This means $\mathbf{K}' = \mathbf{I}_{k+1}$, $r_{\mathbf{K}'} = k+1$, $\mathbf{m} = \mathbf{0}$, $\mathbf{K}'\hat{\boldsymbol{\beta}} - \mathbf{m} = \hat{\boldsymbol{\beta}}$, and $[\mathbf{K}'(\mathbf{X}'\mathbf{X})^{-1}\mathbf{K}]^{-1} = \mathbf{X}'\mathbf{X}$ and so

$$F(H) = \frac{\hat{\boldsymbol{\beta}}'\mathbf{X}'\mathbf{X}\hat{\boldsymbol{\beta}}}{\hat{\sigma}^2(k+1)} = \frac{\hat{\boldsymbol{\beta}}'\mathbf{X}'\mathbf{y}}{\hat{\sigma}^2(k+1)} = \frac{\text{SSR}}{(k+1)\hat{\sigma}^2}. \tag{14.74}$$

Degrees of freedom for $F(H)$ are $k+1$ and $N - k - 1$.

(ii) $H: \boldsymbol{\beta} = \boldsymbol{\beta}_+$. This differs from (i) only in that $\mathbf{m}$ of $H: \mathbf{K}'\boldsymbol{\beta} = \mathbf{m}$ is $\boldsymbol{\beta}_+$. Thus

$$F(H) = (\hat{\boldsymbol{\beta}} - \boldsymbol{\beta}_+)'\mathbf{X}'\mathbf{X}(\hat{\boldsymbol{\beta}} - \boldsymbol{\beta}_+) \big/ \left[(k+1)\hat{\sigma}^2\right], \tag{14.75}$$

with $k+1$ and $N - k - 1$ degrees of freedom.

(iii) $H: \boldsymbol{\lambda}'\boldsymbol{\beta} = m$. Here, $\mathbf{K}' = \boldsymbol{\lambda}'$ with $r_{\mathbf{K}'} = 1$, $\mathbf{m} = m$ and

$$F(H) = \frac{(\boldsymbol{\lambda}'\hat{\boldsymbol{\beta}} - m) \left[\boldsymbol{\lambda}(\mathbf{X}'\mathbf{X})^{-1}\boldsymbol{\lambda}\right]^{-1} (\boldsymbol{\lambda}'\hat{\boldsymbol{\beta}} - m)}{\hat{\sigma}^2}.$$

Because $\boldsymbol{\lambda}'$ is a vector, this can be rewritten as

$$F(H) = \frac{(\boldsymbol{\lambda}'\hat{\boldsymbol{\beta}} - m)^2}{\boldsymbol{\lambda}'(\mathbf{X}'\mathbf{X})^{-1}\boldsymbol{\lambda}\hat{\sigma}^2}; \tag{14.76}$$

it has 1 and $N - k - 1$ degrees of freedom and is the square of a t-statistic having $N - k - 1$ degrees of freedom.

14.10 Predicting and Forecasting

The estimator $\hat{\beta}$ can be used for estimating the expected value $E(y) = \beta_0 + \beta_1 x_1 + \cdots + \beta_k x_k$ in the form

$$\widehat{E(y)} = \hat{\beta}_0 x_0 + \hat{\beta}_1 x_1 + \hat{\beta}_2 x_2 + \cdots + \hat{\beta}_k x_k$$

with $x_0 = 1$. This can be done for any set of x-values

$$\mathbf{x}'_0 = [1 \quad x_{01} \quad \cdots \quad x_{0k}] \tag{14.77}$$

for which we wish to estimate the corresponding value of $E(y)$, in the form

$$\widehat{E(y_0)} = \hat{\beta}_0 + \hat{\beta}_1 x_{01} + \cdots + \hat{\beta}_k x_{0k} = \mathbf{x}'_0 \hat{\beta}. \tag{14.78}$$

This is called the *estimated expected value of y* corresponding to the values $1, x_{01}, \ldots, x_{0k}$.

a. The traditional predictor

When $\mathbf{x}'_0$ of (14.77) is a row of $\mathbf{X}$ then (14.78) is an element of $\mathbf{X}\hat{\beta}$, and so as N special cases of (14.78) we have $\hat{\mathbf{y}} = \mathbf{X}\hat{\beta}$ of (14.44). The values in $\hat{\mathbf{y}}$ correspond to those in the data. They are sometimes called *fitted y-values*, or estimated y-values, names which can be misleading because $\hat{\mathbf{y}}$ and (14.78) are both estimates of expected values of y. They correspond, in (14.78), to any set of predetermined xs in $\mathbf{x}'_0$ of (14.77) and in (14.44) to the observed xs in $\mathbf{X}$.

Variances of the estimators (14.77) and (14.78) are readily obtained using $\text{var}(\hat{\beta})$ of (14.39). Thus

$$\text{var}\left[\widehat{E(y_0)}\right] \;=\; \mathbf{x}'_0 (\mathbf{X}'\mathbf{X})^{-1} \mathbf{x}_0 \sigma^2 \tag{14.79}$$

and

$$\text{var}(\hat{\mathbf{y}}) \;=\; \mathbf{X}(\mathbf{X}'\mathbf{X})^{-1}\mathbf{X}'\sigma^2 = \mathbf{H}\sigma^2. \tag{14.80}$$

Another variance of interest is that of $\mathbf{y} - \hat{\mathbf{y}}$, the observed "error" in $\hat{\mathbf{y}}$ or, put more carefully, the deviation of the predicted $\hat{\mathbf{y}}$ from the data $\mathbf{y}$. This variance is

$$\text{var}(\mathbf{y} - \hat{\mathbf{y}}) = \text{var}\left[(\mathbf{I} - \mathbf{H})\mathbf{y}\right] = (\mathbf{I} - \mathbf{H})\sigma^2 \tag{14.81}$$

on using the idempotence and symmetry of $(\mathbf{I} - \mathbf{H})$.

b. Forecasting

An important application of (14.78) is to predict or forecast the value of y corresponding to some future or potential value of $\mathbf{x}_0$, call it $\mathbf{x}_f$. Then the best available prediction or forecast of y_f, corresponding to x_f is $\tilde{y}_f = \mathbf{x}'_f \hat{\boldsymbol{\beta}}$. Thus $\mathbf{x}'_f \hat{\boldsymbol{\beta}}$ can be used both as a forecast of a potential y-value corresponding to $\mathbf{x}_f$, as well as its more customary use as an estimator of the expected value $E(y_f)$ corresponding to $\mathbf{x}_f$. As a forecast of y_f it is interesting to consider the difference of $\hat{y}_f$ from y_f:

$$y_f - \hat{y}_f = y_f - \mathbf{x}'_f \hat{\boldsymbol{\beta}} = \mathbf{x}'_f(\boldsymbol{\beta} - \hat{\boldsymbol{\beta}}) + e_f \qquad (14.82)$$

where, by the underlying model equation (14.14), we have $y_f = \mathbf{x}'_f \boldsymbol{\beta} + e_f$. Hence (14.82) gives

$$\text{var}(y_f - \hat{y}_f) = \left[\mathbf{x}'_f(\mathbf{X}'\mathbf{X})^{-1}\mathbf{x}_f + 1\right]\sigma^2. \qquad (14.83)$$

In comparing (14.81) and (14.83) it is clear that there is a noticeable difference between $\text{var}(\mathbf{y} - \hat{\mathbf{y}})$ and $\text{var}(y_f - \hat{y}_f)$. The underlying reason is that with the former, $\mathbf{y}$ and $\hat{\mathbf{y}}$ are correlated, whereas with the latter y_f and $\hat{y}_f$ are independent due to assuming y_f to be independent of $\mathbf{y}$, and hence of $\hat{\boldsymbol{\beta}}$ in $\hat{y}_f$.

A warning here is essential: $\mathbf{x}_f$ should not be very much outside the range of values of the rows of $\mathbf{X}$. Interpolation is acceptable, but extrapolation beyond available data can be risky.

14.11 Analysis of Variance

Partitioning the total sum of squares as shown in Section 14.7 is the basis for analysis of variance. Assuming that observations y_i are normally distributed, as in (14.69), leads to sums of squares in the partitioning of SST having χ^2-distributions, and thence to ratios of mean squares having F-distributions from which come familiar tests of hypotheses based on F-statistics. Derivation of these results is now sketched. The reader wanting details is referred to Section 3.5 of Searle (1997).

The basic distributional assumption is that the error terms in $\mathbf{e}$ of the model $\mathbf{y} = \mathbf{X}\boldsymbol{\beta} + \mathbf{e}$ are normally distributed with zero means and variance-covariance matrix $\sigma^2\mathbf{I}$; i.e., using the notation following (14.69), we write

$$\mathbf{e} \sim \mathcal{N}(\mathbf{0}, \sigma^2\mathbf{I}).$$

Distribution properties of $\mathbf{y}$ and of functions of $\mathbf{y}$ follow at once:

(i) $\mathbf{y} \sim \mathcal{N}(\mathbf{X}\boldsymbol{\beta}, \sigma^2\mathbf{I})$.

(ii) $\hat{\boldsymbol{\beta}} \sim \mathcal{N}[\boldsymbol{\beta}, (\mathbf{X}'\mathbf{X})^{-1}\sigma^2]$.

(iii) $\hat{\boldsymbol{\beta}}_* \sim \mathcal{N}[\boldsymbol{\beta}_*, (\mathbf{X}'_*\mathbf{C}\mathbf{X}_*)^{-1}\sigma^2]$.

(iv) $\hat{\boldsymbol{\beta}}$ and $\hat{\sigma}^2$ are independent.

(v) SSE/σ^2 has a χ^2-distri- bution on $N - r_{\mathbf{X}}$ degrees of freedom.

(vi) SSR/σ^2, SSM/σ^2 and SSR_m/σ^2 each has a (noncentral) χ^2-distribution on $k + 1$, 1, and k degrees of freedom, respectively, and each is independent of SSE.

Application of (v) and (vi) to the sums of squares in (14.58) yields the analysis of variance of Table 14.4.

Table 14.4. Analysis of Variance for Regression on k x-Variables

Source of Variation[a]	d.f.[b]	Sum of Squares	Mean Square
Mean	1	$\text{SSM} = N\bar{y}^2$	$\text{MSM} = \dfrac{\text{SSM}}{1}$
Regression (c.f.m.)	k	$\text{SSR}_m = \hat{\boldsymbol{\beta}}_*\mathbf{X}'_*\mathbf{C}\mathbf{y}$	$\text{MSR}_m = \dfrac{\text{SSR}_m}{k}$
Residual	$N - r_{\mathbf{X}}$	$\text{SSE} = \mathbf{y}'\mathbf{y} - N\bar{y}^2 - \hat{\boldsymbol{\beta}}_*\mathbf{X}'_*\mathbf{C}\mathbf{y}$	$\text{MSE} = \dfrac{\text{SSE}}{N - r_{\mathbf{X}}} = \hat{\sigma}^2$
Total	N	$\text{SST} = \mathbf{y}'\mathbf{y}$	

[a]c.f.m. = corrected for the mean.
[b]d.f. = degrees of freedom: $r_{\mathbf{X}} = k + 1$ for k x-variables.

Results (14.76) and (14.73) can now be used (see E 14.9) to show that in Table 14.4

$$F(M) = \frac{\text{MSM}}{\text{MSE}} \qquad \text{tests} \qquad H: \mathbf{1}'\mathbf{X}\boldsymbol{\beta} = 0, \text{ i.e., } H: \text{E}(\bar{y}) = 0$$

and

$$F(R_m) = \frac{\text{MSR}_m}{\text{MSE}} \qquad \text{tests} \qquad H: \boldsymbol{\beta}_* = 0, \text{ i.e., } \beta_i = 0 \,\forall\, i > 0.$$

Note also that $F(M)$, because it has numerator degrees of freedom of unity, is the square of a t-statistic:

$$F(M) = \frac{\text{MSM}}{\text{MSE}} = \frac{N\bar{y}^2}{\hat{\sigma}^2} = \left(\frac{\bar{y}}{\hat{\sigma}/\sqrt{N}}\right)^2,$$

where $\bar{y}/(\hat{\sigma}/\sqrt{N})$ is a t-statistic with $N - k - 1$ degrees of freedom. Also, $F(R_m)$ can be calculated using the multiple correlation squared. Using (14.64) yields

$$
\begin{aligned}
F(R_m) &= \frac{\text{MSR}_m}{\text{MSE}} = \frac{\text{SSR}_m}{k} \frac{(N - k - 1)}{\text{SSE}} = \frac{N - k - 1}{k} \frac{\text{SSR}_m}{\text{SST}_m - \text{SSR}_m} \\
&= \frac{N - k - 1}{k} \frac{R^2}{1 - R^2} = \frac{(N - 1)/k - 1}{1/R^2 - 1}.
\end{aligned} \tag{14.84}
$$

Illustration (*continued*)

The analysis of variance Table 14.4 is shown for the example in Table 14.5. Then $F(M)$ tested against the tabulated $\mathcal{F}$-distribution on 1 and 2 degrees of freedom tests $H: \beta_0 + 5\beta_1 + 7\beta_2 = 0$. This is equivalent to both

$$H: \mathbf{1'X\beta} = 0 \quad \text{and} \quad H: \text{E}(\bar{y}.) = 0.$$

And $F(R_m) = 0.2$ tested against the tabulated F-distribution on 2 and 2 degrees of freedom tests $H: \beta_1 = 0 = \beta_2$.

Demonstration of (14.84) is also readily available. From the end of Section 14.8, $R^2 = 187\frac{1}{2}/1026$, and so (14.84) gives the same value as Table 14.5:

$$F(R_m) = (4/2 - 1)/(1026/187\tfrac{1}{2} - 1) = 187\tfrac{1}{2}/838\tfrac{1}{2} = 0.2.$$

Table 14.5. Analysis of Variance for Example

Source of Variation	d.f.	Sum of Squares	Mean Square	F-statistic
Mean	1	SSM $= 12{,}500$	MSM $= 12{,}500$	$F(M) = 12{,}500/419\tfrac{1}{4} = 29.8$
Regression (c.f.m.)	2	SSR$_m = 187\tfrac{1}{2}$	MSR$_m = 93\tfrac{3}{4}$	$F(R_m) = 93\tfrac{3}{4}/419\tfrac{1}{4} = 0.2$
Residual error	2	SSE $= 838\tfrac{1}{2}$	MSE $= 419\tfrac{1}{4}$	
Total	5	SST $= 13{,}526$		

14.12 Confidence Intervals

With assuming $\mathbf{y} \sim \mathcal{N}(\mathbf{X}\boldsymbol{\beta}, \sigma^2 \mathbf{I})$, we have $\hat{\boldsymbol{\beta}}_* \sim \mathcal{N}\left[\boldsymbol{\beta}_*, (\mathbf{X}'_* \mathbf{C} \mathbf{X}_*)^{-1} \sigma^2\right]$ and so $(\hat{\beta}_i - \beta_i)/\sqrt{a^{ii}\sigma^2} \sim \mathcal{N}(0, 1)$, where a^{ii} is the ith diagonal element of $(\mathbf{X}'_* \mathbf{C} \mathbf{X}_*)^{-1}$. Therefore

$$(\hat{\beta}_i - \beta_i)/\sqrt{a^{ii}\hat{\sigma}^2} \sim t_{N-r_{\mathbf{X}}},$$

where t_{N-r} represents the t-distribution on $N - r$ degrees of freedom, with $r = r_{\mathbf{X}}$. On defining t_α and z_α by

$$P\{t_{N-r} > t_\alpha\} = \tfrac{1}{2}\alpha \quad \text{and} \quad (2\pi)^{-\frac{1}{2}} \int_{z_\alpha}^{\infty} e^{-\frac{1}{2}u^2} du = \tfrac{1}{2}\alpha,$$

the $100(1 - \alpha)\%$ symmetric confidence interval on β_i is

$$\hat{\beta}_i \pm \hat{\sigma} t_\alpha \sqrt{a^{ii}} \quad \text{or} \quad \hat{\beta}_i \pm \hat{\sigma} z_\alpha \sqrt{a^{ii}},$$

the former when degrees of freedom are small ($N - r < 100$, say) and the latter otherwise. Derivation of these intervals and of non-symmetric intervals is given in Searle (1997, p. 107). In the case of β_0, these intervals are used with

$$a^{00} = 1/N + \bar{\mathbf{x}}'(\mathbf{X}'_* \mathbf{C} \mathbf{X}_*)^{-1}\bar{\mathbf{x}},$$

in accord with (14.43), which is $\text{var}(\hat{\beta}) = a^{00}\sigma^2$. a^{00} is not, of course, part of $(\mathbf{X}'_* \mathbf{C} \mathbf{X}_*)^{-1}$.

14.13 The No-Intercept Model

When all xs are zero in the above models, $\widehat{\text{E}(y)} = \hat{\beta}_0$. Thus for $x_1 = 0 = x_2$ in the preceding example the estimated value $\widehat{\text{E}(y)}$ is $\hat{\beta}_0 = 37$. Models of this nature are called *intercept models*; the intercept is β_0, the value of $\text{E}(y)$ when all xs are zero.

Sometimes it is appropriate to have no term β_0 in the model, in which case the model is called a *no-intercept model*. The matrix $\mathbf{X}$ then has no vector $\mathbf{1}$ in it [it is $\mathbf{X}_*$ of (14.25)], and $\mathbf{X}'\mathbf{X}$ is then the matrix of sums of squares and products of the observations, without the first row and column of totals seen in (14.24).

Example (*continued*)

For the no-intercept model, omitting from (14.34) the first element and from $\mathbf{X}'\mathbf{X}$ in (14.35) the first row and column

$$\mathbf{X}'\mathbf{X} = \begin{bmatrix} 155 & 175 \\ 175 & 325 \end{bmatrix} \quad \text{and} \quad \mathbf{X}'\mathbf{y} = \begin{bmatrix} 1265 \\ 1870 \end{bmatrix},$$

for which solution (14.22) is

$$\hat{\boldsymbol{\beta}} = \begin{bmatrix} \hat{\beta}_1 \\ \hat{\beta}_2 \end{bmatrix} = (\mathbf{X}'\mathbf{X})^{-1}\mathbf{X}'\mathbf{y} = \tfrac{1}{790}\begin{bmatrix} 3355 \\ 2739 \end{bmatrix} = \begin{bmatrix} 4.25 \\ 3.47 \end{bmatrix}. \quad (14.85)$$

The no-intercept model thus has $E(y)$ being estimated from these data as

$$\widehat{E(y)} = 4.25x_1 + 3.47x_2.$$

The partitioning of $\boldsymbol{\beta}'$ as $[\beta_0 \;\; \boldsymbol{\beta}'_*]$ does not exist in the no-intercept model. $\boldsymbol{\beta}'$ is itself the vector of the βs corresponding to the x-variables and $\hat{\boldsymbol{\beta}} = (\mathbf{X}'\mathbf{X})^{-1}\mathbf{X}'\mathbf{y}$ is based on uncorrected sums of squares and products as exemplified in (14.85). Using corrected sums of squares and products is therefore not appropriate for the no-intercept model. Similarly, the only appropriate partitioning of SST is that of (14.54).

For the example, (14.85) yields

$$\text{SSE} = \hat{\boldsymbol{\beta}}'\mathbf{X}'\mathbf{y} = \tfrac{1}{790}[3355 \;\; 2739]\begin{bmatrix} 1265 \\ 1870 \end{bmatrix} = 11,855.7025$$

so that the partitioning of SST is

$$\begin{aligned} \text{SSR} &= 11,855.7025 \\ \text{SSE} &= 1,670.2975 \\ \hline \text{SST} &= 13,526 \end{aligned}$$

Analogous to (14.64), the multiple correlation R in the no-intercept model is defined without reference to $\bar{y}$ and so is

$$R^2 = (\Sigma y_i \hat{y}_i)^2 / \big[(\Sigma y_i^2)(\Sigma \hat{y}^2)\big] = \text{SSR/SST}.$$

Its value for the example is therefore $11,855.7025/13,526 = .877$.

14.14 Simultaneous Equations

a. Model

Economic models that involve a set of relationships designed to explain the behavior of model variables are called *simultaneous equation systems*. There are several types of variables in a simultaneous equation system. The variables explained by the model equations are called *endogenous variables*. *Predetermined variables* are those that assist in the explanation. Predetermined variables can be *exogenous* (those determined completely outside the system), or *lagged endogenous* (those variables that are the past values of the endogenous variables).

Illustration   (Market model)

The following equations make up a simple market model with a demand and supply equation, respectively.

$$Q_t = \alpha_1 + \alpha_2 P_t + \alpha_3 Y_t + \epsilon_{1t} \quad \text{and} \quad Q_t = \beta_1 + \beta_2 P_t + \epsilon_{2t}, \quad (14.86)$$

where α_1, α_2, α_3, β_1 and β_2 are parameters, ϵ_{1t} and ϵ_{2t} are stochastic disturbances, Q_t and P_t are endogenous variables representing quantity and price, and Y_t is an exogenous variable representing income. This is a set of simultaneous equations since both relationships are needed to explain the two endogenous variables. The subscript t suggests that the model variables and stochastic disturbances vary over time.

The single most important characteristic of a simultaneous equation system is that endogenous variables appear on the right-hand side of some of the model equations. These model equations, known as the *structural form*, can be solved as they stand, for the endogenous variables.

In general, the structural form of a simultaneous equation system can be described in matrix notation as

$$\mathbf{B}\mathbf{y}_t + \mathbf{\Gamma}\mathbf{x}_t = \mathbf{u}_t, \quad (14.87)$$

where

$$\mathbf{y}_t = \begin{bmatrix} y_{1t} \\ y_{2t} \\ \vdots \\ y_{Gt} \end{bmatrix}, \quad \mathbf{x}_t = \begin{bmatrix} x_{1t} \\ x_{2t} \\ \vdots \\ x_{Kt} \end{bmatrix}, \quad \mathbf{u}_t = \begin{bmatrix} u_{1t} \\ u_{2t} \\ \vdots \\ u_{Gt} \end{bmatrix}, \quad (14.88)$$

$$\mathbf{B} = \begin{bmatrix} \beta_{11} & \beta_{12} & \cdots & \beta_{1G} \\ \beta_{21} & \beta_{22} & \cdots & \beta_{2G} \\ \vdots & \vdots & \ddots & \vdots \\ \beta_{G1} & \beta_{G2} & \cdots & \beta_{GG} \end{bmatrix} = \left\{ {}_m \, \beta_{ij} \right\}_{i,j=1}^{G},$$

$$\mathbf{\Gamma} = \begin{bmatrix} \gamma_{11} & \gamma_{12} & \cdots & \gamma_{1K} \\ \gamma_{21} & \gamma_{22} & \cdots & \gamma_{2K} \\ \vdots & \vdots & \ddots & \vdots \\ \gamma_{G1} & \gamma_{G2} & \cdots & \gamma_{GK} \end{bmatrix} = \left\{ {}_m \, \gamma_{ir} \right\}_{i=1 \; r=1}^{G \quad K}.$$

The **y**s are endogenous variables (G of them), the **x**s are predetermined variables (K of them), and the **u**s are stochastic disturbances analogous to the ϵs in the simple market model example. The βs and the γs are the *structural coefficients*.

The stochastic disturbances satisfy the assumption of the classical normal linear regression model, namely

$$\mathbf{u}_t \sim N(0, \, \boldsymbol{\phi})$$
$$\mathrm{E}(u_t, \, u_s') = 0 \quad \text{for} \quad t \neq s \tag{14.89}$$

where $\boldsymbol{\phi}$ is the variance-covariance matrix of the stochastic disturbances of the structural model.

A reduced form of the model shows how the endogenous variables are jointly determined by the predetermined variables and the stochastic disturbances of the structural model. The reduced form of the system can be obtained by solving the structural form equations for the values of the endogenous variables. Hence, $\mathbf{y}_t$ will be expressed in terms of $\mathbf{x}_t$ and $\mathbf{u}_t$. The model is

$$\mathbf{y}_t = \Pi \mathbf{x}_t + \mathbf{v}_t, \tag{14.90}$$

where Π is the matrix of the *reduced form coefficients* and $\mathbf{v}_t$ is the vector of the reduced form disturbances, which are linear functions of all the disturbances of the structural model. By comparing (14.90) with (14.87) we see that

$$\Pi = -\mathbf{B}^{-1}\mathbf{\Gamma} \tag{14.91}$$

and

$$\mathbf{v}_t = \mathbf{B}^{-1}\mathbf{u}_t. \tag{14.92}$$

The variance-covariance matrix of $\mathbf{v}_t$ is

$$\Psi = \mathbf{B}^{-1}\boldsymbol{\phi}(\mathbf{B}^{-1})'. \tag{14.93}$$

Illustration (*continued*)

The reduced form of the simple market model is

$$\mathbf{y}_t = \Pi \mathbf{x}_t + \mathbf{v}_t \tag{14.94}$$

where

$$\mathbf{y}_t = \begin{bmatrix} Q_t \\ P_t \end{bmatrix}, \qquad \mathbf{x}_t = \begin{bmatrix} 1 \\ Y_t \end{bmatrix},$$

$$\Pi = -\begin{bmatrix} \dfrac{1}{\alpha_2 - \beta_2} \end{bmatrix} \begin{bmatrix} \alpha_1\beta_2 - \alpha_2\beta_1 & \alpha_3\beta_2 \\ \alpha_1 - \beta_1 & \alpha_3 \end{bmatrix},$$

$$\mathbf{v}_t = \begin{bmatrix} \dfrac{1}{\alpha_2 - \beta_2} \end{bmatrix} \begin{bmatrix} -\beta_2\epsilon_{1t} + \alpha_2\epsilon_{2t} \\ -\epsilon_{1t} + \epsilon_{2t} \end{bmatrix}.$$

b.　Identification

Least squares estimators of the structural coefficients of a simultaneous equation system are not consistent because there are endogenous variables included as explanatory variables. In other words, endogenous variables are on the right-hand side of the structural equations. The explanatory variables in the reduced form equations are represented by the predetermined variables. Hence ordinary least squares estimates of the reduced form coefficients are consistent. The goal then is to identify estimates of the structural coefficients from estimates of the reduced form coefficients. In our example we want to obtain the $\mathbf{B}$ and the Γ in the structural model from the Π in the reduced form model. If there is a unique solution for the structural equation coefficients from the reduced form coefficients there is exact identification. Over-identification suggests that there is no unique solution. There are more reduced form coefficients than are needed to obtain the structural coefficients. Under-identification occurs if the number of reduced form coefficients is less than that necessary to obtain the coefficients of the structural equation. For more details on identification, see Judge et al. (1988, Chapters 14 and 15), Greene (1997, Chapter 16), or Kmenta (1986, Chapter 13).

Example

Suppose the estimate of Π is $\hat{\Pi} = \begin{bmatrix} f & g \\ h & k \end{bmatrix}$ where f, g, h and k represent numbers. We want to derive $\hat{\mathbf{B}}$ and $\hat{\Gamma}$ from $\hat{\Pi} = -\hat{\mathbf{B}}^{-1}\hat{\Gamma}$, based on (14.91). We can see that

$$\hat{\Gamma} = -\hat{B}\hat{\Pi}$$

$$\begin{bmatrix} -\hat{\alpha}_1 & -\hat{\alpha}_3 \\ -\hat{\beta}_1 & 0 \end{bmatrix} = -\begin{bmatrix} 1 & -\hat{\alpha}_2 \\ 1 & -\hat{\beta}_2 \end{bmatrix}\begin{bmatrix} f & g \\ h & k \end{bmatrix}$$

$$\begin{aligned} \hat{\alpha}_1 &= f - \hat{\alpha}_2 h & \hat{\alpha}_3 &= g - \hat{\alpha}_2 k \\ \hat{\beta}_1 &= f - \hat{\beta}_2 h & 0 &= g - \hat{\beta}_2 k \end{aligned} \qquad (14.95)$$

From (14.95) we can derive $\hat{\beta}_2 = g/k$ and $\hat{\beta}_1 = f - gh/k$. But notice that we are unable to determine $\hat{\alpha}_1$, $\hat{\alpha}_2$ and $\hat{\alpha}_3$. This clearly shows a case of under-identification for the demand equation [the first equation in (14.86)], and exact identification for the supply equation [the second equation in (14.86)].

c. Methods of estimation

We have discussed how the ordinary least squares method of estimation applied to simultaneous equation systems often leads to inconsistent estimates. There are two sets of methods to estimate structural equations when they are exactly or over-identified. The first set of methods estimates a single structural equation with a limited reference to other parts of the system. This method includes such estimation techniques as indirect least squares, instrumental variable methods, two-stage least squares or limited information maximum likelihood estimator.

The second set of methods for estimating structural equations are those where all equations of a system are estimated simultaneously. These system methods of estimation overcome the problem of single equation estimation wherein the correlation between the disturbances of different structural equations are not taken into consideration. The simplest method of estimation is three-stage least squares. The full information maximum likelihood estimation technique is also often used in system estimation techniques. Details of these methods of system estimation are fairly straightforward but lengthy. Students of economics are often not introduced to the specifics of these methods until a high-level graduate course in econometrics. Hence the details are not covered here. The reader is referred to Judge et al. (1988), Chapter 15, or Greene (1997), Sections 16.5, 16.6 and 16.7, for a complete presentation of the methods.

14.15 Appendix

Consider the regression model

$$E(y_i) = \alpha + \beta_1 x_i + \beta_2 z_i. \tag{14.96}$$

Redefine y_i, x_i and z_i as being connected for their observed means; e.g., y_i becomes $y_i - \bar{y}$ for $\bar{y} = \sum_{i=1}^{n} y_i / n$. With those redefinitions let

$$s_1^2 = \sum x_i^2, \quad s_2^2 = \sum z_i^2 \quad s_{12} = \sum x_i z_i$$
$$p_1 = \sum x_i y_i, \quad p_2 = \sum z_i y_i \tag{14.97}$$

where all summations are over $i = 1, 2, \ldots, n$. Then for the model

$$E(y_i) = \mu + \beta x_i \tag{14.98}$$

the estimator of β is

$$\hat{\beta} = \frac{p_1}{s_1^2}. \tag{14.99}$$

And for the model (14.96) the estimator of β_1 is given by

$$\begin{pmatrix} \hat{\beta}_1 \\ \hat{\beta}_2 \end{pmatrix} = \begin{pmatrix} s_1^2 & s_{12} \\ s_{12} & s_2^2 \end{pmatrix}^{-1} \begin{pmatrix} p_1 \\ p_2 \end{pmatrix} \tag{14.100}$$

$$= \begin{bmatrix} (s_2^2 p_1 - s_{12} p_2)/\Delta \\ (s_2^1 p_2 - s_{12} p_1)/\Delta \end{bmatrix} \quad \text{for} \quad \Delta = s_1^2 s_2^2 - s_{12}^2. \tag{14.101}$$

Therefore the occurrence of $\hat{\beta} = \hat{\beta}_1$ noted for the example at the end of Section 14.3 arises when

$$\frac{p_1}{s_1^2} = \frac{s_2^2 p_1 - s_{12} p_2}{\Delta}. \tag{14.102}$$

On using Δ from (14.101) we find that (14.102) reduces to

$$\frac{p_1}{p_2} = \frac{s_1^2}{s_{12}}. \tag{14.103}$$

This, of course, cannot occur if $s_{12} = 0$. But then from (14.101) $\hat{\beta}_1 = s_2^2 p_1 / s_1^2 s_2^2 = p_1 / s_1^2 = \hat{\beta}$. Thus we have the result that

$$s_{12} = 0 \text{ is a sufficient condition for having } \hat{\beta} = \hat{\beta}_1$$

and for

$s_{12} \neq 0$ a NSC for having $\hat{\beta} = \hat{\beta}_1$ is (14.103),

namely

$$\frac{\sum(x_i - \bar{x})(y_i - \bar{y})}{\sum(z_i - \bar{z})(y_i - \bar{y})} = \frac{\sum(x_i - \bar{x})^2}{\sum(x_i - \bar{x})(z_i - \bar{z})}, \tag{14.104}$$

where x_i, y_i and z_i are the actual data (not corrected for their means). It can also be noted from (14.101) that when (14.103) is true then $\hat{\beta}_2 = 0$.

14.16 Exercises

E 14.1. Show that equations (14.6) and (14.7) can be solved as

$$\begin{bmatrix} \hat{\alpha} \\ \hat{\beta} \end{bmatrix} = \begin{bmatrix} N & x. \\ x. & x_i^2 \end{bmatrix}^{-1} \begin{bmatrix} y. \\ \Sigma x_i y_i \end{bmatrix}.$$

E 14.2. Show details of differentiating Δ of (14.18) to derive (14.21).

E 14.3. For $\mathbf{X} = [\mathbf{1} \quad \mathbf{X}_*]$ of (14.25), show that $\hat{\boldsymbol{\beta}} = (\mathbf{X}'\mathbf{X})^{-1}\mathbf{X}'\mathbf{y}$ of (14.22) yields $\hat{\boldsymbol{\beta}}_*$ of (14.30) and then from (14.33)

$$\hat{\beta}_0 = \bar{y}. - \hat{\boldsymbol{\beta}}'_*\bar{\mathbf{x}}$$

where $\bar{\mathbf{x}}.$ is the vector of x-means $\bar{x}_{.j}$ for $j = 1, \ldots, k$.

E 14.4. Derive (14.47) from (14.46).

E 14.5. For $\mathbf{H}$ of (14.44) reduce (14.45) to (14.46).

E 14.6. Use $\text{var}(\hat{\boldsymbol{\beta}}) = (\mathbf{X}'\mathbf{X})^{-1}\sigma^2$ to derive (14.40), (14.41) and (14.43).

E 14.7. (a) Use $\mathbf{HX} = \mathbf{X}$ to show that $\mathbf{H1} = \mathbf{1}$.
(b) Derive (14.62) and (14.63).

E 14.8. (a) Solve equations (14.6) and (14.7).

(b) Show that equations (14.6) and (14.7) are a special case of the normal equations (14.21).

(c) Show that the solutions to equations (14.6) and (14.7) are of the same form as (14.22).

E 14.9. (a) Show that $\lambda' = 1'X$ and $m = 0$ in (14.76) gives $F(M)$ of Table 14.2.

(b) Show that $K_* = I_k$ and $m_* = 0$ in (14.73) gives $F(R_m)$ of Table 14.2.

[*Hint*: Recall that $X(X'X)^-X'1 = 1$ from E 14.7(a).]

E 14.10. Derive F-statistics for

$$H: 1'X\beta = 0 \qquad \text{and} \qquad H: \beta_* = 0.$$

E 14.11. Use the example of Table 14.1 to:

(a) Multiply every y_i and x_i by 1.1 and recalculate the regression. Why is your value of $\hat{\beta}$ the same as the example? Why does $\hat{\alpha}$ equal 52.25?

(b) Add 5 to every y_i and 10 to every x_i. Recalculate the regression. How are the new $\hat{\beta}$ and $\hat{\alpha}$ compared to the $\hat{\beta}$ and $\hat{\alpha}$ of Section 14.1c?

(c) For a simple regression explain the relationship between the estimated coefficients of the regression of y on x to the estimated coefficients of the regression of $cy + K_1$ on $cx + K_2$. Use (a) and (b) to illustrate your results.

E 14.12. Derive Π and v_t which follow (14.94).

E 14.13. (Simple regression) The market value of a home is often influenced by its size. In fact, one could express this relationship as a linear model $P_t = \alpha + \beta S_t + \epsilon_t$ where P_t are house sales in a small city in a particular month and S_t is the size of the house measured in square footage. Data on these variables are shown in the following table.

Observation	House Price ($1,000)	Size (100 sq. feet)
1	75	12
2	125	20
3	150	18
4	136	19
5	95	17
6	200	35
7	112	20
8	85	14
9	99	16
10	124	23

(a) What is your expectation of the magnitude and signs of α and β?

(b) Estimate α and β from these data. (You may find it helpful to use a spreadsheet for this work.)

(c) Determine the R^2 for this model.

(d) Determine the F-statistic for the hypothesis $H: \beta = 0$.

E 14.14. (Demand) Microeconomic theory indicates that the quantity demanded for a product is affected by its own price, the prices of substitutes and complements, and income. The following data are available from the 1999 Economic Report of the President and the 1999 Agricultural Statistics.

Year	QA	PA	PG	PCE
1989	9,916.8	11.65	351.70	650.1
1990	9,656.8	15.96	311.77	662.9
1991	9,706.7	18.25	318.17	659.6
1992	10,568.5	13.60	306.00	660.0
1993	10,685.1	12.71	328.21	675.3
1994	11,500.9	12.55	312.35	687.9
1995	10,578.4	16.35	332.82	689.5
1996	10,381.9	14.99	404.41	692.6
1997	10,323.8	14.30	397.44	699.3
1998	10,943.6	10.79	414.81	714.0

where QA is the quantity of apples (million pounds), PA is the

price of apples (1992 cents/pound), PG is the price of grapes
(1992 $/ton) and Y is income (1992 billion $).

(a) Write a single equation for a multiple regression model
for the quantity demanded of apples in the United States.
Use quantity demanded as the dependent variable.

(b) What is your hypothesis about the signs of the coefficients in your model? Explain your reasoning.

(c) Estimate the model coefficients using least squares. (You
may find it helpful to use a spreadsheet for this work.)

(d) What is the coefficient of determination?

E 14.15. Show that the following values of x, z and y satisfy the conditions of the Appendix.

x	z	y
7	140	62
14	140	60
11	120	57
8	10	48
10	140	23

Chapter 15

LINEAR STATISTICAL MODELS

The general subject of linear statistical models is a large one and is presented extensively in many texts such as Myers and Milton (1991), Stapleton (1995), Searle (1997) and Scheffé (1999), to name but a few. Linear models are of great importance in data analysis, because of their close association with the widely used methods of analysis of variance. They are also very amenable to presentation in matrix technology which, in turn, brings considerable clarity to the subject. Nevertheless, because the topic is so vast, we give only a brief outline here, with illustration of the general results using a single example which has easy arithmetic. Great reliance is placed on results on regression in the preceding chapter, and considerable use is made of the generalized inverse of a matrix discussed in Chapter 9 and its uses in solving linear equations given in Chapter 10.

15.1   General Description

A simple illustration serves to introduce the general idea of linear models and the matrix notation applicable to them.

a.   Illustration   (Linear model)

Suppose Internet providers in cities of three Ivy League schools have revenues as shown in Table 15.1. The cities are Cambridge, Massachusetts (Harvard), Ithaca, New York (Cornell), and New Haven,

Connecticut (Yale).

Table 15.1. Revenue of Six Firms

	Cambridge	Ithaca	New Haven	
	101	84	32	
	105	88		
	94			
Totals	300	172	32	Grand total $= 504$

b.   Model

For the entries in the table let y_{ij} denote the revenue (in million dollars) of the jth firm in the ith city, i taking values 1,2,3 for the three cities, and $j = 1, \ldots, n_i$, where n_i is the number of firms in the ith city in Table 15.1. The feature of interest is the extent to which cities differ—and for this purpose we want to estimate the effect of city on revenue. To do this we assume that the average or expected value of an observation y_{ij}, which we denote by $E(y_{ij})$, is the sum of two parts:

$$E(y_{ij}) = \mu + \alpha_i, \tag{15.1}$$

where μ represents an overall mean and α_i is the effect of city on revenue. Now in general, no observation is the same as its expected value, and we define the difference to be a random error,

$$y_{ij} - E(y_{ij}) = e_{ij}. \tag{15.2}$$

Therefore

$$y_{ij} = E(y_{ij}) + e_{ij} = \mu + \alpha_i + e_{ij}. \tag{15.3}$$

Equation (15.3) is known as a *model equation*. The model is this equation and description of its terms.

μ and α_i have already been described. They are taken as being fixed (unknown) constants which we would like to gain information about from the data. We now describe e_{ij}. It is a random variable, and from its definition in (15.2) it is easily seen to have expected value zero: $E(e_{ij}) = 0$. Being a random variable it has a variance, and to this we attribute the value σ^2, the same for all i and j:

$$\text{var}(e_{ij}) = \sigma^2 \; \forall \, i \text{ and } j. \tag{15.4}$$

Furthermore, we assume the e_{ij}s to be independent of each other so that

$$\text{cov}(e_{ij}, e_{rs}) = 0 \text{ except for } i = r \text{ and } j = s,$$

and hence the variance-covariance matrix of the vector of e-terms is $\sigma^2 \mathbf{I}$. Thus is the model specified, in this case the model of the general *1-way classification*; it is clearly a linear model since it is based on the assumption that y_{ij} is the simple sum of its three parts μ, α_i and e_{ij}.

The problem is to estimate μ and the α_is, and also σ^2, the variance of the error terms. We will find that not all the terms μ and α_i can be estimated satisfactorily, only certain linear functions of them can. At first thought this may seem to be a matter for concern; sometimes it is, but on many occasions it is not, because the number of linear functions that can be estimated is large and frequently includes those in which we are interested; e.g., differences between the effects, such as $\alpha_1 - \alpha_2$. Sometimes, however, functions of interest cannot be estimated, because of a paucity of data. On all occasions, though, a method is needed for ascertaining which functions can be estimated and which cannot. This is provided in what follows in Section 15.10 dealing with estimable functions.

To develop the method of estimation we write down the six observations in terms of equation (15.3) of the model:

$$
\begin{array}{rcccccccc}
101 & = & y_{11} & = & \mu & + & \alpha_1 & & & & + & e_{11} \\
105 & = & y_{12} & = & \mu & + & \alpha_1 & & & & + & e_{12} \\
94 & = & y_{13} & = & \mu & + & \alpha_1 & & & & + & e_{13} \\
84 & = & y_{21} & = & \mu & & & + & \alpha_2 & & + & e_{21} \\
88 & = & y_{22} & = & \mu & & & + & \alpha_2 & & + & e_{22} \\
32 & = & y_{31} & = & \mu & & & & & + & \alpha_3 + & e_{31}
\end{array}
$$

These equations are easily written in matrix form as

$$
\begin{bmatrix} 101 \\ 105 \\ 94 \\ 84 \\ 88 \\ 32 \end{bmatrix} = \begin{bmatrix} y_{11} \\ y_{12} \\ y_{13} \\ y_{21} \\ y_{22} \\ y_{31} \end{bmatrix} = \begin{bmatrix} 1 & 1 & 0 & 0 \\ 1 & 1 & 0 & 0 \\ 1 & 1 & 0 & 0 \\ 1 & 0 & 1 & 0 \\ 1 & 0 & 1 & 0 \\ 1 & 0 & 0 & 1 \end{bmatrix} \begin{bmatrix} \mu \\ \alpha_1 \\ \alpha_2 \\ \alpha_3 \end{bmatrix} + \begin{bmatrix} e_{11} \\ e_{12} \\ e_{13} \\ e_{21} \\ e_{22} \\ e_{31} \end{bmatrix}, \qquad (15.5)
$$

which is precisely the form

$$
\mathbf{y} = \mathbf{Xb} + \mathbf{e} \qquad (15.6)
$$

where $\mathbf{y}$ is the vector of observations, $\mathbf{X}$ is the matrix of 0s and 1s, and $\mathbf{b}$ is the vector of parameters to be considered,

$$
\mathbf{b}' = \begin{bmatrix} \mu & \alpha_1 & \alpha_2 & \alpha_3 \end{bmatrix}, \qquad (15.7)
$$

and **e** is the vector of error terms.

Note that (15.6) is exactly the same as (14.16) of Chapter 14 except for using **b** in place of β. In both cases **y** and **e** are vectors of observations and errors, respectively. β in Chapter 14 is a vector of regression coefficients β_j whereas **b** in this chapter is a vector of effects such as μ and the αs in (15.7), and in general can include effects for several factors. Both β and **b** are vectors of parameters whose estimates we seek. Also, the properties of **e** are the same: $E(\mathbf{e}) = \mathbf{0}$, so that $E(\mathbf{y}) = \mathbf{Xb}$, and $E(\mathbf{ee'}) = \sigma^2 \mathbf{I}$. The only difference is in the form of **X**: in regression it is a matrix of observations on x-variables, whereas in (15.6) its elements are either 0 or 1, depending on the absence or presence of particular terms of the model in each y_{ij} observation. But, with respect to applying the principle of least squares for estimating the elements of **b**, there is no difference at all between (15.6) of this chapter and (14.16) of Chapter 14. Hence we can go directly to (14.21) to obtain an equation for $\hat{\mathbf{b}}$, the least squares estimator of **b**:

$$\mathbf{X'X\hat{b}} = \mathbf{X'y}. \tag{15.8}$$

In regression analysis the solution $\hat{\mathbf{b}}$ was obtained from this as $\hat{\mathbf{b}} = (\mathbf{X'X})^{-1}\mathbf{X'y}$, where $\mathbf{X'X}$ was non-singular and therefore had an inverse. Now, however, $\mathbf{X'X}$ has no inverse and we have to use the methods of generalized inverses to obtain a solution.

15.2 Normal Equations

a. General form

Equations (15.8) are called the *normal equations*. Before discussing their solution let us look briefly at their form. First, the vector of parameters, **b**: it is the vector of all the elements of the model, in this case the elements μ, α_1, α_2 and α_3. And this is so in general; for example, if data can be arranged in rows and columns according to two different classifications, elements of the vector **b** can be μ and terms representing, say, *a* row effects and *b* column effects.

Second, the matrix **X**: it is called the *model matrix*. For data from designed experiments it is the *design matrix*, and for model equations such as here, e.g., (15.7), it can be called an *incidence matrix*, because the location of the 0s and 1s throughout its elements represents the incidence of terms of the model among the observations—and hence of the classifications in which the observations lie. This is particularly evident if one writes $\mathbf{X}$ as a 2-way table with the parameters as headings to the columns and the observations as labels for the rows, as illustrated in Table 15.2.

Table 15.2.   **X** as a 2-Way Table (Data of Table 15.1)

Observations	\multicolumn{4}{c}{Parameters of Model}
	μ	α_1	α_2	α_3
y_{11}	1	1	0	0
y_{12}	1	1	0	0
y_{13}	1	1	0	0
y_{21}	1	0	1	0
y_{22}	1	0	1	0
y_{31}	1	0	0	1

In Table 15.2, as in equations (15.5), it is clear that the sum of the last three columns equals the first column. (This is so because every y_{ij} observation contains μ and so the first column of $\mathbf{X}$ is all 1s; and every y_{ij} also contains just one α and so the sum of the last three columns is also all 1s.) Thus $\mathbf{X}$ is not of full column rank.

Now consider the normal equations (15.8). They involve $\mathbf{X'X}$, which is obviously square and symmetric. Its elements are the inner products of the columns of $\mathbf{X}$ with each other, as in equation (3.1) of Chapter 3. Hence

$$\mathbf{X'X} = \begin{bmatrix} 6 & 3 & 2 & 1 \\ 3 & 3 & 0 & 0 \\ 2 & 0 & 2 & 0 \\ 1 & 0 & 0 & 1 \end{bmatrix}. \tag{15.9}$$

Furthermore, because $\mathbf{X}$ does not have full column rank, $\mathbf{X'X}$ is not of full rank.

The normal equations (15.8) also involve the vector $\mathbf{X'y}$; its elements are the inner products of the columns of $\mathbf{X}$ with the vector $\mathbf{y}$, and since the only non-zero elements of $\mathbf{X}$ are ones, the elements of $\mathbf{X'y}$ are various sums of elements of $\mathbf{y}$; e.g., from (15.5) and (15.6),

$$\mathbf{X'y} = \begin{bmatrix} 1 & 1 & 1 & 1 & 1 & 1 \\ 1 & 1 & 1 & 0 & 0 & 0 \\ 0 & 0 & 0 & 1 & 1 & 1 \\ 0 & 0 & 0 & 0 & 0 & 1 \end{bmatrix} \begin{bmatrix} y_{12} \\ y_{12} \\ y_{13} \\ y_{21} \\ y_{22} \\ y_{31} \end{bmatrix} = \begin{bmatrix} y_{..} \\ y_{1.} \\ y_{2.} \\ y_{3.} \end{bmatrix} = \begin{bmatrix} 504 \\ 300 \\ 172 \\ 32 \end{bmatrix}. \tag{15.10}$$

This is often the nature of $\mathbf{X'y}$ in linear models—a vector of various subtotals of the observations in $\mathbf{y}$.

b. Many solutions

Whenever $\mathbf{X'X}$ is not of full rank, as in (15.5), the normal equations (15.8) cannot be solved with one solitary solution $\hat{\mathbf{b}} = (\mathbf{X'X})^{-1}\mathbf{X'y}$ as in equation (14.22). Many solutions are available. To emphasize this we write the normal equations as

$$\mathbf{X'Xb}^0 = \mathbf{X'y}, \tag{15.11}$$

using the symbol $\mathbf{b}^0$ to distinguish those many solutions from the solitary solution that exists when $\mathbf{X'X}$ has full rank.

Illustration (*continued*)

The normal equations (15.11) for the data of Table 15.1 are, from (15.9) and (15.10),

$$\begin{bmatrix} 6 & 3 & 2 & 1 \\ 3 & 3 & 0 & 0 \\ 2 & 0 & 2 & 0 \\ 1 & 0 & 0 & 1 \end{bmatrix} \begin{bmatrix} \mu^0 \\ \alpha_1^0 \\ \alpha_2^0 \\ \alpha_3^0 \end{bmatrix} = \begin{bmatrix} y_{..} \\ y_{1.} \\ y_{2.} \\ y_{3.} \end{bmatrix} = \begin{bmatrix} 504 \\ 300 \\ 172 \\ 32 \end{bmatrix}. \tag{15.12}$$

15.3 Solving the Normal Equations

We seek a solution for $\mathbf{b}^0$ to the normal equations (15.11): $\mathbf{X'Xb}^0 = \mathbf{X'y}$. Since $\mathbf{X'X}$ is singular, it has no inverse, and so we obtain a solution for $\mathbf{b}^0$ using a generalized inverse of $\mathbf{X'X}$.

a. Generalized inverses of $\mathbf{X'X}$

Let $\mathbf{G}$ be a generalized inverse of $\mathbf{X'X}$ satisfying

$$\mathbf{X'XGX'X} = \mathbf{X'X} \tag{15.13}$$

in accord with Section 9.2. Then, on defining

$\mathbf{X}$ as having order $N \times p$ with $p < N$, and rank $r_{\mathbf{X}} = r \le p,$ (15.14)

the theorem in Section 9.5c gives:

(i) $\mathbf{X'XGX'X} = \mathbf{X'X}.$ (ii) $\mathbf{XGX'X} = \mathbf{X}.$

(iii) $\mathbf{XGX'}$ is invariant to $\mathbf{G}$. (iv) $\mathbf{XGX'}$ is symmetric.

$$\tag{15.15}$$

b. Solutions

Equation (10.15) indicates that

$$\mathbf{b}^0 = \mathbf{GX'y} + (\mathbf{GX'X} - \mathbf{I})\mathbf{z} \tag{15.16}$$

for arbitrary $\mathbf{z}$ is a solution for $\mathbf{b}^0$ of the normal equations (15.11). And properties (i) and (ii) at the end of Section 10.4 tell us that all possible solutions can be obtained this way, or that they can be obtained using $\mathbf{GX'y}$ for all possible generalized inverses $\mathbf{G}$. For our purposes, only one solution is needed, and we take it as $\mathbf{GX'y}$ for some generalized inverse $\mathbf{G}$. Thus we use

$$\mathbf{b}^0 = \mathbf{GX'y} \tag{15.17}$$

as a solution for the normal equations (15.11).

The notation $\mathbf{b}^0$ in (15.17) for a solution to the normal equations (15.11) emphasizes that what is derived by solving (15.11) is *only* a solution to the normal equations and *not* an estimator of $\mathbf{b}$. This fact cannot be overemphasized. In a general discussion of linear models that are not of full rank, it is essential to realize that what is obtained as a solution of the normal equations is just that, a solution and *nothing more*. It is misleading and in most cases wrong for $\mathbf{b}^0$ to be termed an estimator of $\mathbf{b}$. It is true that $\mathbf{b}^0$ is, as will be shown, an estimator of something, but not of $\mathbf{b}$, and indeed the expression it estimates depends entirely upon which generalized inverse of $\mathbf{X'X}$ is used in obtaining $\mathbf{b}^0$. For this reason $\mathbf{b}^0$ is always referred to as a solution and not an estimator.

Any particular $\mathbf{b}^0$ depends on which particular generalized inverse of $\mathbf{X'X}$ is used as $\mathbf{G}$ in calculating $\mathbf{b}^0 = \mathbf{GX'y}$. There is therefore an infinite number of solutions $\mathbf{b}^0$—except when $\mathbf{X'X}$ has full rank, whereupon $\mathbf{G} = (\mathbf{X'X})^{-1}$ and $\mathbf{b}^0 = \hat{\mathbf{b}} = (\mathbf{X'X})^{-1}\mathbf{X'y}$ is then the only solution. Nevertheless, we find that the properties of $\mathbf{G}$ given in (15.15) ensure that this lack of uniqueness in $\mathbf{b}^0$ is less traumatic than might be anticipated.

Illustration (*continued*)

For (15.12) a generalized inverse of

$$\mathbf{X'X} = \begin{bmatrix} 6 & 3 & 2 & 1 \\ 3 & 3 & 0 & 0 \\ 2 & 0 & 2 & 0 \\ 1 & 0 & 0 & 1 \end{bmatrix} \quad \text{is} \quad \mathbf{G} = \begin{bmatrix} 0 & 0 & 0 & 0 \\ 0 & \frac{1}{3} & 0 & 0 \\ 0 & 0 & \frac{1}{2} & 0 \\ 0 & 0 & 0 & 1 \end{bmatrix} \tag{15.18}$$

for which the solution (15.17) is

$$
\mathbf{b}^0 = \mathbf{GX'y} =
\begin{bmatrix}
0 & 0 & 0 & 0 \\
0 & \frac{1}{3} & 0 & 0 \\
0 & 0 & \frac{1}{2} & 0 \\
0 & 0 & 0 & 1
\end{bmatrix}
\begin{bmatrix}
504 \\
300 \\
172 \\
32
\end{bmatrix}
=
\begin{bmatrix}
0 \\
100 \\
86 \\
32
\end{bmatrix}.
\tag{15.19}
$$

Another generalized inverse and corresponding value of $\mathbf{b}^0$ is

$$
\dot{\mathbf{G}} =
\begin{bmatrix}
\frac{1}{2} & -\frac{1}{2} & 0 & -\frac{1}{2} \\
-\frac{1}{2} & \frac{5}{6} & 0 & \frac{1}{2} \\
0 & 0 & 0 & 0 \\
-\frac{1}{2} & \frac{1}{2} & 0 & 1\frac{1}{2}
\end{bmatrix}
\quad \text{and} \quad
\dot{\mathbf{b}}^0 = \dot{\mathbf{G}}\mathbf{X'y} =
\begin{bmatrix}
86 \\
14 \\
0 \\
-54
\end{bmatrix}.
\tag{15.20}
$$

15.4 Expected Values and Variances

A general statement of the expected values discussed following (15.1) is

$$
E(\mathbf{e}) = \mathbf{0} \quad \text{and} \quad E(\mathbf{y}) = \mathbf{Xb}.
\tag{15.21}
$$

Hence with $\mathbf{b}^0 = \mathbf{GX'y}$,

$$
E(\mathbf{b}^0) = \mathbf{GX'}E(\mathbf{y}) = \mathbf{GX'Xb}.
\tag{15.22}
$$

Thus $\mathbf{b}^0$ has expected value $\mathbf{GX'Xb}$, and so is an unbiased estimator of $\mathbf{GX'Xb}$ but *not* of $\mathbf{b}$.

With $E(\mathbf{y}) = \mathbf{Xb}$ of (15.21), the variance-covariance matrix of $\mathbf{y}$ is

$$
\text{var}(\mathbf{y}) = E(\mathbf{y} - \mathbf{Xb})(\mathbf{y} - \mathbf{Xb})' = E(\mathbf{ee'}) = E(\mathbf{e} - \mathbf{0})(\mathbf{e} - \mathbf{0})' = \text{var}(\mathbf{e}).
$$

Also, as indicated following (15.4), we assume the error terms to be uncorrelated with the variance of each being σ^2, so that by assuming $\text{var}(\mathbf{e}) = \sigma^2\mathbf{I}$,

$$
\text{var}(\mathbf{y}) = \text{var}(\mathbf{e}) = \sigma^2\mathbf{I}.
\tag{15.23}
$$

And then

$$
\text{var}(\mathbf{b}^0) = \text{var}(\mathbf{GX'y}) = \mathbf{GX'}\sigma^2\mathbf{IXG'} = \mathbf{GX'XG'}\sigma^2.
\tag{15.24}
$$

Although this is not the analogue of its counterpart $(\mathbf{X'X})^{-1}\sigma^2$ in regression [see (14.39)] as would be $\mathbf{G}\sigma^2$, we find that (15.24) does, in fact, create no difficulties in applications. And in using a symmetric reflexive, generalized inverse $\mathbf{G}^* = \mathbf{GX'XG'}$ (see E 9.9) in place of $\mathbf{G}$ we get (15.24) as $\mathbf{G}^*\sigma^2$, should that direct analogue with $(\mathbf{X'X})^{-1}\sigma^2$ be required.

15.5 Predicted y-Values

Just as in equation (14.44) dealing with predicted y-values in regression, so here: we denote by $\hat{\mathbf{y}}$ the vector $\mathbf{X}\mathbf{b}^0$ of predicted y-values corresponding to the vector of observations $\mathbf{y}$:

$$\hat{\mathbf{y}} = \mathbf{X}\mathbf{b}^0 = \mathbf{X}\mathbf{G}\mathbf{X}'\mathbf{y}. \tag{15.25}$$

Because $\mathbf{G}$ occurs in $\hat{\mathbf{y}}$ only in the form $\mathbf{X}\mathbf{G}\mathbf{X}'$, we use (iii) of (15.15) to observe that $\hat{\mathbf{y}}$ is invariant to $\mathbf{G}$. This means that it does not matter what generalized inverse of $\mathbf{X}'\mathbf{X}$ is used for $\mathbf{G}$: for each and every $\mathbf{G}$, the value $\hat{\mathbf{y}} = \mathbf{X}\mathbf{G}\mathbf{X}'\mathbf{y}$ of (15.25) is the same. Moreover, because for any solution $\mathbf{b}^0$ there is a corresponding $\mathbf{G}$, whether we know its value or not, $\mathbf{X}\mathbf{b}^0$ is always invariant to $\mathbf{b}^0$.

Illustration (*continued*)

$\mathbf{X}$ is the matrix of 0s and 1s in (15.5). Using $\mathbf{b}^0$ of (15.19) in $\mathbf{X}\mathbf{b}^0$ of (15.25) therefore gives

$$\hat{\mathbf{y}} = \begin{bmatrix} 1 & 1 & 0 & 0 \\ 1 & 1 & 0 & 0 \\ 1 & 1 & 0 & 0 \\ 1 & 0 & 1 & 0 \\ 1 & 0 & 1 & 0 \\ 1 & 0 & 0 & 1 \end{bmatrix} \begin{bmatrix} 0 \\ 100 \\ 86 \\ 32 \end{bmatrix} = \begin{bmatrix} 100 \\ 100 \\ 100 \\ 86 \\ 86 \\ 32 \end{bmatrix} \tag{15.26}$$

and using $\dot{\mathbf{b}}^0$ of (15.20) for $\mathbf{b}^0$ in (15.25) gives the same value for $\hat{\mathbf{y}}$:

$$\hat{\mathbf{y}} = \begin{bmatrix} 1 & 1 & 0 & 0 \\ 1 & 1 & 0 & 0 \\ 1 & 1 & 0 & 0 \\ 1 & 0 & 1 & 0 \\ 1 & 0 & 1 & 0 \\ 1 & 0 & 0 & 1 \end{bmatrix} \begin{bmatrix} 86 \\ 14 \\ 0 \\ -54 \end{bmatrix} = \begin{bmatrix} 100 \\ 100 \\ 100 \\ 86 \\ 86 \\ 32 \end{bmatrix}.$$

15.6 Estimating the Residual Variance

a. Error sum of squares

Having obtained $\hat{\mathbf{y}} = \mathbf{X}\mathbf{b}^0$ as the vector of predicted values corresponding to the vector of observations $\mathbf{y}$, we can now calculate the sum of squares of the deviations of the observed ys from their predicted values. This is the residual (or error) sum of squares, SSE, and in regression analysis [see (14.45)] we found it to be expressible as SSE = $\mathbf{y}'\mathbf{y} - \hat{\boldsymbol{\beta}}'\mathbf{X}'\mathbf{y}$. Essentially the same result holds for the general linear model of this chapter. As before,

$$\text{SSE} = \sum_i \sum_j (y_{ij} - \hat{y}_{ij})^2 = (\mathbf{y} - \hat{\mathbf{y}})'(\mathbf{y} - \hat{\mathbf{y}}),$$

and this reduces (see E 15.3) to

$$\text{SSE} = \mathbf{y}'(\mathbf{I} - \mathbf{XGX}')\mathbf{y}. \tag{15.27}$$

The occurrence of $\mathbf{G}$ in (15.27) only in the form $\mathbf{XGX}'$ has the same consequence as it does in $\hat{\mathbf{y}}$ of (15.25): the residual sum of squares SSE is invariant to whatever generalized inverse of $\mathbf{X}'\mathbf{X}$ is used for $\mathbf{G}$ and, equivalently, is invariant to whatever solution of the normal equations $\mathbf{X}'\mathbf{X}\mathbf{b}^0 = \mathbf{X}'\mathbf{y}$ is used for $\mathbf{b}^0$. This is especially useful because from (15.27) a computing formula is obtainable that is based on $\mathbf{b}^0$:

$$\text{SSE} = \mathbf{y}'\mathbf{y} - \mathbf{b}^{0'}\mathbf{X}'\mathbf{y}. \tag{15.28}$$

This is exactly the same as in regression: $\mathbf{y}'\mathbf{y}$ is the total sum of squares of the observed ys; and $\mathbf{b}^{0'}\mathbf{X}'\mathbf{y}$ is the sum of products of the solutions in $\mathbf{b}^{0'}$ multiplied by the corresponding elements of the right-hand sides of the normal equations $\mathbf{X}'\mathbf{X}\mathbf{b}^0 = \mathbf{X}'\mathbf{y}$ from which $\mathbf{b}^0$ is derived.

Illustration (*continued*)

From (15.5) the total sum of squares, SST, is

$$\text{SST} = \mathbf{y}'\mathbf{y} = 101^2 + 105^2 + 94^2 + 84^2 + 88^2 + 32^2 = 45886; \tag{15.29}$$

and using (15.19) in $\mathbf{b}^{0'}\mathbf{X}'\mathbf{y}$, with $\mathbf{X}'\mathbf{y}$ being the right-hand side of (15.12), gives

$$\mathbf{b}^{0'}\mathbf{X}'\mathbf{y} = 0(504) + 100(300) + 86(172) + 32(32) = 45816. \tag{15.30}$$

Similarly, using $\dot{\mathbf{b}}^0$ of (15.20) gives the same value:

$$\dot{\mathbf{b}}^{0'}\mathbf{X}'\mathbf{y} = 86(504) + 14(300) + 0(172) - 54(32) = 45816. \tag{15.31}$$

Hence, on using either solution (15.19) or (15.20), calculation of SSE from (15.28) is

$$\text{SSE} = \mathbf{y}'\mathbf{y} - \mathbf{b}^{0'}\mathbf{X}'\mathbf{y} = 45886 - 45816 = 70. \tag{15.32}$$

b. Solutions obtained without using a G

We have now seen how two functions involving $\mathbf{b}^0$ are invariant to $\mathbf{b}^0$, namely $\mathbf{Xb}^0 = \mathbf{XGX'y}$ and SSE $= \mathbf{y'y} - \mathbf{b}^{0'}\mathbf{X'y}$. This invariance is true for every solution $\mathbf{b}^0$ of the normal equations $\mathbf{X'Xb}^0 = \mathbf{X'y}$. Thus if we calculate a solution $\mathbf{b}^0$ without having its $\mathbf{G}$, we still have $\mathbf{Xb}^0$ and $\mathbf{b}^{0'}\mathbf{X'y}$, each being their own unique invariant values.

The preceding invariance is important because sometimes solutions are obtained without calculating a generalized inverse. The common procedure is to supplement the normal equations with equations of such nature as to produce a single solution. For the example of (15.12) only one supplemental equation is needed, and the easiest one is $\alpha_3^0 = 0$ which yields [32 68 54 0] as a solution. This equation and extensions of it are used by some computer software packages, although some of those also print the corresponding generalized inverse if requested. Another procedure (if appropriate to the model at hand) is, for our example, to define the α-effects as adding to zero ($\alpha_1 + \alpha_2 + \alpha_3 = 0$) so that we want the solution to satisfy $\alpha_1^0 + \alpha_2^0 + \alpha_3^0 = 0$. For our data this gives $\left[72\frac{2}{3} \quad 27\frac{1}{3} \quad 13\frac{1}{3} \quad -40\frac{2}{3}\right]$ as a solution. Yet one further form of supplemental equation is $\sum_i n_i \alpha_i^0 = 0$, which for our example is $3\alpha_1^0 + 2\alpha_2^0 + \alpha_3^0 = 0$ and this yields [84 16 2 -52] as a solution.

For all of these solutions (and all others) $\hat{\mathbf{y}} = \mathbf{Xb}^0$ is exactly as in (15.26) and $\mathbf{b}^{0'}\mathbf{X'y}$ is 45,816 as in (15.30).

c. Expected value

From (15.21) and (15.23), the mean and variance of $\mathbf{y}$ are, respectively, $E(\mathbf{y}) = \mathbf{Xb}$ and $\text{var}(\mathbf{y}) = \sigma^2 \mathbf{I}$. And in (15.27), SSE is a quadratic form (see Section 3.6) in $\mathbf{y}$. Therefore Section 15.14a can be used to derive the expected value of SSE as

$$E(\text{SSE}) = E[\mathbf{y'}(\mathbf{I} - \mathbf{XGX'})\mathbf{y}] = \text{tr}\left[(\mathbf{I} - \mathbf{XGX'})\mathbf{I}\sigma^2\right] + \mathbf{b'X'}(\mathbf{I} - \mathbf{XGX'})\mathbf{Xb}.$$

On using the results given in (15.15) and (15.19) this simplifies to

$$
\begin{aligned}
E(\text{SSE}) &= \sigma^2\left[\text{tr}(\mathbf{I}) - \text{tr}(\mathbf{XGX'})\right] + \mathbf{b'X'}(\mathbf{X} - \mathbf{XGX'X})\mathbf{b} \\
&= \sigma^2\left[N - \text{tr}\mathbf{GX'X}\right] + 0 = \sigma^2(N - r_{\mathbf{GX'X}}) \\
&= (N - r_{\mathbf{X}})\sigma^2. \tag{15.33}
\end{aligned}
$$

d.   Estimation

Equation (15.33) immediately gives an unbiased estimator of σ^2 as

$$\hat{\sigma}^2 = \frac{\text{SSE}}{N - r_{\mathbf{X}}}. \qquad (15.34)$$

Again we see a similarity with regression, since (15.34) is the same form as (14.50), for which $k+1 = r_{\mathbf{X}}$. Now, though, the importance of using $r_{\mathbf{X}}$ in the expectation is clear, because $\mathbf{X}$ is not of full column rank and its rank is therefore not equal to its number of columns. In fact, the rank of $\mathbf{X}$ depends on the nature of the data available.

Illustration   (*continued*)

$$\hat{\sigma}^2 = \frac{70}{6 - 3} = 23\tfrac{1}{3}. \qquad (15.35)$$

15.7   Partitioning the Total Sum of Squares

Partitioning the total sum of squares as shown in Section 14.7 for regression takes place in the same way for the general linear model being dealt with in this chapter. One difference is that the corrected sums of squares and products of x-variables are of neither use nor interest (because every x-variable is just 0 or 1, representing the incidence of a term of the model in the data). Nevertheless, use is still made of $\text{SST}_m = \mathbf{y}'\mathbf{y} - N\bar{y}^2$, the corrected sum of squares of the y-observations. The three forms of partitioning the sums of squares are shown in Table 15.3.

Table 15.3.   Partitioning Sums of Squares

	$\text{SSM} = N\bar{y}^2$	
$\text{SSR} = \mathbf{y}'\mathbf{XGX}'\mathbf{y}$	$\text{SSR}_m = \mathbf{y}'\mathbf{XGX}'\mathbf{y} - N\bar{y}^2$	$\text{SSR}_m = \mathbf{y}'\mathbf{XGX}'\mathbf{y} - N\bar{y}^2$
$\text{SSE} = \mathbf{y}'(\mathbf{I} - \mathbf{XGX}')\mathbf{y}$	$\text{SSE} = \mathbf{y}'(\mathbf{I} - \mathbf{XGX}')\mathbf{y}$	$\text{SSE} = \mathbf{y}'(\mathbf{I} - \mathbf{XGX}')\mathbf{y}$
$\text{SST} = \mathbf{y}'\mathbf{y}$	$\text{SST} ,= \mathbf{y}'\mathbf{y}$	$\text{SST}_m = \mathbf{y}'\mathbf{y} - N\bar{y}^2$

The three columns in Table 15.3 correspond to the three partitionings shown in (14.54), (14.58) and (14.60). The first column shows

$$\text{SSR} = \text{SST} - \text{SSE} = \mathbf{y}'\mathbf{XGX}'\mathbf{y} = \mathbf{b}^{0'}\mathbf{X}'\mathbf{y}, \qquad (15.36)$$

the sum of squares attributable to fitting the model $\mathbf{y} = \mathbf{Xb} + \mathbf{e}$, similar to the sum of squares for regression in Section 14.7. In the second column,

$$\text{SSM} = N\bar{y}^2 \tag{15.37}$$

is the sum of squares due to fitting a general mean, and

$$\text{SSR}_m = \text{SSR} - \text{SSM} = \text{SSR} - N\bar{y}^2 \tag{15.38}$$

is the sum of squares for fitting the model, corrected for the mean. The third column is identical to the second except that SSM has been deleted from the body of the table and subtracted from SST to give

$$\text{SST}_m = \text{SST} - \text{SSM} = \sum_{i,j} y_{ij}^2 - N\bar{y}^2 \tag{15.39}$$

as the total sum of squares corrected for the mean. In all three columns the error sum of squares is the same, SSE of (15.28).

Table 15.3 forms the basis of traditional analysis of variance tables, as is shown in Section 15.9.

Illustration (*continued*)

With $\bar{y} = 504/6$ from Table 15.1, we have $N\bar{y}^2 = 42,336$; and with SST $= 45,886$ and SSE $= 70$ from (15.29) and (15.32), respectively, Table 15.3 for the example is as shown in Table 15.4.

Table 15.4.   Partitioning Sums of Squares as in Table 15.3
(Data of Table 15.1)

	SSM $= 42,336$	
SSR $= 45,816$	SSR$_m = 3,480$	SSR$_m = 3,480$
SSE $= 70$	SSE $= 70$	SSE $= 70$
SST $= 45,886$	SST $= 45,886$	SST$_m = 3,550$

15.8   Coefficient of Determination

The predicted values of y corresponding to the observations $\mathbf{y}$ are the elements of $\hat{\mathbf{y}}$ given in (15.25). The product-moment correlation between the observed ys and the corresponding elements of $\hat{\mathbf{y}}$ is, when squared, commonly referred to as the *coefficient of determination*, or the *multiple correlation coefficient*. Since the usual linear model contains μ, which corresponds to the intercept in regression, we define

$$R^2 \;=\; \text{coefficient of determination}$$

$$=\; \frac{[\Sigma(y_i - \bar{y})(\hat{y}_i - \bar{\hat{y}})]^2}{\Sigma(y_i - \bar{y}^2 \Sigma(\hat{y}_i - \bar{\hat{y}})^2}, \tag{15.40}$$

similar to (14.61). In simplifying (15.40), use is made of $\mathbf{X'XGX'} = \mathbf{X'}$ obtainable from (15.15), which, because the first row of $\mathbf{X'}$ is $\mathbf{1'}$, also means that

$$\mathbf{1'XGX'} = \mathbf{1'}. \tag{15.41}$$

This leads to the equality $\bar{\hat{y}} = \bar{y}$: hence, as in equation (14.64) we can get

$$R^2 = \frac{(\text{SSR}_m)^2}{\text{SST}_m(\text{SSR}_m)} = \frac{\text{SSR}_m}{\text{SST}_m}. \tag{15.42}$$

Illustration   (*continued*)

From Table 15.4, $R^2 = 3480/3550 = .98$.

15.9   Analysis of Variance

Analysis of variance is based on partitioning sums of squares in the manner of Tables 15.3 and 15.4. The right-most column of those tables is the partitioning seen in most analysis of variance tables, with the total sum of squares being that corrected for the mean, $\text{SST}_m = \mathbf{y'y} - N\bar{y}^2$. In contrast, we concentrate attention on the middle column of Tables 15.3 and 15.4, which has the term $\text{SSM} = N\bar{y}^2$ in it, as well as SSR_m and SSE. The analysis of variance based on this column is shown in Table 15.5, first in terms of the sums of squares in Table 15.3, and then for the data, using Table 15.4.

Table 15.5 is just a summary of calculations. To utilize the F-statistics therein, in the usual manner, we assume that the observations in $\mathbf{y}$ are normally distributed with mean $E(\mathbf{y}) = \mathbf{Xb}$ and variance-covariance matrix $\text{var}(\mathbf{y}) = \sigma^2\mathbf{I}$, already specified in (15.21) and (15.23), respectively, and write

$$\mathbf{y} \sim N(\mathbf{Xb}, \sigma^2\mathbf{I}). \tag{15.43}$$

Applying subsections b, c and e of Section 15.14 to the sums of squares SSM, SSR_m and SSE then establishes that the F-statistics in Table 15.5 have F-distributions with degrees of freedom corresponding to those of the mean squares in their numerator and denominator. Establishing these results is beyond the scope of this book. Details are available in many of the texts mentioned at the beginning of this chapter (e.g., Searle, 1997, Section 5.3).

The hypothesis tested by $F(M)$ in Table 15.5a is $H\colon \text{E}(\bar{y}) = 0$; and $F(M)$ is the square of the t-statistic $\left[\bar{y}/(\hat{\sigma}\sqrt{N})\right]^2$ with $N - r_{\mathbf{x}}$ degrees of freedom.

The hypothesis tested by $F(R_m)$ in Table 15.5a is described by first considering an F-statistic not shown in Table 15.5, namely $F(R)$ based on recombining SSM and SSR_m into SSR and calculating MSR $=$ $\text{SSR}/r_{\mathbf{x}}$ and $F(R) = \text{MSR}/\text{MSE}$.

Table 15.5.   Analysis of Variance

Source of Variation	d.f.[a]	Sum of Squares	Mean Square	F-statistic
(a) Based on Table 15.3				
Mean	1	SSM	$\text{MSM} = \dfrac{\text{SSM}}{1}$	$F(M) = \dfrac{\text{MSM}}{\text{MSE}}$
Model, a.f.m.[b]	$r_{\mathbf{x}} - 1$	SSR_m	$\text{MSR}_m = \dfrac{\text{SSR}_m}{r_{\mathbf{x}} - 1}$	$F(R_m) = \dfrac{\text{MSR}_m}{\text{MSE}}$
Error	$N - r_{\mathbf{x}}$	SSE	$\text{MSE} = \dfrac{\text{SSE}}{N - r_{\mathbf{x}}}$	
Total	N	SST		
(b) Sums of squares from Table 15.4 used in Table 15.5				
Mean	1	42,336	$\dfrac{42,336}{1} = 42,336$	$F(M) = \dfrac{42,336}{23\frac{1}{3}} = 1,814\frac{2}{7}$
Model,	2	3,480	$\dfrac{3,480}{2} = 1,740$	$F(R_m) = \dfrac{1,740}{23\frac{1}{3}} = 74\frac{4}{7}$
Error	3	70	$\dfrac{70}{3} = 23\frac{1}{3}$	
Total	6	45,886		

[a]d.f. $\equiv$ degrees of freedom.
[b]a.f.m. = adjusted for mean.

The results given in Section 15.14 lead to also establishing $F(R)$ as having an F-distribution. This is just like the $F(H)$ of regression which, as in (14.74), provides a test of $H: \boldsymbol{\beta} = \mathbf{0}$. In contrast, for the linear model $E(\mathbf{y}) = \mathbf{Xb}$, where $\mathbf{X}$ does not have full column rank (as is being dealt with in this chapter) $F(H)$ tests $H: \mathbf{Xb} = \mathbf{0}$. This is *not* equivalent to $H: \mathbf{b} = \mathbf{0}$ because, with $\mathbf{X}$ having less than full column rank, $\mathbf{Xb} = \mathbf{0}$ does not imply $(\not\Rightarrow)$ $\mathbf{b} = \mathbf{0}$. [At all times, $\mathbf{Xb} = \mathbf{0} \Rightarrow \mathbf{X'Xb} = \mathbf{0}$, and when $\mathbf{X}$ does have full column rank, $\mathbf{X'X}$ is non-singular, as in regression (see Chapter 14) and so then $\mathbf{Xb} = \mathbf{0} \Rightarrow \mathbf{X'Xb} = \mathbf{0} \Rightarrow \mathbf{b} = \mathbf{0}$.]

When $F(H)$ is significant we say that there is concordance of the data with the model $E(\mathbf{y}) = \mathbf{Xb}$; i.e., the model accounts for a significant portion of the variation in the y-variable. This does not mean that the model used is necessarily the most suitable model: there may be a subset of its elements that is as significant as the whole set; or there may be other elements (factors) which, when used alone or in combination with some or all of those already used, are significantly better than those used; or there may be non-linear models that are at least as suitable as the model used. None of these contingencies is inconsistent with $F(H)$ being significant and the ensuing conclusion that the data are in concordance with the model $E(\mathbf{y}) = \mathbf{Xb}$.

Just as $F(H)$ provides a test of the model $E(\mathbf{y}) = \mathbf{Xb}$, so does $F(R_m)$ of Table 15.5(a) provide a test of the model $E(\mathbf{y}) = \mathbf{Xb}$ over and above the model $E(\mathbf{y}) = \mu\mathbf{1}$. Since the latter can be considered as fitting a general mean, we look upon $F(R_m)$ as providing a test of the model $E(\mathbf{y}) = \mathbf{Xb}$ over and above the mean. When $F(R_m)$ is significant we conclude that the model satisfactorily accounts for variation in the y-variable. This is not to be taken as evidence that all elements of $\mathbf{b}$ other than μ are non-zero but only that at least one of them, or one linear combination of them, may be. If $F(M)$ has first been found significant, then $F(R_m)$ being significant indicates that a model with terms in it additional to a mean explains significantly more of the variation in the y-variable than does the model $E(\mathbf{y}) = \mu\mathbf{1}$.

Illustration   (*continued*)

The particular hypotheses tested by $F(M)$ and $F(R_m)$ in Table 15.5b are as follows:

$$F(M) \text{ tests } H: \mu + \tfrac{1}{6}(3\alpha_1 + 2\alpha_2 + \alpha_3) = 0 \qquad (15.44)$$

and

$$F(R_M) \text{ tests } H\colon \alpha_1 = \alpha_2 = \alpha_3. \tag{15.45}$$

The latter is the familiar hypothesis of equality of the effects of city on revenue (see Table 15.1).

15.10 Estimable Functions

Equation (15.22) shows that $\mathbf{b}^0$ is not an unbiased estimator of $\mathbf{b}$. Nevertheless, unbiased estimators of certain linear combinations of elements of $\mathbf{b}$ can be obtained from $\mathbf{b}^0$, and they are also estimators that are invariant to $\mathbf{b}^0$; i.e., invariant to whatever generalized inverse of $\mathbf{X}'\mathbf{X}$ is used for $\mathbf{G}$. The situation is as follows.

A solution to the normal equations $\mathbf{X}'\mathbf{X}\mathbf{b}^0 = \mathbf{X}'\mathbf{y}$ is $\mathbf{b}^0 = \mathbf{G}\mathbf{X}'\mathbf{y}$ for $\mathbf{X}'\mathbf{X}\mathbf{G}\mathbf{X}'\mathbf{X} = \mathbf{X}'\mathbf{X}$. From this we can notice for $\mathbf{q}' = \mathbf{t}'\mathbf{X}$ for any $\mathbf{t}'$ that $\mathbf{q}'\mathbf{b}^0 = \mathbf{t}'\mathbf{X}\mathbf{G}\mathbf{X}'\mathbf{y}$ is invariant to $\mathbf{G}$ because $\mathbf{X}\mathbf{G}\mathbf{X}'$ is. Thus for

$$\mathbf{q}' = \mathbf{t}'\mathbf{X} \tag{15.46}$$

$\mathbf{q}'\mathbf{b}^0$ is invariant to $\mathbf{b}^0$; i.e., $\mathbf{q}'\mathbf{b}^0$ (for given $\mathbf{q}' = \mathbf{t}'\mathbf{X}$) is the same no matter what $\mathbf{b}^0$ is. The corresponding expression $\mathbf{q}'\mathbf{b}$ is called an *estimable function* and $\mathbf{q}'\mathbf{b}^0$ is an unbiased estimator of $\mathbf{q}'\mathbf{b}$.

More than that, $\mathbf{q}'\mathbf{b}^0$ is the *b.l.u.e.* (best, linear, unbiased estimator) of $\mathbf{q}'\mathbf{b}$. This means that not only is it unbiased and linear in the elements of $\mathbf{y}$ (because $\mathbf{q}'\mathbf{b}^0 = \mathbf{q}'\mathbf{G}\mathbf{X}'\mathbf{y}$), but it is also "best" in the sense that among all linear (in $\mathbf{y}$) unbiased estimators of $\mathbf{q}'\mathbf{b}$ it is the one with smallest variance (see, for example, Searle, 1997, p. 182). This variance has a very simple form. Denoting the variance of the scalar $\mathbf{q}'\mathbf{b}^0$ by $v(\mathbf{q}'\mathbf{b}^0)$, its value is, using (15.24),

$$v(\mathbf{q}'\mathbf{b}^0) = \mathbf{q}'\mathbf{G}\mathbf{q}\sigma^2. \tag{15.47}$$

Notice that with $\mathbf{q}' = \mathbf{t}'\mathbf{X}$ the variance in (15.47) is also $v(\mathbf{q}'\mathbf{b}^0) = \mathbf{t}'\mathbf{X}\mathbf{G}\mathbf{X}'\mathbf{t}\sigma^2$, which is invariant to $\mathbf{G}$. This is the kind of result alluded to following (15.24). Whereas $\mathbf{b}^0$ is not invariant to $\mathbf{G}$ and is accordingly of little interest in itself, $\mathbf{q}'\mathbf{b}^0$ for $\mathbf{q}'\mathbf{b}$ being estimable is not only invariant to $\mathbf{G}$, it is also of interest, being the b.l.u.e. of $\mathbf{q}'\mathbf{b}$, and its variance $v(\mathbf{q}'\mathbf{b}^0) = \mathbf{q}'\mathbf{G}\mathbf{q}\sigma^2$ is also invariant to $\mathbf{G}$.

We use the traditional "hat" over an expression to indicate the b.l.u.e. of an estimable function, and so when, for some $\mathbf{t}'$,

$$\mathbf{q}' = \mathbf{t}'\mathbf{X}, \tag{15.48}$$

then

$$\mathbf{q'b} \text{ is an estimable function} \qquad (15.49)$$

and

$$\text{the b.l.u.e. of } \mathbf{q'b} \quad \text{is} \quad \widehat{\mathbf{q'b}} = \mathbf{q'b^0} \qquad (15.50)$$

with variance

$$v(\mathbf{q'b^0}) = \mathbf{q'Gq}\sigma^2. \qquad (15.51)$$

Two further aspects of estimable functions are very useful in application:

$$\begin{array}{c} \text{The expected value of any observation} \\ \text{is an estimable function;} \end{array} \qquad (15.52)$$

$$\begin{array}{c} \text{Linear combinations of estimable} \\ \text{functions are estimable functions.} \end{array} \qquad (15.53)$$

These, as available alternates to (15.48), provide ready opportunity for developing estimable functions.

Details of many aspects of estimable functions are given in Searle (1997, Section 5.4).

Illustration (*continued*)

The model is $\mathrm{E}(y_{ij}) = \mu + \alpha_i$ of (15.1). Therefore, by (15.52), $\mu + \alpha_i$ is estimable and so by (15.50)

$$\widehat{\mu + \alpha_i} = \mu^0 + \alpha_i^0.$$

Examples from (15.19) are

$$\widehat{\mu + \alpha_1} = 0 + 100 = 100 \quad \text{and} \quad \widehat{\mu + \alpha_2} = 0 + 86 = 86.$$

The reader should use $\dot{\mathbf{G}}$ of (15.20) to obtain the same values and so demonstrate the invariance of $\mu^0 + \alpha_i^0$ to the value used for $\mathbf{b^0}$. Similarly, since the estimable function $\mu + \alpha_1$ is

$$\mu + \alpha_1 = [1 \quad 1 \quad 0 \quad 0]\,\mathbf{b},$$

using $\mathbf{G}$ of (15.18) in (15.51) gives

$$v(\widehat{\mu + \alpha_1}) = [1 \quad 1 \quad 0 \quad 0]\begin{bmatrix} 0 & 0 & 0 & 0 \\ 0 & \frac{1}{3} & 0 & 0 \\ 0 & 0 & \frac{1}{2} & 0 \\ 0 & 0 & 0 & 1 \end{bmatrix}\begin{bmatrix} 1 \\ 1 \\ 0 \\ 0 \end{bmatrix}\sigma^2 = \tfrac{1}{3}\sigma^2.$$

The reader, using $\dot{\mathbf{G}}$ of (15.20), should get the same result.

Further, from (15.53), because $\mu + \alpha_1$ and $\mu + \alpha_2$ are estimable, so is $(\mu + \alpha_1) - (\mu + \alpha_2) = \alpha_1 - \alpha_2$; i.e., $\alpha_1 - \alpha_2$ is estimable. Therefore by (15.50) the b.l.u.e. of $\alpha_1 - \alpha_2$ is

$$\widehat{\alpha_1 - \alpha_2} = \alpha_1^0 - \alpha_2^0 = 100 - 86 = 14.$$

15.11   Testing Linear Hypotheses

We now consider the calculations required for testing linear hypotheses, a linear hypothesis being a hypothesis that some linear function of the parameters equals some arbitrary constant. The discussion is confined to linear hypotheses involving estimable functions, since these are the only ones that can be tested. Thus in the preceding section, the function $\alpha_1 - \alpha_2$ is estimable and the hypothesis that $\alpha_1 - \alpha_2$ equals some constant, m_1 say, can be tested. We would write this hypothesis as

$$H: \alpha_1 - \alpha_2 = m_1.$$

More generally we might want to test the hypothesis

$$H: \begin{cases} \alpha_1 - \alpha_2 = m_1 \\ \alpha_2 - \alpha_3 = m_2 \end{cases} \tag{15.54}$$

which can also be written as

$$H: \begin{bmatrix} 0 & 1 & -1 & 0 \\ 0 & 0 & 1 & -1 \end{bmatrix} \begin{bmatrix} \mu \\ \alpha_1 \\ \alpha_2 \\ \alpha_3 \end{bmatrix} = \begin{bmatrix} m_1 \\ m_2 \end{bmatrix}. \tag{15.55}$$

This is a particular case of a general way of writing a linear hypothesis as

$$H: \mathbf{K'b} = \mathbf{m}, \tag{15.56}$$

where $\mathbf{b}$ is, of course, the vector of parameters in the model $E(\mathbf{y}) = \mathbf{Xb}$, and $\mathbf{m}$ is the vector of known constants that form part of the hypothesis. For example, if (15.54) were the hypothesis

$$H: \begin{cases} \alpha_1 - \alpha_2 = 6 \\ \alpha_2 - \alpha_3 = 5 \end{cases},$$

then on rewriting this as (15.56), $\mathbf{m}$ would be $\mathbf{m'} = \begin{bmatrix} 6 & 5 \end{bmatrix}$.

The matrix $\mathbf{K}'$ in the general statement (15.56) of a linear hypothesis must satisfy two simple requirements. Each is stated mathematically but has important practical ramifications. First, $\mathbf{K}'$ must be such that each element of $\mathbf{K}'\mathbf{b}$ is an estimable function. This is required because, in essence, the only linear hypotheses that can be tested are those which involve estimable functions. Second, $\mathbf{K}'$ must have full row rank. This is a very practical requirement. It means that hypotheses such as

$$H_1: \begin{matrix} \alpha_1 & - & \alpha_2 & = & 6 \\ 2\alpha_1 & - & 2\alpha_2 & = & 12 \end{matrix} \quad \text{or} \quad H_2: \begin{matrix} \alpha_1 & - & \alpha_2 & = & 6 \\ \alpha_2 & - & \alpha_3 & = & 5 \\ \alpha_1 & - & \alpha_3 & = & 11 \end{matrix}$$

cannot be considered. This is very reasonable, because in each of these there is a redundant statement that need not be there. In H_1 only the first statement, $\alpha_1 - \alpha_2 = 6$, need be considered, and in H_2 just the first two statements will suffice, for the third one is the sum of the other two.

Provided $\mathbf{K}'$ meets the two requirements just specified, and on assuming $\mathbf{y} \sim N(\mathbf{Xb}, \sigma^2\mathbf{I})$, the F-statistic for testing

$$H: \mathbf{K}'\mathbf{b} = \mathbf{m} \tag{15.57}$$

is calculated as follows. First, calculate the sum of squares

$$Q = (\mathbf{K}'\mathbf{b}^0 - \mathbf{m})'(\mathbf{K}'\mathbf{GK})^{-1}(\mathbf{K}'\mathbf{b}^0 - \mathbf{m}). \tag{15.58}$$

Then

$$F(H) = Q/\hat{\sigma}^2 r_{\mathbf{K}} \tag{15.59}$$

is the F-statistic (Section 15.14e) with $r_{\mathbf{K}}$ and $N - r_{\mathbf{X}}$ degrees of freedom. Details of deriving this F-statistic are beyond the scope of this book and are to be found in Searle (1997, Section 5.5b, with numerous ramifications considered in succeeding sections). It suffices to state here that the requirements placed on $\mathbf{K}'$ in the preceding paragraph ensure that $r_{\mathbf{K}}$, the rank of $\mathbf{K}$, equals the number of rows in $\mathbf{K}'$, that the inverse $(\mathbf{K}'\mathbf{GK})^{-1}$ in (15.58) always exists, and that (15.58), and hence (15.59), are invariant to $\mathbf{G}$.

The importance of (15.57), (15.58) and (15.59) is their broad generality, limited only by the practical need for $\mathbf{K}'$ to be of full row rank, with every element of $\mathbf{K}'\mathbf{b}$ being an estimable function. Within this framework, (15.57), (15.58) and (15.59) provide the entire calculation procedure for the F-statistic for testing *any* linear hypothesis about

$\mathbf{b}$ in the model $\mathbf{y} \sim N(\mathbf{Xb}, \sigma^2 \mathbf{I})$. This includes analysis of variance models with either equal or unequal numbers of observations in the subclasses of the data, be they survey data or data from well-designed and executed experiments. It also includes regression and analysis of variance in the presence of covariates. Thus the applicability of (15.57), (15.58) and (15.59) is very broad indeed.

Illustration (*continued*)

We have seen that $\alpha_1 - \alpha_2$ and $\alpha_2 - \alpha_3$ are both estimable functions. We consider testing

$$H: \begin{cases} \alpha_1 - \alpha_2 = 10 \\ \alpha_2 - \alpha_3 = 50 \end{cases}, \text{ i.e., } H: \begin{bmatrix} 0 & 1 & -1 & 0 \\ 0 & 0 & 1 & -1 \end{bmatrix} \mathbf{b} = \begin{bmatrix} 10 \\ 50 \end{bmatrix},$$
(15.60)

where the 2×4 matrix in (15.60) is $\mathbf{K}'$, having $r_\mathbf{K} = 2$. Hence Q of (15.58) has

$$\mathbf{K}'\mathbf{b}^0 - \mathbf{m} = \begin{bmatrix} 0 & 1 & -1 & 0 \\ 0 & 0 & 1 & -1 \end{bmatrix} \begin{bmatrix} 0 \\ 100 \\ 86 \\ 32 \end{bmatrix} - \begin{bmatrix} 10 \\ 50 \end{bmatrix} = \begin{bmatrix} 4 \\ 4 \end{bmatrix}$$

and for $\mathbf{G}$ of (15.18) simple arithmetic yields

$$(\mathbf{K}'\mathbf{GK})^{-1} = \tfrac{1}{6} \begin{bmatrix} 9 & 3 \\ 3 & 5 \end{bmatrix}.$$
(15.61)

Therefore (15.58) is

$$Q = [4 \ \ 4] \tfrac{1}{6} \begin{bmatrix} 9 & 3 \\ 3 & 5 \end{bmatrix} \begin{bmatrix} 4 \\ 4 \end{bmatrix} = (16/6)(9 + 6 + 5) = 53\tfrac{1}{3}$$
(15.62)

and so from (15.59), using $\hat{\sigma}^2 = 23\tfrac{1}{3}$ from (15.35),

$$F(H) = 53\tfrac{1}{3}/[2(23\tfrac{1}{3})] = 8/7.$$

This has an $\mathcal{F}$-distribution with 2 and 2 degrees of freedom and is obviously not significant.

Numerous aspects of (15.57), (15.58) and (15.59) could be commented on, and are discussed in Searle (1997, Section 5.5, 5.6 and 5.7). Attention is directed here to but three of them.

(i) Any hypothesis (a) which consists of linearly independent linear combinations of the equations of $H: \mathbf{K}'\mathbf{b} = \mathbf{m}$ and (b) which has the same number of equations as H has the same Q and the same F-statistic as H.

(ii) The hypothesis

$$H:\mathbf{K'b} = \mathbf{0} \quad \text{has} \quad Q = \mathbf{b}^{0\prime}\mathbf{K}(\mathbf{K'GK})^{-1}\mathbf{K'b}^0. \tag{15.63}$$

(iii) When in (15.63) the matrices $\mathbf{K}$ and $\mathbf{X}$ have the same rank, $r_{\mathbf{K}} = r_{\mathbf{X}}$, then $Q = $ SSR. Further, provided $\mathbf{b}$ includes μ, a general mean (in which case $\mathbf{X}$ has $\mathbf{1}$ for one of its columns), and if $r_{\mathbf{K}} = r_{\mathbf{X}} - 1$ and $\mathbf{K'1} = \mathbf{0}$, then $Q = $ SSR$_m$.

Illustration (*continued*)

(i) Consider the hypothesis

$$H^*:\begin{cases} 5\alpha_1 - 6\alpha_2 + \alpha_3 = 0 \\ \alpha_1 \qquad\quad - \alpha_3 = 60 \end{cases}. \tag{15.64}$$

Its first equation is five times the first equation of (15.60) minus the second; and the second equation in (15.64) is the sum of the two in (15.60). Then, for (15.64)

$$\mathbf{K'} = \begin{bmatrix} 0 & 5 & -6 & 1 \\ 0 & 1 & 0 & -1 \end{bmatrix} \quad \text{and} \quad \mathbf{m} = \begin{bmatrix} 0 \\ 60 \end{bmatrix},$$

and so for Q of (15.58)

$$\mathbf{K'b}^0 - \mathbf{m} = \begin{bmatrix} 5(100) - 6(86) + 32 \\ 100 - 32 \end{bmatrix} - \begin{bmatrix} 0 \\ 60 \end{bmatrix} = \begin{bmatrix} 16 \\ 8 \end{bmatrix}$$

and it is easy to calculate

$$(\mathbf{K'GK})^{-1} = \begin{bmatrix} 27\frac{1}{3} & \frac{2}{3} \\ \frac{2}{3} & 1\frac{1}{3} \end{bmatrix}^{-1} = \frac{1}{36} \begin{bmatrix} 1\frac{1}{3} & -\frac{2}{3} \\ -\frac{2}{3} & 27\frac{1}{3} \end{bmatrix}.$$

Therefore the Q for H^* of (15.64) is

$$\tfrac{1}{36}\left[16^2(1\tfrac{1}{3}) + 8^2(27\tfrac{1}{3}) - 2(16)8(\tfrac{2}{3})\right] = 53\tfrac{1}{3} = Q \text{ of (15.62)}.$$

(ii) The hypothesis $H:\mu + \alpha_1 = 0$ has $\mathbf{K'} = \begin{bmatrix} 1 & 1 & 0 & 0 \end{bmatrix}$, $\mathbf{K'b}^0 = 100$ and $(\mathbf{K'GK})^{-1} = \tfrac{1}{3}$, for which $Q = (100)(\tfrac{1}{3})^{-1}(100) = 30000$.

(iii) The hypothesis

$$H:\begin{cases} \alpha_1 - \alpha_2 = 0 \\ \alpha_2 - \alpha_3 = 0 \end{cases} \quad \text{has} \quad \mathbf{K'b}^0 = \begin{bmatrix} 14 \\ 54 \end{bmatrix}$$

and the same $(\mathbf{K'GK})^{-1}$ as in (15.61). Therefore from (15.63)

$$Q = [14 \quad 54]\tfrac{1}{6} \begin{bmatrix} 9 & 3 \\ 3 & 5 \end{bmatrix} \begin{bmatrix} 14 \\ 54 \end{bmatrix} = 3240 = \text{SSR}_m$$

of Tables 15.4 and 15.5(b).

15.12 Confidence Intervals

Since it is only estimable functions that have useful estimators (b.l.u.e.s) that are invariant to the solution of the normal equations, they are the only functions for which establishing confidence intervals is valid. On assuming $\mathbf{y} \sim N(\mathbf{Xb}, \sigma^2\mathbf{I})$ we then have, from the unbiasedness of $\mathbf{q'b}^0$ and from (15.51), that for an estimable function $\mathbf{q'b}$ its b.l.u.e. is $\mathbf{q'b}^0 \sim N(\mathbf{q'b}, \mathbf{q'Gq}\sigma^2)$. Therefore, akin to Section 14.12, a $100(1-\alpha)\%$ symmetric confidence interval on $\mathbf{q'b}$ is

$$\mathbf{q'b}^0 \pm \hat{\sigma} t_\alpha \sqrt{\mathbf{q'Gq}} \quad \text{or} \quad \mathbf{q'b}^0 \pm \hat{\sigma} z_\alpha \sqrt{\mathbf{q'Gq}},$$

the former when degrees of freedom are small ($N - r_\mathbf{X} < 100$, say) and the latter otherwise. t_α and z_α are points on the t- and normal-distributions, respectively, defined in Section 14.12.

15.13 The Illustration Generalized

The model equation (15.3) for the example of Table 15.1 is $y_{ij} = \mu + \alpha_i + e_{ij}$ for $i = 1, 2, 3$ for $j = 1, 2, \ldots, n_i$ with $n_1 = 3$, $n_2 = 2$ and $n_3 = 1$. We here show the generalization of this for $i = 1, 2, \ldots, a$ where a $(= 3$, the number of cities in Table 15.1) is, in general terms, the number of groups or classes. The general form of the normal equations (15.12) is

$$\begin{bmatrix} N & n_1 & n_2 & \cdots & n_a \\ n_1 & n_1 & 0 & \cdots & 0 \\ n_2 & 0 & n_2 & \cdots & 0 \\ \vdots & & & & \\ n_a & 0 & \cdots & 0 & n_a \end{bmatrix} \begin{bmatrix} \mu^0 \\ \alpha_1^0 \\ \alpha_2^0 \\ \vdots \\ \alpha_a^0 \end{bmatrix} = \begin{bmatrix} y_{..} \\ y_{1.} \\ y_{2.} \\ \vdots \\ y_{a.} \end{bmatrix} \qquad (15.65)$$

where

$$y_{i.} = \sum_{j=1}^{n_i} y_{ij} \quad \text{and} \quad \bar{y}_{i.} = y_{i.}/n_i \quad \text{and} \quad y_{..} = \sum_{i=1}^{a} y_{i.} .$$

The solution to the normal equations corresponding to (15.19) is

$$\mu^0 = 0 \quad \text{and} \quad \alpha_i^0 = \bar{y}_{i\cdot\cdot} \qquad (15.66)$$

The basic estimable function from (15.52) is $\mu + \alpha_i$, and from (15.50) its b.l.u.e. is

$$\widehat{\mu + \alpha_i} = \mu^0 + \alpha_i^0 = 0 + \bar{y}_{i\cdot} = \bar{y}_{i\cdot\cdot} \qquad (15.67)$$

Any linear combination of the $(\mu + \alpha_i)$-expressions is, from (15.53), also estimable, with its b.l.u.e. being that same linear combination of the $\bar{y}_{i\cdot}$.s.

Of many features of this model (as considered, for example, in Searle, 1997, Section 6.2), only two will be mentioned, and but briefly. First, from (15.67) and the sentence that follows it, μ is not estimable. In contrast, three other commonly used functions, which are estimable and could each be taken as a definition of some kind of overall mean, are shown in Table 15.6. The first is the simple mean of means, the second is the weighted mean using the numbers of observations in the groups as weights, and the third is a general weighted mean using weights $\lambda_i / \Sigma\lambda_i$. Each function in the table is followed by an example.

From the first function of Table 15.6 can be seen the effect of using "convenient restraint" $\alpha_1 + \alpha_2 + \alpha_3 = 0$ as part of the model: it makes μ be estimable with b.l.u.e. $(1/a)\Sigma\bar{y}_{i\cdot}$, the mean of sample means. But if a different "convenient restraint" $\Sigma n_i\alpha_i = 0$ is used, as in the second function, then again μ becomes estimable, but with a different b.l.u.e., namely $\bar{y}_{\cdot\cdot}$, the average of all observations. This apparent occurrence of two different b.l.u.e.'s for μ arises because μ in the two cases is not the same: it is μ in two different models. In the first case it is a model that includes $\Sigma\alpha_i = 0$, and in the second it is a different model, one that includes $\Sigma n_i\alpha_i = 0$.

Overall, this model is known by statisticians as the *1-way classification* model; it is the easiest of all linear models (and a widely used and important one). Readers wanting to read about more complicated models should refer to the books mentioned at the beginning of this chapter.

Table 15.6 Estimable Functions in the 1-Way Classification
That Could Be Used for Estimating an Overall Mean

Function of class means	Estimable Function[a] $q'b$	b.l.u.e. of $q'b$ $\widehat{q'b} = q'b^0$	Variance of b.l.u.e. $v(q'b^0) = q'Gq\sigma^2$
Mean of means	$\mu + \frac{1}{a}\Sigma\alpha_i$	$\frac{1}{a}\Sigma\bar{y}_{i\cdot}$	$\frac{\sigma^2}{a}\Sigma(1/n_i)$
Example	$\mu + \frac{1}{3}(\alpha_1 + \alpha_2 + \alpha_3)$	$\frac{1}{3}(100 + 86 + 32) = 72\frac{2}{3}$	$\frac{1}{9}\left(\frac{1}{3} + \frac{1}{2} + \frac{1}{1}\right)\sigma^2 = \frac{11}{27}\sigma^2$
Weighted mean	$\mu + \frac{1}{N}\Sigma n_i\alpha_i$	$\frac{1}{N}\Sigma n_i\bar{y}_{i\cdot} = \bar{y}_{\cdot\cdot}$	$\frac{1}{N}\sigma^2$
Example	$\mu + \frac{1}{6}(3\alpha_1 + 2\alpha_2 + \alpha_3)$	$\frac{1}{6}(504) = 84$	$\frac{1}{6}\sigma^2$
General weighted mean	$\mu + \Sigma\lambda_i\alpha_i/\Sigma\lambda_i$	$(\Sigma\lambda_i\bar{y}_{i\cdot})/\Sigma\lambda_i$	$(\Sigma\lambda_i^2/n_i)\sigma^2/(\Sigma\lambda_i)^2$
Example	$\mu + \dfrac{(\lambda_1\alpha_1 + \lambda_2\alpha_2 + \lambda_3\alpha_3)}{\lambda_1 + \lambda_2 + \lambda_3}$	$\dfrac{100\lambda_1 + 86\lambda_2 + 32\lambda_3}{\lambda_1 + \lambda_2 + \lambda_3}$	$\dfrac{\frac{1}{3}\lambda_1^2 + \frac{1}{2}\lambda_2^2 + \lambda_3^2}{(\lambda_1 + \lambda_2 + \lambda_3)^2}\sigma^2$

[a] All summations are over $i = 1,\ldots,a$ and $N = \Sigma n_i$.

15.14   Appendix: Results on Quadratic Forms

We here list (without proof) a few results from mathematical statistics which bear directly on applying linear model methodology to data. To do so it is to be emphasized that any sum of squares of elements of a data vector $\mathbf{y}$ can always be expressed as a quadratic form (Section 3.6) $\mathbf{y'Ay}$ for $\mathbf{A} = \mathbf{A'}$. With this notation we have the following very useful results. For proofs, each result is accompanied by a page reference to Searle (1997, abbreviated as LM).

a.   Expected value

For $\mathbf{y} \sim (\boldsymbol{\mu}, \mathbf{V})$,   [LM 55]

$$E(\mathbf{y'Ay}) = \text{tr}(\mathbf{AV}) + \boldsymbol{\mu}'\mathbf{A}\boldsymbol{\mu}.$$

b.   Chi-square distributions

For $\mathbf{y} \sim \mathcal{N}(\boldsymbol{\mu}, \mathbf{V})$ with $\mathbf{V}$ non-singular,   [LM 58]

$\mathbf{y'Ay}$ is distributed as $\chi^2_{r_\mathbf{A}}$ iff $\mathbf{AV}$ is idempotent and $\mathbf{A}\boldsymbol{\mu} = \mathbf{0}$.

c.   Independence

For $\mathbf{y}$ as in b,   [LM 59]

$\mathbf{y'Ay}$ and $\mathbf{y'By}$ are independent if $\mathbf{AVB} = \mathbf{0}$.

d.   Independence with linear forms

For $\mathbf{y}$ as in b,   [LM 59]

$\mathbf{y'Ay}$ and $\mathbf{By}$ are independent if $\mathbf{BVA} = \mathbf{0}$.

e. $\mathcal{F}$-distributions

If $u_1 \sim \chi^2_{n_1}$ is independent of $u_2 \sim \chi^2_{n_2}$, then   [LM 48]

$$\frac{u_1/n_1}{u_2/n_2} \sim \mathcal{F}_{n_1, n_2}.$$

15.15 Exercises

E 15.1. (Linear model) In the pre-Christmas shopping week suppose that the number of Pokemon sales in Chicago in four Woolworth stores was 53, 43, 52 and 47; in two K-marts it was 71 and 73, in three Walmarts it was 62, 61 and 57, and in two Toys R Us shops it was 52 and 62.

    (a) Write a model equation for these data so that differences between the four makes of stores can be estimated.

    (b) What is the estimate of the residual error variance?

    (c) Test the hypothesis that Woolworth's sales do not differ from those of Toys R Us.

    (d) Calculate three different values that could be used as estimators of the overall store mean for Chicago. Suppose the square footage of all Woolworth stores is 11,000, of K-mart it is 10,000, of Walmarts it is 4000 and of Toys R Us it is 15,000.

E 15.2. Ignoring the arithmetic, why in equation (15.26) are three entries all the same and two other entries are also?

E 15.3. (a) Derive (15.27) from the preceding expression for SSE.
(b) Derive (15.28) from (15.27).

E 15.4. Show and explain each step that leads from $E[\mathbf{y}'(\mathbf{I} - \mathbf{XGX}'\mathbf{y}]$ to the result in (15.33).

E 15.5. Show that SSR and SSE of Table 15.3 are independent for $\mathbf{y}$ of (15.43).

E 15.6. Derive (15.42) from (15.40), using (after proving) $\bar{\hat{y}} = \bar{y}$.

E 15.7. For E 15.1, list 8 estimable functions and their b.l.u.e.s.

E 15.8. For calculating Q of (15.58) one person uses the same $\mathbf{G}$ in $\mathbf{K}'\mathbf{GK}$ as in $\mathbf{b}^0 = \mathbf{GX}'\mathbf{y}$. Another person uses one $\mathbf{G}$ in

K′GK and another in **b**^{0}. Explain the relationship between their two values of Q.

E 15.9. Show that Q of (15.58) is the same for H_1: **LK′b** = **Lm** where **L** is non-singular as it is for H_2: **K′b** = **m**.

E 15.10. For estimable **q′b** derive (15.47) giving a reason for each step.

E 15.11. Show that the b.l.u.e. of an estimable function is unbiased.

Chapter 16

LINEAR PROGRAMMING

Mathematical programming is a tool used for solving optimization problems. The most common application is that of allocating scarce resources among various activities to maximize or minimize an objective such as profit or cost. There are two types of mathematical programming used by economists: linear programming and non-linear programming. Linear programming, the focus of this chapter, involves problems where the mathematical relationships are expressed as linear equations. Non-linear programming involves relationships which are non-linear.

16.1 The Maximization Problem

a. Illustration (Profit maximization)

Sarah's Sewing Company designs and produces two types of garments: dresses and skirts. Each product requires time in the design center and the sewing center. Dresses require 12 hours per week in the design center and six hours per week in the sewing center. Sarah is allowed to have only 300 hours in the sewing center each week. Sarah can get all the fabric and thread she needs. Fabric and thread for each dress and for each skirt cost $30 and $27, respectively. Overhead costs $42 per dress and $30 per skirt. Sarah is able to sell all the dresses and skirts she produces each week. Each dress sells for $90. Each skirt sells for $63. Sarah wishes to maximize her weekly profits from the sale of dresses and skirts.

The solution to Sarah's problem can be solved with a technique known as *linear programming*. Each linear programming problem has *decision variables* that capture the decision that must be made. In this case, Sarah must determine

$$x_1 \; = \; \text{number of dresses to produce each week,}$$

and

$$x_2 \; = \; \text{number of skirts to produce each week.}$$

In each linear programming problem the decision maker either wishes to maximize (or minimize) an *objective function*, a linear function of the decision variables. In the illustration, Sarah wishes to maximize the profits from the production of dresses and skirts. Profits for Sarah consist of revenue less costs (fabric, thread and overhead) for the production of dresses and skirts. The profit Sarah can expect from the production of each dress is $\$90 - \$30 - \$42 = \$18$. The profit expected from the production of each skirt is $\$63 - \$27 - \$30 = \$6$. Hence, the *objective function* is

$$\text{Maximize } z = (90 - 30 - 42)x_1 + (63 - 27 - 30)x_2$$

or more simply as

$$\text{Maximize } z = 18x_1 + 6x_2.$$

The coefficient of each variable in the objective function is called the *objective function coefficient*. It describes the contribution of the variable to the overall objective. In this illustration, the objective function coefficient for dresses is 18, suggesting that each dress contributes $18 to the company's profits.

Sarah could choose to make a large number of dresses and skirts. However, the number is limited by certain restrictions, often called *constraints*. Sarah's restrictions are:

Constraint 1: Sarah may not use more than 300 hours in the design center per week, and

Constraint 2: Sarah may not use more than 240 hours in the sewing center each week.

Constraints 1 and 2 must be expressed in terms of the decision variables x_1 and x_2. Constraint 1 can be rewritten as $12x_1 + 3x_2 \leq 300$. Note that the units on the left-hand side of the equation are (design hours per dress) × (dresses made per week) + (design hours per skirt) × (skirts made per week). The units on the right-hand side are design hours per week. Constraint 2 can be expressed as $6x_1 + 3x_2 \leq 240$. Once again the units on the left-hand side of the equation match the units on the right-hand side of the equation.

The coefficients of the variables in the constraints are called the *technological coefficients*. The right-hand side of each constraint represents the resource capacity available for a particular operation or input.

Some additional constraints are often included in linear programming problems. These constraints require that negative production is not feasible. Hence to ensure that Sarah does not produce negative dresses or skirts, the following equations are included in the linear programming problem: $x_1 \geq 0$ and $x_2 \geq 0$.

b. Matrix formulation

Sarah's complete maximization problem can be written as

$$
\begin{aligned}
\text{Maximize } z = 18x_1 + 6x_2 \\
\text{subject to } 12x_1 + 3x_2 &\leq 300 \\
6x_1 + 3x_2 &\leq 240 \\
x_1 &\geq 0 \\
x_2 &\geq 0.
\end{aligned}
$$

There are three essential properties to all linear programming problems:

(i) All relationships are linear, including the objective function.

(ii) Constraints may be equalities or inequalities in either direction.

(iii) Solutions must be non-negative.

All linear programming problems, where the purpose is to maximize the objective function, can be expressed as:

$$
\begin{aligned}
\text{Maximize } z = \mathbf{c}'\mathbf{x} \\
\text{subject to } \mathbf{Ax} &\lessgtr \mathbf{b} \\
\mathbf{x} &\geq 0
\end{aligned}
\tag{16.1}
$$

where

c is an $n \times 1$ vector of objective function coefficients,

x is an $n \times 1$ vector of decision variables,

A is a $m \times n$ matrix of technological coefficients,

b is a $m \times 1$ vector of scarce resources with $b_i \geq 0$, $i = 1, \ldots, m$,

z is the value of the objective function.

In Sarah's illustration

$$\mathbf{c} = \begin{bmatrix} 18 \\ 6 \end{bmatrix}, \quad \mathbf{x} = \begin{bmatrix} x_1 \\ x_2 \end{bmatrix}, \quad \mathbf{A} = \begin{bmatrix} 12 & 3 \\ 6 & 3 \end{bmatrix}, \quad \mathbf{b} = \begin{bmatrix} 300 \\ 240 \end{bmatrix}.$$

c. Graphical solution

A linear programming problem with two variables can easily be illustrated graphically. The first step in this solution is to determine the feasible region for the problem. The *feasible region* is the set of all points (x_1, x_2) satisfying all the constraints $\mathbf{Ax} \lesseqgtr \mathbf{b}$, $\mathbf{x} \geq 0$; and for Sarah, it is represented by the shaded area in Figure 16.1. It lies strictly within the first quadrant of the (x_1, x_2) plane. The problem is to find the point (x_1, x_2) in the feasible region that maximizes the value of the objective function.

Begin by choosing a point in the feasible region, say $(5, 0)$. The profit for this point, determined from the objective function $z = 18x_1 + 6x_2$, is 90. This point lies on a line, $90 = 18x_1 + 6x_2$, where all points along the line yield equal profits for Sarah. This line is called an *isoprofit line*, as is every other line of the form $18x_1 + 6x_2 = $ constant, all with the same slope. These lines represent combinations of dress and skirt production that yield equal profit for Sarah. The isoprofit line $270 = 18x_1 + 6x_2$ can also be seen in Figure 16.1.

To simulate maximizing the objective function, one can place a ruler on an isoprofit line and push it away from the origin, remembering to keep the ruler parallel to the dashed lines. The ruler represents the objective function $z = 18x_1 + 6x_2$ where Sarah's profits, z, are increasing as you push the ruler further from the origin. Continue to push the ruler out until it touches the feasible region at the outermost point, point A in Figure 16.1, remembering to keep the ruler parallel to the dashed lines. This point, $(10, 60)$, at the intersection of the first and second constraints, is the optimal solution to the linear programming problem. It can be found by simultaneously solving the equations, $12x_1 + 3x_2 = 300$ and $6x_1 + 3x_2 = 240$. These are the first and second

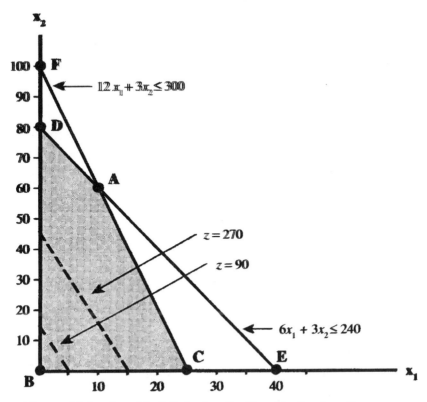

Figure 16.1: Graphical Solution for Sarah's Sewing Company

constraints expressed as equalities. The optimal value of z, Sarah's profits, can be found by substituting the values of x_1 and x_2 into the objective function. At point A Sarah's profits from the production of 10 dresses and 60 skirts are \$540. There is no other combination of dresses and skirts in the feasible region that will give Sarah higher profits.

Notice that the optimal solution to Sarah's problem is at the intersection of two constraints. These constraints are considered binding and prevent Sarah from producing more dresses and skirts. A constraint is *binding* (*non-binding*) if the left-hand side and the right-hand side of the constraint are (not) equal when the optimal values of the decision variables are substituted in the constraint.

d.   Extreme points

The feasible region for Sarah's problem is considered a convex set since a line segment joining any pair of points in the feasible set is contained completely in the feasible set. Solutions to linear programming problems use the concept of a convex set and a related concept of extreme points to determine the optimal solution.

An *extreme point* of any n-dimensional convex set in a linear programming problem is the n-element vector $\mathbf{x}' = [x_1 \quad x_2 \quad \cdots \quad x_n]$ which is the unique solution to the set of consistent equations formed by selecting n of the constraints defining the feasible region and requiring them to hold as equalities. If an extreme point also satisfies the remaining constraints, it is called an extreme point of the feasible region or a *feasible extreme point*.

Illustration   (*continued*)

In Sarah's illustration four extreme points, seen in Figure 16.1, solved from the constraints are within the feasible region. The sets of equations formed from the constraints and their related extreme points are shown in Table 16.1. Two other extreme points can be determined: E $(40, 0)$ and F $(0, 100)$. They are not within the feasible region. Hence, they are infeasible solutions to Sarah's problem.

Table 16.1. Sarah's Extreme Points and
Related Constraints

Extreme Point		Constraints
Label	Coordinates	
A	$(10, 60)$	$12x_2 + 3x_2 = 300 \quad 6x_1 + 3x_2 = 240$
B	$(0, 0)$	$x_1 = 0, \quad x_2 = 0$
C	$(25, 0)$	$12x_1 + 3x_2 = 300, \quad x_2 = 0$
D	$(0, 80)$	$6x_1 + 3x_2 = 240, \quad x_1 = 0$

Given that an optimal solution is always an extreme point of the feasible region, solving linear programming problems can easily be simplified. If there are n unknowns and $m + n$ constraints, taking n of the constraints at a time to form an extreme point means there are $(m + n)!/(m!n!)$ extreme points. Computer algorithms, such as the revised simplex method, progressively move from one extreme point to the next so that the values of the objective function form a monotonic non-decreasing series. Thus, not all extreme points are considered in the process of determining the extreme point that maximizes the objective function.

e. Slack variables

Linear programming problems can be rewritten where the inequalities are converted to equalities by inserting *"slack" variables* to take up the slack. Thus the linear programming problem is transformed into standard form and becomes

$$\text{Maximize } z = \mathbf{c}'\mathbf{x}^*$$
$$\text{subject to } \mathbf{A}^*\mathbf{x}^* = \mathbf{b}$$
$$\mathbf{x}^* \geq \mathbf{0} \qquad (16.2)$$

where

$$\mathbf{c}^* = [\mathbf{c}' \ \mathbf{0}']', \text{ a column vector of order } n+m,$$
$$\mathbf{x}^* = [\mathbf{x}' \ \mathbf{x}_s']', \text{ a column vector of order } n+m,$$
$$\mathbf{x}_s = \text{ a column vector of order } m \text{ of slack variables,}$$
$$\mathbf{A}^* = [\mathbf{A} \ \mathbf{I}_m] \text{ is a matrix of order } m \text{ by } (n+m), \text{ and}$$

where $\mathbf{c}$, $\mathbf{x}$, $\mathbf{b}$ and $\mathbf{A}$ are as defined earlier.

By requiring the slack variables to be non-negative and to have zero profit associated with them, we ensure that if $\mathbf{x}^*$ is such that $\mathbf{A}^*\mathbf{x}^* = \mathbf{b}$ then $\mathbf{A}\mathbf{x} \leq \mathbf{b}$ and $z = \mathbf{c}^{*\prime}\mathbf{x}^* = [\mathbf{c}' \ \mathbf{0}'][\mathbf{x}' \ \mathbf{x}_s']' = \mathbf{c}'\mathbf{x}$. Thus the vector $\mathbf{x}$ in $\mathbf{x}^*$ that satisfies these constraints also satisfies the constraints without the slack variables and produces the same value for the objective function. If (16.2) can be solved for its optimal solution $\mathbf{x}_0^* = [\mathbf{x}_0' \ \mathbf{x}_{s0}']'$ then the optimal solution to (16.1) is $\mathbf{x}_0$.

Illustration (*continued*)

Sarah's sewing problem can be rewritten as

$$\text{Maximize} \quad z = [18 \ 6 \ 0 \ 0] \begin{bmatrix} x_1 \\ x_2 \\ x_{s1} \\ x_{s2} \end{bmatrix}$$

$$\text{subject to} \quad \begin{bmatrix} 12 & 3 & 1 & 0 \\ 6 & 3 & 0 & 1 \end{bmatrix} \begin{bmatrix} x_1 \\ x_2 \\ x_{s1} \\ x_{s2} \end{bmatrix} = \begin{bmatrix} 300 \\ 240 \end{bmatrix}$$

and
$$x_1, \quad x_2, \quad x_{s1}, \quad x_{s2} \geq 0.$$

The preceding graphical solution to this maximization problem indicates that the first two constraints are binding and thus $x_{s1} = x_{s2} = 0$. If there were a non-binding constraint the value of the slack variable would be greater than or less than zero to achieve the equality.

f. Basic solution

From the standard form of a linear programming problem, (16.2), we see that $\mathbf{A}^* = [\mathbf{A} \quad \mathbf{I}_m]$. There is at least one set of m linearly independent columns in $\mathbf{A}^*$. In fact, since $\mathbf{A}^*$ has $m+n$ columns there are at most $(m+n)!/(m!n!)$ sets of m linearly independent columns. Each $m \times m$ matrix, with columns consisting of m linearly independent columns of $\mathbf{A}^*$, is a *basis*.

Consider a basis $\mathbf{B}$ consisting of k columns from $\mathbf{A}$ associated with the xs and $m - k$ columns from $\mathbf{I}$ associated with elements of $\mathbf{x}$. From (16.2) we see that
$$\mathbf{A}^*\mathbf{x}^* = \mathbf{b}.$$

Alternatively we can write this as
$$[\mathbf{B} \quad \mathbf{A}_2] \begin{bmatrix} x_B \\ x_2 \end{bmatrix} = \mathbf{b},$$

where $\mathbf{B}$ is a basis, $\mathbf{x}_B$ is an $m \times 1$ vector of basic variables, $\mathbf{A}_2$ has columns of $\mathbf{A}^*$ not included in the basis and $\mathbf{x}_2$ is its corresponding vector of variables. The solution, by partitioning, yields
$$\begin{bmatrix} x_B \\ x_2 \end{bmatrix} = \begin{bmatrix} \mathbf{B}^{-1}\mathbf{b} - \mathbf{B}^{-1}\mathbf{A}_2\mathbf{x}_2 \\ x_2 \end{bmatrix}.$$

Suppose we set the nonbasic variables, $\mathbf{x}_2$, to 0. Then
$$\mathbf{x}^* = \begin{bmatrix} x_B \\ 0 \end{bmatrix} = \begin{bmatrix} \mathbf{B}^{-1}\mathbf{b} \\ 0 \end{bmatrix} \qquad (16.3)$$

is a *basic solution* to $\mathbf{A}^*\mathbf{x}^* = \mathbf{b}$. Since a certain number of elements of $\mathbf{x}$, say k, are set equal to 0, k of the inequalities $\mathbf{A}\mathbf{x} \leq \mathbf{b}$ are forced to hold as equalities. In addition, $n - k$ of the xs are set equal to zero, forcing the associate non-negativity constraints to hold as inequalities.

Then the real variables in $\mathbf{x}^*$ represent the solution to n equations formed by forcing n of the constraints, $\mathbf{Ax} \leq \mathbf{b}$, $\mathbf{x} \geq \mathbf{0}$, to hold as equalities. The unique solution to this set of equations is by definition an extreme point. Any basis $\mathbf{B}$ has associated with it a unique extreme point which may or may not be in the feasible region, depending on whether or not it satisfies the remaining constraints. A *basic feasible solution* is a basic solution where all the variables are non-negative.

Illustration (*continued*)

In Sarah's problem there are six basic solutions. Table 16.2 shows their relation to extreme points and whether or not they are considered a basic feasible solution. Note that only the four extreme points A, B, C and D are basic feasible solutions, because they lie within the feasible region. Extreme points E and F lie outside that region, as seen in Figure 16.1, and so are not basic feasible solutions.

Table 16.2. Sarah's Basic Solutions and Related Extreme Points

Basis	Basic Variables	Nonbasic Variables	Basic Solution	Is Solution a Basic Feasible Solution?	Extreme Point (Figure 16.1)
$\begin{bmatrix} 12 & 3 \\ 6 & 3 \end{bmatrix}$	x_1, x_2	x_{s1}, x_{s2}	$x_1 = 10 \quad x_2 = 60$ $x_{s1} = 0 \quad x_{s2} = 0$	Yes	A
$\begin{bmatrix} 1 & 0 \\ 0 & 1 \end{bmatrix}$	x_{s1}, x_{s2}	x_1, x_2	$x_1 = 0 \quad x_2 = 0$ $x_{s1} = 300 \quad x_{s2} = 240$	Yes	B
$\begin{bmatrix} 12 & 0 \\ 6 & 1 \end{bmatrix}$	x_1, x_{s2}	x_2, x_{s1}	$x_1 = 25 \quad x_2 = 0$ $x_{s1} = 0 \quad x_{s2} = 9$	Yes	C
$\begin{bmatrix} 3 & 1 \\ 3 & 0 \end{bmatrix}$	x_2, x_{s1}	x_1, x_{s2}	$x_1 = 0 \quad x_2 = 80$ $x_{s1} = 60 \quad x_{s2} = 0$	Yes	D
$\begin{bmatrix} 12 & 1 \\ 6 & 0 \end{bmatrix}$	x_1, x_{s1}	x_2, x_{s2}	$x_1 = 40 \quad x_2 = 0$ $x_{s1} = -180 \quad x_{s2} = 0$	No	E
$\begin{bmatrix} 3 & 0 \\ 3 & 1 \end{bmatrix}$	x_2, x_{s2}	x_1, x_{s1}	$x_1 = 0 \quad x_2 = 100$ $x_{s1} = 0 \quad x_{s2} = -60$	No	F

16.2 The Minimization Problem

a. Matrix formulation

Minimization problems often occur in situations where the objective is to attain a specified level of performance at a minimum cost. The matrix form of the minimization problem of linear programming is

$$\text{Minimize } z = \mathbf{c}'\mathbf{x}$$
$$\text{subject to } \mathbf{Ax} \gtrless \mathbf{b}$$
$$\mathbf{x} \geq 0. \qquad (16.4)$$

b. Illustration

A typical minimization problem is one where the objective is to minimize the cost of a diet while satisfying requirements related to nutritional intake. In fact, the first linear programming problems to be solved by computer, proposed by Stigler (1945), included 77 types of food with 10 nutritional requirements. The optimal solution consisted of a diet with cornmeal, wheat flour, evaporated milk, peanut butter, lard, beef, liver, potatoes, spinach and cabbage. Current applications of the diet problem are used in hospitals and institutions for menu planning. The diet problem is also used in agricultural operations requiring a controlled diet for efficient growth of animals. We shall consider a simplified case.

Suppose you were responsible for determining the grams of food mix A and food mix B that should be purchased to meet the dietary requirements of your feedlot. The animals in your feedlot require 70 grams of protein and 150 grams of carbohydrate. Feed mix A costs .2 cent per gram and contains .5 gram of protein and .2 gram of carbohydrate per gram. Feed mix B costs .03 cent per gram and contains .02 grams of protein and .6 gram of carbohydrate per gram.

The problem can be formulated as (16.4), where

$$\mathbf{c}' = [.2 \quad .03], \quad \mathbf{A} = \begin{bmatrix} .5 & .02 \\ .2 & .6 \end{bmatrix}, \quad \mathbf{b} = \begin{bmatrix} 70 \\ 150 \end{bmatrix}, \quad \mathbf{x}' = [x_1 \quad x_2].$$

As illustrated in Figure 16.2, this problem has a feasible region above the equations formed by the constraints. Since we are minimizing cost the optimal solution will be near the origin. To obtain a graphical

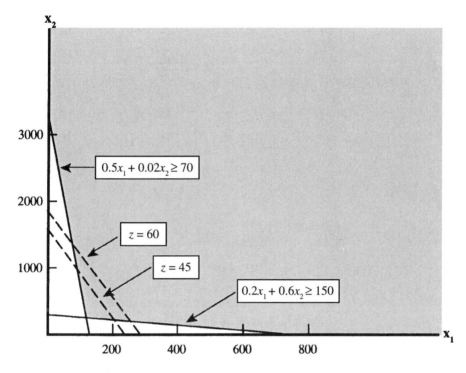

Figure 16.2: Graphical Solution for Feedlot Diet Problem

solution, move along the *isocost lines*, the lines of decreasing values of the objective function, toward the origin until the line touches the extreme point at the farthest point of the feasible region. There are three feasible extreme points $(750, 0)$, $(0, 3500)$ and $(131.8, 206.1)$. The latter solution, which minimizes the cost of the diet while satisfying the nutritional constraints, gives the minimum cost as 32.54 cents.

c. Surplus variables

When constraints in a linear programming problem are expressed with "greater than or equal to" signs, we write the equalities in (16.2) using slack variables with a negative coefficient. They are then called *surplus variables*.

The diet problem becomes

$$
\text{Minimize} \qquad z = [.2 \quad .03 \quad 0 \quad 0]
\begin{bmatrix}
x_1 \\
x_2 \\
x_{s1} \\
x_{s2}
\end{bmatrix}
$$

$$
\text{subject to} \qquad
\begin{bmatrix}
.5 & .02 & -1 & 0 \\
.2 & .6 & 0 & -1
\end{bmatrix}
\begin{bmatrix}
x_1 \\
x_2 \\
x_{s1} \\
x_{s2}
\end{bmatrix}
=
\begin{bmatrix}
70 \\
150
\end{bmatrix}
$$

and

$$
x_1, \quad x_2, \quad x_{s1}, \quad x_{s2} \geq 0.
$$

16.3 Simplex Method

The revised simplex method is often used by computers to find the optimal solutions to linear programming problems. The method begins with a basis corresponding to the extreme point representing the origin. A variable is selected to enter the basis according to which of the nonbasic variables in $\mathbf{x}^*$ will increase the value of the objective function. A column is selected for removal from the existing basis so that feasibility is maintained. A new basis and corresponding new basic solution are obtained. A new feasible extreme point is also found such that the value of the objective function at the new extreme point is no lower than the value of the objective function at the previous extreme point. The process continues as long as a new extreme point can be found that will increase (or at least not decrease) the objective function. When no other substitutions are possible the current extreme point is the optimal solution. Since there is only a finite number of feasible extreme points and no extreme point is found more than once, this method obtains the optimal solution in a finite number of iterations.

We are not going to delve into the specifics of the method. However, several references are available that provide the details. The reader may wish to refer to Dantzig (1963), Luenberger (1984), Gass (1985), Bazaraa and Jarvis (1990), or Winston (1995).

16.4 Related Topics

a. Changing a minimization problem to a maximization problem

It is often easier to deal with minimization problems if they could be rewritten as maximization problems. To change a minimization problem to a maximization problem, simply multiply the objective function by -1 and change the word minimize to maximize; everything else remains unchanged. The problem

$$
\begin{aligned}
\text{Minimize} \quad & \mathbf{c}'\mathbf{x} \\
\text{subject to} \quad & \mathbf{A}\mathbf{x} \lesseqgtr \mathbf{b} \\
& \mathbf{x} \geq \mathbf{0}
\end{aligned}
$$

is equivalent to

$$
\begin{aligned}
\text{Maximize} \quad & -\mathbf{c}'\mathbf{x} \\
\text{subject to} \quad & \mathbf{A}\mathbf{x} \lesseqgtr \mathbf{b} \\
& \mathbf{x} \geq \mathbf{0}.
\end{aligned}
$$

To change a maximization problem to a minimization problem, simply do the opposite.

b. Mixed constraints

The linear programming problems described in previous sections either had all constraints "less than or equal to" or all "greater than or equal to". Linear programming problems can, and often do, have a mix of constraints. These, too, are simple to solve by combining the procedures discussed earlier. That is, when the constraints are written in inequality form, add slack variables with coefficients of $+1$ to the less-than constraints and surplus variables with coefficients of -1 to the greater-than constraints.

c. Changing the direction of an inequality

The elements of $\mathbf{b}$ must be non-negative for the revised simplex method. If any b_i, $i = 1, \ldots, m$, is negative when the problem is initially formulated, the corresponding constraint can be multiplied through by -1, thus changing the signs of the elements of the constraint matrix

A associated with that constraint and the direction of the inequality, as well as the sign of **b**, as desired. This does not affect the meaning of the constraint at all; the feasible region is unchanged. For example, a greater-than constraint multiplied by -1 becomes a less-than constraint, but the problem and the resulting solution are unchanged.

d. Unconstrained variables

Linear programming procedures are designed under the assumption that all variables are non-negative. If the problem to be solved is such that one or more of the variables can be negative, i.e., these variables are unconstrained, linear programming procedures can still be utilized. For every unconstrained x_i, substitute two new variables $x_i^+ - x_i^- = x_i$ into the original formulation of the problem and proceed to solve the reformulated problem in terms of the difference between the non-negative variables $(x_i^+ - x_i^-)$ rather than the unrestricted x_i. When the solution has been obtained, replace $x_i^+ - x_i^-$ by x_i.

e. The dual problem

For any linear programming problem written in the form

$$\text{Maximize } z = \mathbf{c'x}$$
$$\text{subject to } \mathbf{Ax} \leq \mathbf{b}$$
$$\mathbf{x} \geq \mathbf{0}, \tag{16.5}$$

there exists a corresponding problem

$$\text{Minimize } z = \mathbf{b'y}$$
$$\text{subject to } \mathbf{A'y} \geq \mathbf{c}$$
$$\mathbf{y} \geq \mathbf{0}. \tag{16.6}$$

The first of these problems is called the *primal problem* and the second is called the *dual problem*. If the original problem is (16.5), its dual is (16.6); if the original is (16.6), its dual is (16.5). The dual of the dual is the primal. The dual problem occupies an important position in linear programming theory and applications; it is discussed fully by Dantzig (1963) and Luenberger (1984). The importance of the dual arises from the fact that the optimal solution to the dual problem, y_0, an $m \times 1$ column vector if **A** is $m \times n$, has as its elements the marginal value of each of the constraint limits, b_i. In most business and economic situations the b_is represent scarce resources at the decision-maker's disposal; knowledge of how much profits would increase or costs would decrease with an additional unit of each of the various scarce resources is valuable information.

The dual variables are also known as shadow prices or simplex multipliers. The shadow price for the ith constraint identifies the amount by which the optimal z value is improved—decreased in a minimization problem or increased in a maximization problem—if the right-hand side of the ith constraint is increased by 1. This holds provided the current basis remains optimal.

Illustration (Profit maximization)

The maximization problem for Sarah's Sewing Company in the first illustration was

$$\text{Maximize } z = 18x_1 + 6x_2$$
$$\text{subject to} \quad 12x_1 + 3x_2 \leq 300 \quad\quad (16.7)$$
$$6x_1 + 3x_2 \leq 240$$
$$x_1, \ x_2 \geq 0. \quad\quad (16.8)$$

The dual of this problem is

$$\text{Minimize } z = 300y_1 + 240y_2$$
$$\text{subject to} \quad 12y_1 + 6y_2 \geq 18 \quad\quad (16.9)$$
$$3y_1 + 3y_2 \geq 6$$
$$y_1, \ y_2 \geq 0, \quad\quad (16.10)$$

where y_1 and y_2 indicate the amount by which the profits of Sarah's Sewing Company would increase if the constraints on design hours and sewing hours are relaxed by one hour.

Illustration (Diet cost minimization)

The minimization of the diet illustration was

$$\text{Minimize } z = 0.2x_1 + 0.03x_2$$
$$\text{subject to} \quad 0.5x_1 + 0.02x_2 \geq 70,$$
$$0.2x_1 + 0.6 \ x_2 \geq 150,$$
$$x_1 \geq 0,$$
$$x_2 \geq 0.$$

The dual of this problem is

$$\begin{aligned}
\text{Maximize } z \;=\;\;& 70y_1 + 150y_2 \\
\text{subject to} \quad\;\;& 0.5\,y_1 + 0.2\,y_2 \le 0.2, \\
& 0.02y_1 + 0.6y_2 \le 0.03, \\
& y_1 \ge 0, \\
& y_2 \ge 0.
\end{aligned}$$

The variable y_1 identifies how much the optimal cost of the diet would decrease if the constraint on protein were relaxed by 1 gram. The variable y_2 identifies how much the optimal cost of the diet would decrease if the constraint on carbohydrates were relaxed by 1 gram.

Let's look at the economic interpretation of the dual problem using Sarah's Sewing Company as an example. Suppose an entrepreneur wants to purchase all of Sarah's resources. Then the entrepreneur must determine the price she is willing to pay for a unit of each resource. Hence, one must determine

$$y_1 \;=\; \text{the price paid for one hour of design work}$$

and

$$y_2 \;=\; \text{the price paid for one hour of sewing work.}$$

The total cost of purchasing these resources is what is to be minimized and is expressed by the objective function of the dual problem.

These resource prices must be set high enough to induce Sarah to sell. For example, the entrepreneur must offer at least $18 for a combination of the resources that include 12 hours of design time and three hours of sewing time because Sarah could use these resources to make a dress that would net a profit of $18. Similar reasoning suggests that the entrepreneur must offer at least $6 for 3 hours of design time and 3 hours of sewing time because Sarah could use these resources to make a skirt that would net a profit of $6.

When we include the objective function and the constraints we see that the solution yields the value of the design and sewing time for Sarah. The solution to Sarah's dual problem yields an optimal value of the objective function of $540. The optimal values of y_1 and y_2 are 1 and 1, respectively. Hence, the marginal value of an hour of design work is $1. Similarly, the marginal value of an hour of sewing work is $1.

The fundamental relation between the primal and the dual problems is that if $\mathbf{x}_0^*$ and $\mathbf{y}_0^*$ are optimal solutions to these problems, respectively, then

$$\mathbf{c}'\mathbf{x}_0 = \mathbf{c}^{*\prime}\mathbf{x}_0^* = z_0 = \mathbf{b}^{*\prime}\mathbf{y}_0^* = \mathbf{b}'\mathbf{y}_0; \qquad (16.11)$$

i.e., at their respective optima, the objective functions of the primal and the dual have identical values. Let us rearrange the elements of $\mathbf{c}^*$, $\mathbf{x}^*$, $\mathbf{b}^*$ and $\mathbf{y}^*$ such that $\mathbf{x}_0^{*\prime} = [\mathbf{x}_B' \quad \mathbf{x}']$, $\mathbf{c}^{*\prime} = [\mathbf{c}_B' \quad \mathbf{0}]$ and $\mathbf{y}^{*\prime} = [\mathbf{y}_0' \quad \mathbf{y}_S']$ where $\mathbf{x}_B$ is the basis variables, $\mathbf{x}$ is the non-basis variables (thus equal to zero), $\mathbf{y}_0$ is the ordinary variables in the dual, $\mathbf{y}_S$ is the surplus variables associated with the dual and $\mathbf{c}^*$ and $\mathbf{b}^*$ have been altered in a corresponding fashion. Then $\mathbf{c}_B'\mathbf{x}_B = \mathbf{b}'\mathbf{y}_0$ by (16.11); but $\mathbf{x}_B = \mathbf{B}^{-1}\mathbf{b}$ from (16.3). Hence $\mathbf{c}_B'\mathbf{B}^{-1}\mathbf{b} = \mathbf{b}'\mathbf{y}_0$, which is true for all non-negative $\mathbf{b}$ so that $\mathbf{y}_0' = \mathbf{c}_B'\mathbf{B}^{-1}$. Thus the value of the dual variables associated with the dual at its optimum can be obtained directly from the optimal basis $\mathbf{B}$ available from the revised simplex solution.

The ordinary dual variables $\mathbf{y}_0' = \mathbf{c}_B'\mathbf{B}^{-1}$, as mentioned above, can be interpreted as the marginal value (*shadow price*) of an additional unit of each of the resources denoted by b_i. If the optimal solution to the primal problem contains some slack variables, the value of additional units of those resources should be zero. Suppose the ith constraint is non-binding; i.e., that slack variable is positive at the optimum. Suppose further that the column associated with that slack variable is the ith column of $\mathbf{B}$. That column will have all zeros except for the ith element, which is equal to 1, and it can be shown that the ith column of $\mathbf{B}^{-1}$ will be the same. Hence the ith element of $\mathbf{c}_B'\mathbf{B}^{-1}$ is c_{Bi}. But since the variable is a slack variable, its cost coefficient is zero, so that y_i in $\mathbf{y}_0' = \mathbf{c}_B'\mathbf{B}^{-1}$ is zero. Hence the dual variable y_i associated with a non-binding constraint (constraint i) is zero.

16.5 Applications of Linear Programming

There are numerous applications of linear programming in economics. In this chapter we have examined a profit maximization problem (Sarah's Sewing Company) and a blending problem (feedlot diet) when optimal inputs are determined for the production of a product. Linear programming can also be used for (1) work scheduling problems where the minimum cost method for satisfying workforce requirements are determined, (2) financial planning such as capital budgeting and cash flow analysis and (3) transportation problems where the optimal path between origins and destinations are determined. We provide an additional illustration of the last type of problem.

Illustration   (Transportation)

Upperville Electric and Gas has two electric plants, A and B, that supply the power needs of Midtown, Downtown and Uptown. The plants can supply 125 million and 100 million kilowatt hours (kwh) of electricity, respectively. The peak power demands of the three towns occur at the same time and are 75 million, 125 million and 25 million kwh, respectively. The cost of delivering 1 million kwh of electricity from a plant to a town depends on the distance the electricity must travel according to Table 16.3.

Table 16.3. Shipping Costs for Upperville Electric and Gas
($/million kwh)

FROM	TO		
	Midtown	Downtown	Uptown
Plant A	13	9	8
Plant B	11	12	12

Upperville Electric and Gas wants to minimize the cost of delivering electricity to the three towns yet still meet the towns' peak demand.

Upperville Electric and Gas' objective function is

Minimize $z = 13x_{AM} + 9x_{AD} + 8x_{AU} + 11x_{BM} + 12x_{BD} + 12x_{BU}$

when x represents the quantity of electricity shipped from one plant to a town. The first subscript on x represents the plant (either A or B), the second subscript represents the town (i.e., M for Midtown, D for Downtown and U for Uptown).

Upperville Electric and Gas faces three types of constraints. The first is that the total power supplied by each plant cannot exceed the plant's capacity. These constraints are expressed as

$$x_{AM} + x_{AD} + x_{AU} \leq 125$$
$$x_{BM} + x_{BD} + x_{BU} \leq 100.$$

The second type of constraint is that the demand by each town will be met by the plants. These demand constraints are

$$x_{AM} + x_{BM} \geq 75$$

$$x_{AD} + x_{BD} \geq 125$$
$$x_{AU} + x_{BU} \geq 25.$$

The third type of constraint ensures that negative power is not shipped from a plant to a town. Hence the following constraints are added to the problem:

$$\mathbf{x}_{ij} \geq \mathbf{0} \quad (i = A, B \text{ and } j = M, D, U).$$

Notice that the total supply of electricity from all of Upperville Electricity and Gas' plants equals the total demand of the three towns. When this occurs the problem becomes a balanced transportation problem and the supply and demand constraints must be binding. Clearly, if one of the supply constraints was not binding, the remaining available electricity would not be sufficient to meet all three towns' needs. In a balanced transportation problem the supply and demand constraints are written as equalities.

Combining the objective function, the supply, demand and non-negativity constraints and expressing them in matrix notation yields the following linear programming formulation to Upperville Electric and Gas' power distribution problem.

$$
\begin{aligned}
\text{Minimize} \quad &= \mathbf{c}'\mathbf{x} \\
\text{subject to} \quad &\mathbf{Ax} = \mathbf{b} \\
&\mathbf{x} \geq \mathbf{0}
\end{aligned}
$$

where

$$\mathbf{c}' = [13 \ 9 \ 8 \ 11 \ 12 \ 12] \qquad \mathbf{x} = \begin{bmatrix} x_{AM} \\ x_{AD} \\ x_{AU} \\ x_{BM} \\ x_{BD} \\ x_{BU} \end{bmatrix}$$

$$\mathbf{A} = \begin{bmatrix} 1 & 1 & 1 & 0 & 0 & 0 \\ 0 & 0 & 0 & 1 & 1 & 1 \\ 1 & 0 & 0 & 1 & 0 & 0 \\ 0 & 1 & 0 & 0 & 1 & 0 \\ 0 & 0 & 1 & 0 & 0 & 1 \end{bmatrix} \quad \text{and} \quad \mathbf{b} = \begin{bmatrix} 125 \\ 100 \\ 75 \\ 125 \\ 25 \end{bmatrix}.$$

The solution to this problem can be found easily using available computer software. The optimal solution for Upperville Electric and

Gas turns out to be

$$\mathbf{x}' = [0 \quad 100 \quad 25 \quad 75 \quad 25 \quad 0]$$

with a minimum cost of $2225. Hence, Plant A should ship 100 million kwh to Downtown and 25 million kwh to Uptown. Plant A would not ship to Midtown. Plant B would ship 75 million kwh to Midtown and 25 million kwh to Downtown. Plant B would not send any electricity to Uptown.

16.6   Exercises

E 16.1. (Profit maximization) As an economic analyst for a landscape firm you have been asked to determine the optimal mix of trays of annuals and numbers of perennials in a retail outlet so that profits can be maximized. Each tray of annuals contributes $5/day to profit. Each perennial contributes $2/day. Space in the retail outlet is limited to 20 square units. Each perennial requires four square units of space, whereas each tray of annuals requires one square unit. There are 16 hours of labor per week that can be dedicated to the annuals and perennials. Each perennial requires just one hour per week, while a tray of annuals requires three. The manager has indicated she wants no more than four perennials in the retail outlet.

(a) Write the linear programming problem.

(b) Solve the problem graphically. Identify the feasible region, extreme points and optimal solution.

(c) Formulate the dual problem.

(d) Show that the dual of the dual is the original problem.

(e) Using computer software, solve the dual. Interpret the results.

E 16.2. (Diet cost minimization) You operate your own catering business and have been hired to provide *hor d'oeuvres* for a small reception. Your client has requested that you bring vegetable trays and fruit trays for the function. You are required to provide a balanced selection and must have a total of six units of vitamin C and ten units of vitamin B1 in your trays. Each fruit tray contributes one unit of vitamin C and one unit of vitamin B1. Each vegetable tray contributes one unit of vitamin B1. You wish to minimize the costs of providing the trays when each vegetable and fruit tray costs $2 and $4, respectively, to prepare.

(a) Write the linear programming problem.

(b) Solve the problem graphically. Identify the feasible region, extreme points and optimal solution.

(c) Formulate the dual problem.

(d) Verify that the value of the dual objective function is equal to the value of the original objective function.

(e) Show that the dual of the dual is the original problem.

E 16.3. (Financial planning) Beddings, Inc. is a company specializing in the production of pillows and comforters. It wishes to maximize its profit for each month's production. Comforters contribute $40 per month to profit while requiring $75 per month of labor to produce. Pillows require $50 per month of labor to produce and contribute $25 to the monthly profit. Demand for comforters and pillows is large enough to guarantee that everything produced will be sold. Beddings, Inc. has enough raw material each month to produce 80 comforters and 125 pillows. Beddings' cash balance on February 1 is $8500. Beddings' cash balance at the end of February must be $3000. During February, they must pay $1500 in rent and $1000 in loan payments. They will receive $3500 in accounts receivable. Beddings needs to determine how many comforters and pillows to produce in February.

(a) Write and solve the problem and its dual.

(b) Interpret the results of the problem and its dual.

(c) How much more profit can be achieved if Beddings management relaxes (1) the first constraint by one comforter, (2) the second constraint by one pillow and (3) the third constraint by $10.

E 16.4.   (a) Show that if a basis $\mathbf{B}$ has as its jth column the ith
                column of the associated identity matrix $\mathbf{I}_m$ in $\mathbf{A}^*$, then
                $\mathbf{B}^{-1}$ has as its ith column the vector containing all zeros
                except for a 1 in the jth position, as as a consequence
                that y_i in $\mathbf{y}' = \mathbf{c}'_B \mathbf{B}^{-1}$ is zero.

          (b) Show that if $\mathbf{B}$ has as its jth column the ith column of
                the associated identity matrix $\mathbf{I}_m$ in $\mathbf{A}^*$, then $\mathbf{B}^{-1}$ has
                as its ith column the vector containing all zeros except
                for a 1 in the jth position, and as a consequence that y_i
                in $\mathbf{y}' = \mathbf{c}'_B \mathbf{B}^{-1}$ is zero.

Chapter 17

MARKOV CHAIN MODELS

Markov process models are a class of probability models used to study the evolution of a system over time. Transition probabilities are used to identify how a system evolves from one time period to the next. A Markov chain is the behavior of the system over time, as described by the transition probabilities and the probability of the system being in various states. A unique property of a Markov chain is that the probability of going from one state to any other state is independent of how the chain reached its current state. If this property is present in a continuous-state system, the system is a Markov process. If the states are discrete, we refer to the system as a Markov chain. We deal only with Markov chains in this chapter.

17.1 Illustration   (Market Share)

Suppose you have been hired by a large consumer products firm to assess market share and customer loyalty for two kinds of pork and beans, Molasses and Barbeque, the only two pork and beans on the market. We will assume that a customer purchases one type of beans each week. We will refer to the weekly purchases of beans as *trials of the process*. Hence at each trial the customer will buy either Molasses or Barbeque. The *state of the system* is the specific type purchased in a given week, where we define the two states as

State 1:   The customer purchases Molasses

and

State 2:   The customer purchases Barbeque.

Using a Markov chain model we can determine the probability that a customer will purchase each type of beans in any trial. To determine the probabilities of the states occurring at trials of the Markov chain we need information on the probability that a customer purchases the same type of beans each time or switches to a competing product as the process continues from trial to trial. Suppose we collect data and find that 90% of the customers who purchased Molasses in a given period purchased the same beans in a subsequent period, while 10% switched to Barbeque. Data also show that 80% of the customers who purchase Barbeque in a given week purchase it the following week. The other 20% of the customers who purchase Barbeque switch to Molasses. These probabilities, shown in matrix **P**, are the transition probabilities because they are the probabilities that a customer makes a transition from a state in a given period to each state in the following period. The transition probability matrix is

$$\mathbf{P} = \begin{bmatrix} .9 & .1 \\ .2 & 8 \end{bmatrix}.$$

Note that the sum of each row is equal to 1. We will assume that the transition probabilities are the same for every customer and that they will not change over time.

To determine the state of the system in subsequent periods, we must introduce *state probability vector*, $\mathbf{x}'_0$. This vector represents the probability of the system being in a particular state in the initial time period. Hence

$$\mathbf{x}'_0 = [1 \quad 0]$$

identifies that in the initial time period the customer purchased Molasses. We can determine the state probabilities for the next time period by post-multiplying the state vector of the initial period by the transition probability matrix **P**. Hence

$$\mathbf{x}'_1 = \mathbf{x}'_0 \mathbf{P} = [1 \quad 0] \begin{bmatrix} .9 & .1 \\ .2 & .8 \end{bmatrix} = [.9 \quad .1]. \qquad (17.1)$$

The probability that a customer, who purchased Molasses initially, will buy Molasses again in period 1 is .9, while the probability that a customer who purchased Molasses initially will switch to Barbeque in period 1 is .1.

For the second week we see the probability of buying Molasses is .83, while the probability of buying Barbeque is .17. This is because, similar to (17.1),

$$\mathbf{x}_2' = \mathbf{x}_1'\mathbf{P} = \mathbf{x}_0'\mathbf{P}^2 = [.83 \quad .17].$$

A simple extension of this is that the state probabilities for period n are given by

$$\mathbf{x}_n' = \mathbf{x}_0'\mathbf{P}^n.$$

The state probabilities for future periods, beginning initially with the purchase of Molasses, are shown in Table 17.1.

Table 17.1. State Probabilities for the First Five Time Periods
Starting with Purchasing Molasses

Period	Molasses	Barbeque
0	1	0
1	.9	.1
2	.83	.17
3	.78	.22
4	.75	.25
5	.72	.28

17.2 Steady-State Probabilities

An important aspect of a Markov chain is its long-run behavior; i.e., its behavior after the initial effects have worn off. If, after a large number of periods, the state vector $\mathbf{x}_n'$ is equal to the state vector at period $n+1$; i.e., $\mathbf{x}_{n+1}'$, and $\mathbf{x}_n'$ is independent of the initial state vector $\mathbf{x}_0'$, then we have reached $\mathbf{x}' = \mathbf{x}_n' = \mathbf{x}_{n+1}'$, the *steady-state vector* for the Markov chain described by $\mathbf{P}$. The elements of $\mathbf{x}'$ are called the *steady-state probabilities*, given by

$$\mathbf{x}' = \mathbf{x}'\mathbf{P}. \tag{17.2}$$

Illustration (*continued*)

If the customer had purchased Barbeque pork and beans in the initial time period, the state vectors would be

$$\mathbf{x}_0' = [0 \ 1]$$

$$\mathbf{x}_1' = [0 \ 1] \begin{bmatrix} .9 & .1 \\ .2 & .8 \end{bmatrix} = [.2 \ .8]$$

$$\mathbf{x}_2' = [.2 \ .8] \begin{bmatrix} .9 & .1 \\ .2 & .8 \end{bmatrix} = [.34 \ .66].$$

For periods 3 through 10 see Table 17.2. Notice that it appears the steady-state vector will converge to $[2/3 \ \ 1/3]$ regardless of the initial state.

We can calculate the steady-state probabilities quite easily. Given the matrix $\mathbf{P}$ we can write (17.2) as

$$\mathbf{x}' = \mathbf{x}' \begin{bmatrix} .9 & .1 \\ .2 & .8 \end{bmatrix}.$$

Table 17.2. State Probabilities for Customers
Purchasing Pork and Beans

Period (10)	Customer Initially Purchasing Molasses Pork and Beans $\mathbf{x}_n'$		Customers Initially Purchasing Barbeque Pork and Beans $\mathbf{x}_n'$	
0	[1	0]	[0	1]
1	[.9	.1]	[.2	.8]
2	[.83	.17]	[.34	.66]
3	[.781	.219]	[.438	.562]
4	[.747	.253]	[.507	.493]
5	[.723	.277]	[.555	.445]
6	[.706	.294]	[.589	.411]
7	[.694	.306]	[.612	.388]
8	[.686	.314]	[.628	.372]
9	[.680	.320]	[.640	.360]
10	[.676	.324]	[.648	.352]

Assuming $\mathbf{x}_1' = [\pi_1 \ \ \pi_2]$ gives

$$\pi_1 = .9\pi_1 + .2\pi_2$$
$$\pi_2 = .1\pi_1 + .8\pi_2. \tag{17.3}$$

We also know that
$$\pi_1 + \pi_2 = 1. \tag{17.4}$$

Using (17.4), substituting into (17.3) and simplifying yields $\pi_1 = \frac{2}{3}$, $\pi_2 = \frac{1}{3}$.

Alternatively we see that one of the equations in (17.3) is redundant because they both simplify to

$$\pi_1 = (.2/.1)\pi_2 = 2\pi_2. \tag{17.5}$$

We can write (17.5) and (17.4) as

$$\begin{bmatrix} 1 & -2 \\ 1 & 1 \end{bmatrix} \begin{bmatrix} \pi_1 \\ \pi_2 \end{bmatrix} = \begin{bmatrix} 0 \\ 1 \end{bmatrix}, \tag{17.6}$$

which has the solution

$$\begin{bmatrix} \pi_1 \\ \pi_2 \end{bmatrix} = \begin{bmatrix} 1 & -2 \\ 1 & 1 \end{bmatrix}^{-1} \begin{bmatrix} 0 \\ 1 \end{bmatrix} = \begin{bmatrix} 1/3 & 2/3 \\ -1/3 & 1/3 \end{bmatrix} \begin{bmatrix} 0 \\ 1 \end{bmatrix} = \begin{bmatrix} \frac{2}{3} \\ \frac{1}{3} \end{bmatrix}.$$

If we have 1000 customers buying pork and beans the Markov model tells us that two-thirds of those customers, or 667 of them, would purchase Molasses and one-third of the customers or 333 of them would purchase Barbeque. These steady-state probabilities are the market shares for the two products.

This market share information is often helpful in evaluating the impact of firm policies, such as advertising, and can often lead to better decision making. Suppose a consumer goods firm begins an advertising campaign to attract Molasses consumers to purchase more Barbeque. Suppose this firm believes its strategy will increase the probability of Molasses consumers switching to Barbeque from .1 to .3. The new transition probability matrix will then be not the **P** of (17.1) but

$$\mathbf{P} = \begin{bmatrix} .7 & .3 \\ .2 & .8 \end{bmatrix}.$$

The new steady-state probabilities can be found from

$$\begin{aligned} \pi_1 &= .7\pi_1 + .2\pi_2 \\ \pi_2 &= .3\pi_1 + .8\pi_2 \end{aligned}$$

and

$$\pi_1 + \pi_2 = 1.$$

Simplifying and expressing in matrix notation

$$\begin{bmatrix} 1 & -\frac{2}{3} \\ 1 & 1 \end{bmatrix} \begin{bmatrix} \pi_1 \\ \pi_2 \end{bmatrix} = \begin{bmatrix} 0 \\ 1 \end{bmatrix}$$

$$\begin{bmatrix} \pi_1 \\ \pi_2 \end{bmatrix} = \begin{bmatrix} 1 & -\frac{2}{3} \\ 1 & 1 \end{bmatrix}^{-1} \begin{bmatrix} 0 \\ 1 \end{bmatrix} = \begin{bmatrix} .4 \\ .6 \end{bmatrix}.$$

The advertising increased the market share of Barbeque from .33 to .6. With our customer base of 1000, the number of customers purchasing Barbeque would increase from 333 to 600, while the customers purchasing Molasses would decrease from 667 to 400. If Barbeque sold for $3/can and Molasses sold for $2/can, the consumer products firm could expect an increase of revenue of $267. If the advertising strategy cost less than $267, the firm would implement the marketing effort.

17.3   Transient States

A *transient state* in a Markov chain is a state with a steady-state probability of 0. Consider a Markov chain model with the transition probability

$$\mathbf{P} = \begin{bmatrix} .2 & .8 \\ 0 & 1 \end{bmatrix}.$$

Once the system reaches the second state it remains there forever since the $p_{22} = 1$ and $p_{21} = 0$. In this case State 1 is the transient state. The steady-state probability vector is $\mathbf{x}' = \begin{bmatrix} 0 & 1 \end{bmatrix}$. State 2 is called the *trapping state* or an *absorbing state*.

17.4   Periodic or Cycling Behavior

A Markov chain exhibiting periodic or cyclic behavior is said to be *periodic*.

Example

The transition probability matrix

$$\mathbf{P} = \begin{bmatrix} 0 & 1 & 0 \\ 0 & 0 & 1 \\ 1 & 0 & 0 \end{bmatrix}$$

generates a Markov chain exhibiting cyclic behavior. The state probabilities for large n do not tend to specific values independent of the steady state. Hence, steady-state probabilities for this chain do not exist. Suppose

$$\mathbf{x}_0' = [1 \ 0 \ 0].$$

Then

$$\begin{aligned}
\mathbf{x}_1' &= \mathbf{x}_0'\mathbf{P} = [0 \ 1 \ 0] \\
\mathbf{x}_2' &= \mathbf{x}_1'\mathbf{P} = [0 \ 0 \ 1] \\
\mathbf{x}_3' &= \mathbf{x}_2'\mathbf{P} = [1 \ 0 \ 0] = \mathbf{x}_0' \\
\mathbf{x}_4' &= \mathbf{x}_3'\mathbf{P} = [0 \ 1 \ 0] = \mathbf{x}_1' \\
\mathbf{x}_5' &= \mathbf{x}_2' \\
\mathbf{x}_6' &= \mathbf{x}_3'
\end{aligned}$$

and so on. Even though $\mathbf{x}' = [1/3 \ \ 1/3 \ \ 1/3]$ satisfies the equation $\mathbf{x}' = \mathbf{x}'\mathbf{P}$, $\mathbf{x}'$ is not the steady-state probability vector. The state of the system is predictable if one knows the starting point. If one starts in State 1 at the initial time period, we know the system is in State 1 in periods 3, 6, 9, ... and in State 2 in periods 1, 4, 7, ... and in State 3 in period 2, 5, 8, ... The probability vector, $\mathbf{x}'$, for a periodic system identifies the likelihood of the system being in a particular state in a future period if the probabilities of the system starting in each state are equal.

17.5 Recurrent Sets

A Markov chain with one or more transient sets does not always have a trapping state. There may be a set of states such that once a system enters the set it always makes transitions within the set. This type of set is called a *recurrent set*.

Example

For the transition probability matrix

$$\mathbf{P} = \begin{bmatrix} .3 & .7 & 0 & 0 \\ .6 & .4 & 0 & 0 \\ 0 & .2 & .3 & .5 \\ .1 & .2 & .4 & .3 \end{bmatrix},$$

once the system reaches either State 1 or State 2 it remains in one of these states. States 1 and 2 form a recurrent set. The probability of going from either State 1 or State 2 to any other state is zero.

A recurrent set with a single state is a *trapping set*. A recurrent set with more than one state is called a *generalized trapping set*.

A Markov chain is *ergodic* if the entire set of states is a recurrent set. In other words it is possible to go from any state to any other state in a finite number of transitions.

If a Markov chain is not ergodic, there may be more than one recurring set. In these chains, i.e., ones with more than one recurrent set, steady-state probabilities do not exist.

Example

A system, characterized by the transition probability matrix

$$\mathbf{P} = \begin{bmatrix} 0 & 1 & 0 & 0 \\ 1 & 0 & 0 & 0 \\ 0 & .3 & 0 & .7 \\ 0 & 0 & 0 & 1 \end{bmatrix},$$

has one recurrent set (States 1 and 2), a transient state (State 3) and a trapping state (State 4). If the system started in States 1 or 2, it would remain there forever cycling between the two states. If the system started in State 3, it would have a 30% chance of going to the first recurrent set and a 70% chance of being trapped in State 4. If the system started in State 4 it would stay there forever.

17.6 Existence of Steady-State Probabilities

If a system contains more than one recurrent set then steady-state probabilities do not exist. Even if a system is ergodic the steady-state probabilities do not always exist.

A Markov chain is *regular* if there exists some positive integer n such that every element of the matrix $\mathbf{P}^n$ except those relating to transient states is greater than zero. If a chain is regular, it can be shown that as n grows large the matrix $\mathbf{P}^n$ approaches a limiting form (call it $\mathbf{P}^*$) in which all rows are identical and equal to the steady-state probability vector $\mathbf{x}'$. The steady-state probabilities for a system exist if, and only if, the system is regular.

Example

Consider the system described by the matrix

$$\mathbf{P} = \begin{bmatrix} .6 & 0 & .4 \\ .3 & .5 & .2 \\ .7 & 0 & .3 \end{bmatrix}.$$

State 2 is a transient state whose limiting probability is zero. Then **P**, with the probabilities relating to State 2 eliminated from consideration, satisfies the definition of a regular system. The steady-state probabilities for States 1 and 3 do indeed exist. As n grows large the matrix $\mathbf{P}^n$ approaches

$$\mathbf{P}^* = \begin{bmatrix} \frac{7}{11} & 0 & \frac{4}{11} \\ \frac{7}{11} & 0 & \frac{4}{11} \\ \frac{7}{11} & 0 & \frac{4}{11} \end{bmatrix},$$

a matrix with all rows identical and equal to the steady-state probability vector $\mathbf{x}' = \mathbf{x}'\mathbf{P}$.

Example

Consider the matrix

$$\mathbf{P} = \begin{bmatrix} .7 & .3 & 0 \\ 0 & .8 & .2 \\ .4 & .3 & .3 \end{bmatrix};$$

there are no transient states. For $n = 2$

$$\mathbf{P}^2 = \begin{bmatrix} .49 & .45 & .06 \\ .08 & .70 & .22 \\ .40 & .45 & .15 \end{bmatrix}.$$

Note that every element is greater than zero. This Markov chain is regular and steady-state probabilities will exist.

17.7 Rewards in Markov Chains

We can add information on rewards or payoffs that can be obtained from the transitions in a Markov chain. Let r_{ij} be the reward associated with a transition from State i to State j; it may be interpreted as the reward from the transition itself or the reward for being in State i (or State j) during one time period. The first interpretation would be appropriate if the states were locations in a city and the transitions were taxi rides; r_{ij} would be the profit from the fare collected in going from location i to location j. The second interpretation would be appropriate if the states were the alternative conditions of a machine; r_{ij} might be the profit earned from being in State i for the period of time before the transition.

Suppose there are N states in the system. We can define the matrix of rewards, $\mathbf{R}$, to be

$$\mathbf{R} = \begin{bmatrix} r_{11} & r_{12} & \cdots & r_{1N} \\ r_{21} & \cdots & & r_{2N} \\ \vdots & & & \vdots \\ r_{N1} & \cdots & & r_{NN} \end{bmatrix}.$$

Suppose the system starts in State i. Let $v_i(n)$ be the total expected reward after n transitions, beginning from State i. Then

$$[\mathbf{v}(n)]' = [v_1(n) \quad v_2(n) \quad \cdots \quad v_N(n)]$$

is the vector of total expected rewards over n transitions for each of the N possible starting states of the system.

Suppose that at the first transition the system moves to State j. The reward for this first transition is r_{ij}. The expected reward over all n transitions can be expressed as $r_{ij} + v_j(n-1)$, where $v_j(n-1)$ is the expected reward over the remaining $n-1$ transitions when the system begins in State i and first moves to State j. The total expected reward after n transitions, starting in State i, may be expressed as

$$v_i(n) = \sum_{j=1}^{N} p_{ij} r_{ij} + \sum_{j=1}^{N} p_{ij} v_j(n-1), \qquad (17.7)$$

where p_{ij} are elements of $\mathbf{P}$. If $q_i = \sum_{j=1}^{N} p_{ij} r_{ij}$ represents the expected reward from the next transition when i is the current state then

$$v_i(n) = q_i + \sum_{j=1}^{N} p_{ij} v_j(n-1).$$

We can write this in vector form as

$$\mathbf{v}(n) = \mathbf{q} + \mathbf{P}\mathbf{v}(n-1), \qquad (17.8)$$

where $\mathbf{q}' = [q_1 \quad q_2 \quad \cdots \quad q_N]$.

If a Markov chain is regular then steady-state probabilities exist. A steady-state expected reward per period, $\mathbf{g}$ can be expressed as the weighted sum of the expected rewards g_i from transitions from State i weighted by the steady-state probabilities, π_i, of being in State i,

$$\mathbf{g} = \mathbf{x}'\mathbf{q}. \tag{17.9}$$

Illustration   (Machine operation)

A machine is either operating properly (State 1) or in need of adjustment (State 2). If a machine is operating properly today, the probability that it will operate properly tomorrow is 0.7. The probability that it will need adjustment tomorrow is 0.3. If the machine needs adjustment today the probability that it will operate properly tomorrow is 0.6; the probability that it still needs adjustment tomorrow is .4. The transition probability matrix is given by

$$\mathbf{P} = \begin{bmatrix} .7 & .3 \\ .6 & .4 \end{bmatrix}.$$

When the machine operates properly before and after a transition, the profit is \$4. When the machine operates properly but then needs adjustment after a transition, the profit is \$2. When a machine needs adjustment and then operates properly, the profit is \$2. Finally, if a machine is in need of adjustment before and after a transition, a loss of \$2 will be incurred. The reward matrix can be expressed as

$$\mathbf{R} = \begin{bmatrix} 4 & 2 \\ 2 & -2 \end{bmatrix}.$$

We can determine

$$\mathbf{PR}' = \begin{bmatrix} .7 & .3 \\ .6 & .4 \end{bmatrix} \begin{bmatrix} 4 & 2 \\ 2 & -2 \end{bmatrix} = \begin{bmatrix} 3.4 & .8 \\ 3.2 & .4 \end{bmatrix}$$

and from (17.7)

$$\mathbf{q} = [3.4 \quad .4].$$

Therefore

$$\mathbf{v}(n) = \begin{bmatrix} 3.4 \\ .4 \end{bmatrix} + \begin{bmatrix} .7 & .3 \\ .6 & .4 \end{bmatrix} \mathbf{v}(n-1).$$

If $\mathbf{v}(0) = 0$ then

$$\mathbf{v}(1) = \begin{bmatrix} 3.4 \\ .4 \end{bmatrix} + \begin{bmatrix} .7 & .3 \\ .6 & .4 \end{bmatrix} \mathbf{0} = \begin{bmatrix} 3.4 \\ .4 \end{bmatrix}.$$

If the machine starts out in working condition the expected profit after one time period is \$3.40. If the machine is in need of adjustment the expected profit after one time period is only 40 cents. After two time periods we see that the expected profit is

$$\mathbf{v}(2) = \begin{bmatrix} 3.4 \\ .4 \end{bmatrix} + \begin{bmatrix} .7 & .3 \\ .6 & .4 \end{bmatrix} \begin{bmatrix} 3.4 \\ .4 \end{bmatrix} = \begin{bmatrix} 5.9 \\ 2.6 \end{bmatrix}.$$

Note that we can determine the expected profit for any n time periods by using (17.8) recursively. Hence for $n = 5$, the expected profit for five periods is

$$\mathbf{v}(5) = \mathbf{q} + \mathbf{Pq} + \mathbf{P}^2\mathbf{q} + \mathbf{P}^3\mathbf{q} + \mathbf{P}^4\mathbf{q} + \mathbf{P}^5\mathbf{v}_0$$

$$= \begin{bmatrix} 13.11 \\ 9.78 \end{bmatrix}.$$

The steady-state expected award is

$$\mathbf{q} = \mathbf{x}'\mathbf{q} = \begin{bmatrix} 2/3 & 1/3 \end{bmatrix} \begin{bmatrix} 3.4 \\ .4 \end{bmatrix} = 2.40.$$

Thus if the system has been operating for a long period of time and the state of the system is unknown, the expected return per period in the next and every subsequent period is \$2.40.

17.8 Additional Applications of Markov Chains

Markov chains can also be used when decision making is introduced to a system. For example, in our illustration one can introduce the decision to always repair the machine when it needs adjustment. One can decide if this decision rule should be followed or not. The introduction of decision rules in Markov chains may affect both the transition probability matrix and the reward matrix. In addition, one must assume that the optional decisions are made in each of the periods and that the maximum expected reward is achieved. We can determine the decision rule that will maximize the steady-state expected reward per period. Furthermore, we could also take into account the time value of money by discounting future returns. These applications, and additional ones, can be found in books dedicated to the study of Markov chains, such as Puterman (1994), Norris (1997) and Bremaud (1999).

17.9 Exercises

E 17.1. (Commuting) Travel to Washington, DC from Virginia requires crossing the Potomic River. Traffic for commuters can be in one of two states: delay or no delay. Consider one period to be 15 minutes. Assume that the probability of no traffic delay in one period, given no traffic delay in the preceding period, is .85. The probability of finding a traffic delay in one period given a delay in the preceding period is .75. Assume you are ready to begin your commute to Washington and hear on the radio there is a traffic delay.

(a) What is the probability that for the next 30 minutes the traffic will be delayed?

(b) What is the probability that in the long run the traffic will not be delayed?

(c) A key assumption of Markov chain models presented in this chapter is that the transition probabilities are constant. Is this assumption valid in this problem? Why or why not?

E 17.2. (City migration) Three percent of the individuals living within the Saint Louis city limits move to the suburbs each year. During the same time period only 1% of the individuals move from the suburbs to the city. Assume this process is modeled as a Markov chain with two states: city and suburbs.

(a) Show the matrix of transition probabilities.

(b) Compute the steady-state probabilities.

(c) Suppose 35% of the population live in Saint Louis, while 65% live in the suburbs. What steady-state population changes can you project to Saint Louis?

E 17.3. (Travel) Each summer the Willetts take a vacation. They never take the same kind of vacation two years in a row. They flip a coin to choose which of the two types of vacation they didn't take the previous year they will take that year. The three children prefer a visit to grandparents. Mom prefers a camping trip and Dad prefers a visit to the beach. What proportion of the time will the Willetts visit Grandma and Grandpa, go camping, or go to the beach?

E 17.4. (Machine operation) Two machines, A and B, are candidates for leasing at the same cost. The two machines have differing transition probability matrices for changing from an "operating properly" state (State 1) to a "requires adjustment" state (State 2) as follows:

Machine A   Machine B

$$P_A = \begin{bmatrix} .9 & .1 \\ .6 & .4 \end{bmatrix} \qquad P_B = \begin{bmatrix} .8 & .2 \\ .7 & .3 \end{bmatrix}.$$

Derive the steady-state probabilities for each machine. Which machine would be most desirable to lease?

E 17.5. (Weather) The Snowflake Ski Resort has found that after a clear day the probability of stormy weather is .3, while after a stormy day the probability of clear weather is .8. Write down the transition probability matrix. What are the steady-state probabilities for clear and stormy days?

E 17.6. (Travel) The ACME Car Rental Company rents cars from three airports: A, B and C. Customers return cars to each of the airports according to the table below.

TABLE OF TRANSITION PROBABILITIES

To:	A	B	C
From:			
A	.8	.2	0
B	.2	0	.8
C	.2	.2	.6

(a) Calculate the vector $\mathbf{x}'$ which satisfies $\mathbf{x}' = \mathbf{x}'\mathbf{P}$ and $\sum_{i=1}^{3} x_i = 1$. Does this vector represent the steady-state probabilities? Justify your answer.

(b) The ACME Company is planning to build a maintenance facility at one of the three airports. Which airport would you recommend? Why?

E 17.7. Consider the following general two-state transition probability matrix:

$$\mathbf{P} = \begin{bmatrix} p_{11} & p_{12} \\ p_{21} & p_{22} \end{bmatrix}.$$

Derive the elements of the steady-state probability vector $\mathbf{x}' = \begin{bmatrix} \pi_1 & \pi_2 \end{bmatrix}$ in terms of the elements of $\mathbf{P}$. (Recall that the elements in each row of $\mathbf{P}$ sum to 1.)

E 17.8. (Stock market) The behavior of stock market prices exhibits a systematic tendency for transactions in which the price change is in one direction to be followed by transactions in which the price change is in the opposite direction. Suppose the conditional probability of a price increase, given that the previous change was a decrease, is .75, as is the conditional probability of a decrease given a previous increase. Define two relevant states and write down a transition probability matrix for this system. What are the steady-state probabilities?

E 17.9. Suppose the state description in E 17.8 was defined as the current price of the stock. Would this description violate the Markovian property of "no memory"? Describe using the information in E 17.8.

E 17.10. Consider a Markov chain with the following transition matrix:

$$\mathbf{P} = \begin{bmatrix} .7 & .3 & 0 \\ .8 & .2 & 0 \\ 0 & .2 & .8 \end{bmatrix}.$$

(a) If the process starts in State 1 and a very large number of transitions occur, what fraction of these transitions are from State 1 to State 2? *Hint*: First calculate the steady-state probability of being in State 1.

(b) Repeat part (a) assuming the process starts in State 3.

E 17.11. (Travel) A mathematically oriented cab driver has found that when he is in Town 1 there is a .8 probability that his next fare will take him to Town 2; otherwise he will stay in Town

1. When he is in Town 2 there is a .4 probability that his next fare will take him to Town 1; otherwise his next fare will keep him in Town 2. The average profit for each type of trip is as follows.: Between Towns: \$2 (either direction). Within either Town: \$1.

(a) What is the transition probability matrix P for two states?

(b) What is the reward matrix R?

(c) Compute the steady-state probabilities of being in the two towns.

(d) Calculate the vector of expected rewards from the next transition, q.

(e) Calculate the steady-state expected reward per period, g.

E 17.12. For any transition probability matrix P, show for any positive integer n

(a) that the row sums of P^n are 1.0;

(b) that if the elements of x sum to unity then so do the elements of $x'P^n$. *Hint*: Utilize a vector 1 whose every element is unity.

E 17.13. Verify that the matrix $(I - P)$, for P being any transition probability matrix, is singular.

References

Aitken, A. C. (1948) *Determinants and Matrices*, Oliver & Boyd, Edinburgh.

Bazaraa, M. and Jarvis, J. (1990) *Linear Programming and Network Flows*, John Wiley & Sons, New York.

Ben-Israel, A. and Greville, N. E. (1974) *Generalized Inverses: Theory and Applications*, John Wiley Sons, New York.

Bernadelli, H. (1941) Population waves, *Journal of Burma Research Society* **31**, 1–18.

Boullion, T. L. and Odell, P. L. (1971) *Generalized Inverse Matrices*, John Wiley & Sons, New York.

Bremaud, P. (1999) *Markov Chains: Gibbs Fields, Monte Carlo Simulation and Queues*, Springer-Verlag, New York.

Casella, G. and Berger, R. L. (1990) *Statistical Inference*, Springer-Verlag, New York.

Dantzig, G. (1963) *Linear Programming and Extensions*, Princeton University Press, Princeton, N.J.

Doolittle, M. H. (1878) Method employed in the solution of normal equations and the adjustment of triangulation, *U.S. Coast and Geodetic Survey Report 1878*, 59–61.

Draper, N. and Smith, H. (1981) *Applied Regression Analysis*, 2nd ed., John Wiley & Sons, New York.

Ferrar, W. L. (1941) *Algebra: A Textbook of Determinants, Matrices and Algebraic Forms*, Oxford University Press, Oxford.

Gass, S. (1985) *Linear Programming: Methods and Applications*, 5th ed., McGraw-Hill, New York.

Golub, G. H. and Van Loan, C. F. (1989) *Matrix Computations*, 3rd ed., Johns Hopkins University Press, Baltimore.

Greene, W. H. (1997) *Econometric Analysis*, 3rd ed., Prentice-Hall, Upper Saddle River, N.J.

Griffiths, W. E., Hill, R. C. and Judge, G. G. (1993) *Learning and Practicing Econometrics*, John Wiley & Sons, New York.

Gujariti, D. N. (1995) *Basic Econometrics*, 3rd ed., McGraw-Hill, New York.

Harville, D. A. (1997) *Matrix Algebra from a Statistican's Perspective*, Springer-Verlag, New York.

Henderson, H. V. and Searle, S. R. (1979) Vec and vech operators for matrices with some uses in Jacobians and multivariate statistics, *Canadian* Journal of Statistics **7**, 65–81.

Henderson, H. V. and Searle, S. R. (1981a) On deriving the inverse of a sum of matrices, *SIAM Review* **23**, 53–60.

Henderson, H. V. and Searle, S. R. (1981b) The vec-permutation matrix, the vec operator and Kronecker products: a review, *Linear and Multilinear Algebra* **9**, 271–88.

Judge, G. G., Hill, R. C., Griffiths, W. E., Lutkepohl, H. and Lee, T. C. (1988) *Introduction to the Theory and Practice of Econometrics*, 2nd ed., John Wiley & Sons, New York.

Keynes, J. M. (1951) Alfred Marshall: 1842–1924 in *Essays in Biography*, Rubert Hart-Davis, London, 140–41. Preface.

Kmenta, J. (1986) *Elements of Econometrics*, 2nd ed., Macmillan, New York.

Luenberger, D. G. (1984) *Linear and Nonlinear Programming*, 2nd ed., Addison-Wesley, Reading, Mass.

McCullough, B. D. (1998) Statistical computing software reviews assessing the reliability of statistical software: Part I, *The American Statistician* **52**, 358–66.

McCullough, B. D. and Vinod, H. D. (1999) The numerical reliability of econometrics software, *Journal of Economic Literature* **37**, 633–65.

Moore, E. H. (1920) On the reciprocal of the general algebraic matrix, *Bulletin of the American Mathematical Society* **26**, 394–95.

Myers, R. H. and Milton, J. S. (1991) *A First Course in the Theory of*

Linear Statistical Models, PWS-Kent Publishing Company, Boston.

Norris, J. R. (1997) *Markov Chains*, Cambridge University Press, New York.

Penrose, R. A. (1955) A generalized inverse for matrices, *Proceedings, Cambridge Philosophical Society* **51**, 406–13.

Pringle, R. M. and Rayner, A. A. (1971) *Generalized Inverse Matrices with* Applications to Statistics, Griffin, London.

Puterman, M. L. (1994) *Markov Decision Processes: Discrete Stochastic Dynamic Programming*, John Wiley & Sons, New York.

Rao, C. R. (1962) A note on the generalized inverse of a matrix with applications to problems in statistics, *Journal of the Royal Statistical Society (B)* **24**, 152–58.

Rao, C. R. and Mitra, S. K. (1971) *Generalized Inverse of Matrices and Its Applications*, John Wiley & Sons, New York.

Samuelson, P. A. (1939) Interactions between the multiplier analysis and the principle of acceleration, *Review of Economic Statistics* **21**, 75–78.

Scheffé, H. (1999) *The Analysis of Variance*, John Wiley & Sons, New York.

Searle, S. R. (1966) *Matrix Algebra for the Biological Sciences*, John Wiley & Sons, New York.

Searle, S. R. (1971) *Linear Models*, John Wiley & Sons, New York.

Searle, S. R. (1982) *Matrix Algebra Useful for Statistics*, John Wiley & Sons, New York.

Searle, S. R. (1997) *Linear Models*, 2nd ed., John Wiley & Sons, New York.

Seelye, C. J. (1958) Conditions for a positive definite quadratic form established by induction, *American Mathematics Monthly* **65**, 355–56.

Silberberg, E. (1990) *The Structure of Economics: A Mathematical Analysis*, 2nd ed., McGraw-Hill, New York.

Stapleton, J. H. (1995) *Linear Statistical Models*, John Wiley & Sons, New York.

Stewart, G. W. (1973) *An Introduction to Matrix Computations*, Academic Press, New York.

Stigler, G. (1945) The cost of subsistence, *Journal of Farm Economics* **27**, 303–14.

Urquhart, N. S. (1969) The nature of the lack of uniqueness of generalized inverse matrices, *Society for Industrial and Applied Mathematics Review* **11**, 268–71.

U.S. Department of Agriculture, National Agricultural Statistics Service (1999) *Agricultural Statistics*, retrieved from the World Wide Web, April 15, 1999. <http://www.usda.gov/nass/pubs/agstats.htm>

U.S. Government Printing Office (1999) *Economic Report of the President*, retrieved from the World Wide Web, April 15, 1999. <http://w3.access.gpo.gov/usbudget/fy2000/pdf/1999_erp.pdf>

Varian, H. R. (1992) *Microeconomic Analysis*, 3rd ed., W.W. Norton & Co., New York.

Winston, W. L. (1995) *Introduction to Mathematical Programming Application and Algorithms*, Wadsworth Publishing Co., Belmont, Calif.

Index

WILEY SERIES IN PROBABILITY AND STATISTICS
ESTABLISHED BY WALTER A. SHEWHART AND SAMUEL S. WILKS

Wiley Series in Probability and Statistics is well established and authoritative. It covers many topics of current research interest in both pure and applied statistics and probability theory. Written by leading statisticians and institutions, the titles span both state-of-the-art developments in the field and classical methods.

Reflecting the wide range of current research in statistics, the series encompasses applied, methodological and theoretical statistics, ranging from applications and new techniques made possible by advances in computerized practice to rigorous treatment of theoretical approaches.

This series provides essential and invaluable reading for all statisticians, whether in academia, industry, government, or research.

*Now available in a lower priced paperback edition in the Wiley Classics Library.

*Now available in a lower priced paperback edition in the Wiley Classics Library.

*COX · Planning of Experiments

CRESSIE · Statistics for Spatial Data, *Revised Edition*

CSÖRGŐ and HORVÁTH · Weighted Approximations in Probability Statistics

CSÖRGŐ and HORVÁTH · Limit Theorems in Change Point Analysis

DANIEL · Applications of Statistics to Industrial Experimentation

DANIEL · Biostatistics: A Foundation for Analysis in the Health Sciences, *Sixth Edition*

*DANIEL · Fitting Equations to Data: Computer Analysis of Multifactor Data, *Second Edition*

DAVID · Order Statistics, *Second Edition*

*DEGROOT, FIENBERG, and KADANE · Statistics and the Law

DETTE and STUDDEN · The Theory of Canonical Moments with Applications in Statistics, Probability, and Analysis

DEY and MUKERJEE · Fractional Factorial Plans

DILLON and GOLDSTEIN · Multivariate Analysis: Methods and Applications

DODGE · Alternative Methods of Regression

*DODGE and ROMIG · Sampling Inspection Tables, *Second Edition*

*DOOB · Stochastic Processes

DOWDY and WEARDEN · Statistics for Research, *Second Edition*

DRAPER and SMITH · Applied Regression Analysis, *Third Edition*

DRYDEN and MARDIA · Statistical Shape Analysis

DUDEWICZ and MISHRA · Modern Mathematical Statistics

DUNN and CLARK · Applied Statistics: Analysis of Variance and Regression, *Second Edition*

DUNN and CLARK · Basic Statistics: A Primer for the Biomedical Sciences, *Third Edition*

DUPUIS and ELLIS · A Weak Convergence Approach to the Theory of Large Deviations

*ELANDT-JOHNSON and JOHNSON · Survival Models and Data Analysis

ETHIER and KURTZ · Markov Processes: Characterization and Convergence

EVANS, HASTINGS, and PEACOCK · Statistical Distributions, *Third Edition*

FELLER · An Introduction to Probability Theory and Its Applications, Volume I, *Third Edition,* Revised; Volume II, *Second Edition*

FISHER and VAN BELLE · Biostatistics: A Methodology for the Health Sciences

*FLEISS · The Design and Analysis of Clinical Experiments

FLEISS · Statistical Methods for Rates and Proportions, *Second Edition*

FLEMING and HARRINGTON · Counting Processes and Survival Analysis

FREEMAN and SMITH · Aspects of Uncertainty: A Tribute to D. V. Lindley

FULLER · Introduction to Statistical Time Series, *Second Edition*

FULLER · Measurement Error Models

GALLANT · Nonlinear Statistical Models

GHOSH, MUKHOPADHYAY, and SEN · Sequential Estimation

GIFI · Nonlinear Multivariate Analysis

GLASSERMAN and YAO · Monotone Structure in Discrete-Event Systems

GNANADESIKAN · Methods for Statistical Data Analysis of Multivariate Observations, *Second Edition*

GOLDSTEIN and LEWIS · Assessment: Problems, Development, and Statistical Issues

GREENWOOD and NIKULIN · A Guide to Chi-Squared Testing

GROSS and HARRIS · Fundamentals of Queueing Theory, *Third Edition*

GUTTORP · Statistical Inference for Branching Processes

*HAHN · Statistical Models in Engineering

HAHN and MEEKER · Statistical Intervals: A Guide for Practitioners

HALD · A History of Probability and Statistics and their Applications Before 1750

HALD · A History of Mathematical Statistics from 1750 to 1930

HALL · Introduction to the Theory of Coverage Processes

HAMPEL · Robust Statistics: The Approach Based on Influence Functions

*Now available in a lower priced paperback edition in the Wiley Classics Library.

*Now available in a lower priced paperback edition in the Wiley Classics Library.

*Now available in a lower priced paperback edition in the Wiley Classics Library.

*MILLER · Survival Analysis, *Second Edition*

MONTGOMERY, PECK, and VINING · Introduction to Linear Regression Analysis, *Third Edition*

MORGENTHALER and TUKEY · Configural Polysampling: A Route to Practical Robustness

MUIRHEAD · Aspects of Multivariate Statistical Theory

MURRAY · X-STAT 2.0 Statistical Experimentation, Design Data Analysis, and Nonlinear Optimization

MYERS and MONTGOMERY · Response Surface Methodology: Process and Product in Optimization Using Designed Experiments

NELSON · Accelerated Testing, Statistical Models, Test Plans, and Data Analyses

NELSON · Applied Life Data Analysis

NEWMAN · Biostatistical Methods in Epidemiology

OCHI · Applied Probability and Stochastic Processes in Engineering and Physical Sciences

OKABE, BOOTS, and SUGIHARA · Spatial Tesselations: Concepts and Applications of Voronoi Diagrams

OLIVER and SMITH · Influence Diagrams, Belief Nets and Decision Analysis

PANKRATZ · Forecasting with Dynamic Regression Models

PANKRATZ · Forecasting with Univariate Box-Jenkins Models: Concepts and Cases

*PARZEN · Modern Probability Theory and Its Applications

PEÑA, TIAO, and TSAY · A Course in Time Series Analysis

PIANTADOSI · Clinical Trials: A Methodologic Perspective

PORT · Theoretical Probability for Applications

POURAHMADI · Foundations of Time Series Analysis and Prediction Theory

PRESS · Bayesian Statistics: Principles, Models, and Applications

PRESS and TANUR · The Subjectivity of Scientists and the Bayesian Approach

PUKELSHEIM · Optimal Experimental Design

PURI, VILAPLANA, and WERTZ · New Perspectives in Theoretical and Applied Statistics

PUTERMAN · Markov Decision Processes: Discrete Stochastic Dynamic Programming

RACHEV · Probability Metrics and the Stability of Stochastic Models

RAO · Asymptotic Theory of Statistical Inference

RAO · Linear Statistical Inference and Its Applications, *Second Edition*

RAO and SHANBHAG · Choquet-Deny Type Functional Equations with Applications to Stochastic Models

RENCHER · Linear Models in Statistics

RENCHER · Methods of Multivariate Analysis

RENCHER · Multivariate Statistical Inference with Applications

RÉNYI · A Diary on Information Theory

RIPLEY · Spatial Statistics

RIPLEY · Stochastic Simulation

ROBERTSON, WRIGHT, and DYKSTRA · Order Restricted Statistical Inference

ROGERS and WILLIAMS · Diffusions, Markov Processes, and Martingales, Volume I: Foundations, *Second Edition;* Volume II: Îto Calculus

ROHATGI and SALEH · An Introduction to Probability and Statistics, *Second Edition*

ROLSKI, SCHMIDLI, SCHMIDT, and TEUGELS · Stochastic Processes for Insurance and Finance

ROSS · Introduction to Probability and Statistics for Engineers and Scientists

ROUSSEEUW and LEROY · Robust Regression and Outlier Detection

RUBIN · Multiple Imputation for Nonresponse in Surveys

RUBINSTEIN · Simulation and the Monte Carlo Method

RUBINSTEIN and MELAMED · Modern Simulation and Modeling

RUBINSTEIN and SHAPIRO · Discrete Event Systems: Sensitivity Analysis and Stochastic Optimization by the Score Function Method

*Now available in a lower priced paperback edition in the Wiley Classics Library.

RUZSA and SZEKELY · Algebraic Probability Theory
RYAN · Modern Regression Methods
RYAN · Statistical Methods for Quality Improvement, *Second Edition*
SCHEFFE · The Analysis of Variance
SCHIMEK · Smoothing and Regression: Approaches, Computation, and Application
SCHOTT · Matrix Analysis for Statistics
SCHUSS · Theory and Applications of Stochastic Differential Equations
SCOTT · Multivariate Density Estimation: Theory, Practice, and Visualization
*SEARLE · Linear Models
SEARLE · Linear Models for Unbalanced Data
SEARLE · Matrix Algebra Useful for Statistics
SEARLE, CASELLA, and McCULLOCH · Variance Components
SEARLE and WILLETT · Matrix Algebra for Applied Economics
SEBER · Linear Regression Analysis
SEBER · Multivariate Observations
SEBER and WILD · Nonlinear Regression
SENNOTT · Stochastic Dynamic Programming and the Control of Queueing Systems
SERFLING · Approximation Theorems of Mathematical Statistics
SHAFER and VOVK · Probability and Finance: It's Only a Game!
SHORACK and WELLNER · Empirical Processes with Applications to Statistics
SMALL and McLEISH · Hilbert Space Methods in Probability and Statistical Inference
STAPLETON · Linear Statistical Models
STAUDTE and SHEATHER · Robust Estimation and Testing
STOYAN, KENDALL, and MECKE · Stochastic Geometry and Its Applications, *Second Edition*
STOYAN and STOYAN · Fractals, Random Shapes and Point Fields: Methods of Geometrical Statistics
STOYANOV · Counterexamples in Probability
STYAN · The Collected Papers of T. W. Anderson: 1943–1985
TANAKA · Time Series Analysis: Nonstationary and Noninvertible Distribution Theory
THOMPSON · Empirical Model Building
THOMPSON · Sampling
THOMPSON · Simulation: A Modeler's Approach
THOMPSON and SEBER · Adaptive Sampling
TIAO, BISGAARD, HILL, PEÑA, and STIGLER (editors) · Box on Quality and Discovery: with Design, Control, and Robustness
TIERNEY · LISP-STAT: An Object-Oriented Environment for Statistical Computing and Dynamic Graphics
TIJMS · Stochastic Modeling and Analysis: A Computational Approach
TIJMS · Stochastic Models: An Algorithmic Approach
TITTERINGTON, SMITH, and MAKOV · Statistical Analysis of Finite Mixture Distributions
UPTON and FINGLETON · Spatial Data Analysis by Example, Volume 1: Point Pattern and Quantitative Data
UPTON and FINGLETON · Spatial Data Analysis by Example, Volume II: Categorical and Directional Data
VAN RIJCKEVORSEL and DE LEEUW · Component and Correspondence Analysis
VIDAKOVIC · Statistical Modeling by Wavelets
WEISBERG · Applied Linear Regression, *Second Edition*
WELSH · Aspects of Statistical Inference
WESTFALL and YOUNG · Resampling-Based Multiple Testing: Examples and Methods for p-Value Adjustment
WHITTAKER · Graphical Models in Applied Multivariate Statistics
WHITTLE · Systems in Stochastic Equilibrium

*Now available in a lower priced paperback edition in the Wiley Classics Library.

WONNACOTT and WONNACOTT · Econometrics, *Second Edition*
WOODING · Planning Pharmaceutical Clinical Trials: Basic Statistical Principles
WOOLSON · Statistical Methods for the Analysis of Biomedical Data
WU and HAMADA · Experiments: Planning, Analysis, and Parameter Design
  Optimization
YANG · The Construction Theory of Denumerable Markov Processes
*ZELLNER · An Introduction to Bayesian Inference in Econometrics